Space is the Ultimate Luxury

African Social Studies Series

VOLUME 49

The titles published in this series are listed at *brill.com/afss*

Space is the Ultimate Luxury

Capitalists, Conservationists and Ancestral Land in Namibia

By

Bernard C. Moore
Luregn Lenggenhager

BRILL

LEIDEN | BOSTON

'The Open Access publication of this book has been published with the support of the Swiss National Science Foundation'.

Cover illustration: Visiting family graves along the Garib/Orange River, photo: L. Lenggenhager, 2020

The Library of Congress Cataloging-in-Publication Data is available online at https://catalog.loc.gov

Typeface for the Latin, Greek, and Cyrillic scripts: "Brill". See and download: brill.com/brill-typeface.

ISSN 1568-1203
ISBN 978-90-04-71244-7 (paperback)
ISBN 978-90-04-71245-4 (e-book)
DOI 10.1163/9789004712454

This book is printed on acid-free paper and produced in a sustainable manner.

Contents

Preface: Willem Basson and the Historians

FIGURE 1 Willem Basson III, photographed in Karasburg
PHOTO: B.C. MOORE, 2022

This book is a historical and contemporary narrative of disparate people thrust into contact and conflict in a specific place: the far south of Namibia, just near the Lower Orange River. This is today a far-off place by nearly all definitions of remoteness. It is more than a hundred kilometres from the nearest tarred road; nearly one hundred and fifty kilometres from the closest railway station; almost 1,000 kilometres from any international airport. As remote as it was and is, the far south of Namibia has stood at the crossroads of international connections and global capitalism. For decades, its sheep satiated South Africa's appetite for mutton and Europe's desire for luxury lambskin fur coats; its venison and game meat adorned the plates of hotel guests in Cape Town and Paris; its tantalum ores clad nuclear reactors in Europe and ballistic missiles in North America. Today, southern Namibia sits at the forefront of new, private nature conservation schemes, which seek to 'rewild' endangered species for a high-wealth tourism clientele, monetising the region's perceived distance from places of power. In the words of the White Afrikaans-speaking manager of one particular tourism venture: 'Remoteness is what people seek'. In the words of its English South African owner: 'Space is the ultimate luxury'.

This is a book which neither of us intended to write. What started as an amusing curiosity developed into something bordering on obsession, and it first arose from each of our own personal meetings with Willem Basson III, a small-scale communal farmer at the Orange River. When historians begin to pursue a new project, it often originates from big-picture intellectual questions that eventually become rooted in a specific locale. Ours is the exact opposite. This is the product of close, personal relationships with the subjects of this study and of our gradual realisation that this seemingly microscopic situation at the Orange River was actually the outcome of processes so much larger than we had thought. The book starts with Basson and ends with Basson, but as the pages that follow show, the story is so much bigger than the Basson family.

•••

To contextualise how this project evolved from unplanned curiosity to a full book – and to be open about our own positions here – it is relevant to first provide some background about how each of us met the Basson family and what made us realise that much more was at stake.

Bernie Meets Basson

In 2021, I was finalising archival and fieldwork for my PhD dissertation in History, looking at the development of the sheep-farming sector in southern Namibia and the relationship between agriculture and apartheid-era homelands in the country. Since 2015, in my capacity as a researcher at Michigan State University in the USA and as a part-time history lecturer at the University of Namibia, I had been conducting extensive documentary and oral history work on this topic. One day at the archives in Cape Town, I stumbled across some documentation from the 1960s and 1970s about an abandoned proposal for an apartheid-era 'Coloured' homeland along the Orange River. One particular 1971 letter claimed that there were Bondelswarts (Nama) 'squatters' residing on the state-owned farm Hartebeesmund, right at the Orange River where this homeland was planned. Since most of my study had dealt with communal areas farther north, near Karasburg, and the White-owned commercial farms surrounding them, I had never been to this farm. But to write about southern Namibia means writing about the Orange River, so I drove there to see this specific farm with my own eyes.

The whole of southern Namibia is extremely arid, and as you approach the Orange River the level of rainfall drops as the mountains continue to rise. To this day, the 6,500-hectare Hartebeesmund Farm has no owner; it is a remnant of an old system of state-run 'Crown lands' from the early days of colonialism in Namibia. It had been surveyed for use as a private farm, but the apartheid-era government could not find a buyer for this large, rocky piece of land. Indeed, the government's cadastral maps I had in my possession showed no graded roads to the farm at all. Even by Namibian standards, that is very isolated. Friends of mine in Karasburg recommended that I try to get in from the neighbouring farm to the east, KumKum; maybe the owners there could show me a path. Although it may not be the biggest or strongest truck, my little Korean SsangYong 4x4 has not let me down yet. KumKum is about an hour and a half south-east from Karasburg on the gravel roads.

When I pulled up to the farmhouse at KumKum, I was greeted not by a burly, sunburnt Afrikaner – as normally is the case in this part of Namibia – but by a slender young English guy with long hair and a distinguished accent. Also unusual was that there was no sign of sheep, or anything to indicate that there was livestock on the farm at all – like bales of lucerne or kraals/pens filled with manure. It was very weird. He introduced himself as Edward 'Red' Barthorp, and he welcomed me inside for some coffee. After more than a year and a half of COVID-pandemic lockdowns, vaccinations and isolation, he was clearly happy to talk with someone from the outside. He lamented the struggles to get a strong enough internet signal to chat with friends back home, and he was considering asking his employers to upgrade the satellite dishes or else build a repeater to tap into the cellular signal from Warmbad village fifty kilometres away. To be honest, Red looked bored.

'Sorry for barging in on you guys unannounced', I said, 'but I'm trying to get through to Hartebeesmund, and I'm hoping you guys can show me a 4x4 path to get down to the river there'. Red said that from KumKum it was not really possible; maybe once upon a time, but not anymore. 'The only way to get to Hartebeesmund is to go via our business partners at Sandfontein to the west', said Red, pointing to the location on my map. 'I can give you Mrs Erika's phone number. She runs Sandfontein's camping spot there by the river. It's a pretty cool place, used to be a church in the old days'.[1] Red seemed perplexed by my question, though. 'Why do you want to go to that place, anyway?', he asked. 'There's nothing really to see there but sharp rocks that'll puncture your tyres'.

Red said he worked for a company that was developing a nature conservation initiative on KumKum, together with a nearby farm to the south-east called Pelladrift. He explained that it was about outdoor education and 'adventure tourism', alongside other more traditional conservationist goals, like

1 Edward Barthorp, interview with Bernard C. Moore (KumKum farm, 18 September 2021).

protecting threatened or endangered wildlife. He described what they were doing as 'rewilding': making KumKum no longer a farm, as it had been for more than a century, but returning it to its 'natural state'. In addition to bringing in undergraduates and gap-year travellers, Red remarked that he also wanted to build a 'digital nomad' workspace near the Orange River as a way to fund some of their operations.

I explained to Red that I was a historian, and that I wanted to follow up on some records that I had found about Africans living outside the state's purview on Hartebeesmund. I had my laptop computer in the car, and I showed him the letter from the archives. Red could not read Afrikaans, so I quickly translated it for him. According to the letter, the state ethnologist had travelled to Hartebeesmund, where he met the heads of several families, including Gert Witbooi, Abraham August, Matewis Kraai, Piet Botes and Willem Basson II.[2] I said that I wanted to see what remained of these living places, since the letter implied that the residents had been staying there for more than six decades.

'Basson … ', murmured Red. 'There's this guy Basson who farms somewhere at the Orange River who my business partners really don't like. I urged them to just buy him a farm somewhere else', said Red, 'I know they've got the money to do it, but it seems that they've got other plans'. I asked Red what the issue was, and he said that farming and conservation just could not go together; they were ecologically at odds with one another. 'But it's bigger than just that', added Red. 'My clients don't want to see some guy with a herd of goats. It's not what they came to Namibia for'.

At Red's recommendation, I drove the following day to Homsrivier Farm, where the Sandfontein company runs a camping site and small-scale accommodation, in stark contrast to the hyper-luxury lodge that the company has on their other neighbouring property to the north. *Tannie* [auntie] Erika Kotze was sitting on the veranda at the main house. She quickly came out and showed me the right place to park, writing my number plate down in a notebook, as well as a description of the vehicle and of me. Erika was clearly used to monitoring everything very carefully, as though this tiny farm road along the Orange River was anything but tiny. I mentioned to her that I was sent by Red about a history project I was doing, that I was trying to get upriver to Hartebeesmund Farm. 'Oh, you can't go there', said Erika. 'There's nothing to see that side, plus the road is too poor. We've never repaired it since last year's rains.' We sat at the table and were served a cup of tea. 'But if you want to know

2 Western Cape Archives and Records Service (hereafter, Buro-TBK), Department of Coloured Affairs (hereafter, KUS) B-615 File B26/2/1/3 (vol. 2): K.F.R. Budack, Etnologiese Afdeling van BAO to Hoofbantoesakekommissaris van SWA 'Nama en Namasprekende Ovaherero op Hartebeesmund 108, Warmbad' – 11 March 1971.

about history, you have to speak with Katrina here', said Erika, gesturing to her housekeeper who had just served us. 'Katrina grew up here at Homsrivier'.[3]

Katrina Pieters was in her forties and spoke only Afrikaans. She explained a lot about when Homsrivier was a Roman Catholic mission station. Many Nama, Coloured,[4] Damara and Herero people came to the church for Holy Mass, as well as to work on the church's irrigated fields. 'You can still see some of the plough marks and furrows on that side of the property', said Katrina.[5] I told her that I had come to look into an old record from the state archives, about some people staying on Hartebeesmund in the 1970s, and I showed her a copy of the document. We read through it together, and upon reaching the name Willem Basson II, *Tannie* Erika – who was listening from a distance – came closer. '*Is dit dieselfde Basson?*' ['Is it the same Basson?'], she asked Katrina in Afrikaans. '*Nee, dit is sy pa en grootouers*' ['No, his father and elders'], replied Katrina.

Erika quickly read the rest of the document. The tone of the meeting had clearly changed from one of curious interest in what Katrina knew about the region's history to one of suspicion. Erika interrogated me about the document. 'What is this for, and where did you get this?', she asked. I replied that it was from the plans of an old commission from the apartheid years to make a homeland for Coloureds along the Orange River. During the planning, they had learned that there were these river-folk living on Hartebeesmund. For a number of reasons, the homeland did not come to fruition, but all the farms in this area were once slated to be purchased by the government for that reason. 'But they can't do that anymore, right? They can't make a *tuisland* [homeland] here anymore. These are private lands!' Erika said, bordering on anger now. '*Is Willem Basson kleurling? Hy is Herero of Nama, nee?*' ['Is Basson even Coloured? He's Herero or Nama, right?'], Erika asked Katrina (who identifies as Coloured). '*Ek weet nie. Maar in daai jare, miskien was die familie*' ['I don't know, but in those years, maybe his family was'], replied Katrina.

Willem Basson III was clearly well known to the Afrikaner matron, Erika, and my disconnected line of questioning had obviously struck a nerve. It was not going to be possible for me to pass through Homsrivier to the next farm, Girtis (also owned by the same people), and then to Hartebeesmund (owned by the state). Upon returning to Karasburg and the cellular network, I dialled some acquaintances within the Bondelswarts Traditional Authority to see if I

3 Erika Kotze, interview with Bernard C. Moore (Homsrivier farm, 19 September 2021).

4 As we explain in more detail in chapter 7 and chapter 12, 'Coloured' in the context of Namibia and South Africa is a category broader than simply 'mixed race'.

5 Katrina Pieters, interview with Bernard C. Moore (Homsrivier farm, 19 September 2021).

could get in touch with Basson. To my luck, he was due to come into town the following day to sell some goats to the slaughterhouse. I left requests to meet with him to chat about his family's history, and potentially learn a bit about what the heck was going on with these people on the Orange River.

Basson was tall, slender, in his late fifties. He drives a 1997 Nissan Hardbody *bakkie* [pickup], which, based on the stream of swear-words he directed at its engine, had clearly seen better days. I gave Willem copies of the records I had found in the archives, and he smiled, saying that he remembered when the government officials visited Hartebeesmund. 'The Willem Basson mentioned in this document is my father', he said. 'I was about seven years old when they came to the farm to count us'.[6] Basson told me that his people were Bondelswarts people from the Orange River [river-folk], and although the Bassons were the biggest of the families, there were about a dozen extended families who over the years had lived on the banks of the Orange River, coming from time to time to Warmbad, Karasburg as well as other settlements on the South African side of the river. I realised that over the past several years of dissertation research, I had actually met his sister Monika and his aunt Ana, both of whom lived in Warmbad.

Willem Basson said that unfortunately he could not talk for long, because he had to travel north to Windhoek the following morning to deliver some papers. Since I was going in the same direction to return to the university, I offered to take them for him. 'No', said Willem. 'I have to go in person this time to sign some documents at the lawyer's office. It's about those people you met at Homsrivier and KumKum. They've opened another lawsuit against me and my family'. Concerned, I asked him what it was about. 'Eviction. It's always eviction', said Basson.

> Those people there, they say they're doing conservation, and maybe they are, but really they just want to own the Orange River. They don't want to see us African people with our animals. But we're not on their farms, we're on the state lands at the river. That's Bondelswarts ancestral land, and it shouldn't belong to those people.

Willem Basson III departed for the night, but he urged me to visit him the following month, and he would take me to their farming operations at the Orange River, so that I could see it all for myself. When I returned to Windhoek, I called Luregn Lenggenhager in Basel, to update him on what had happened. Luregn

6 Willem Basson III, interview with Bernard C. Moore (Karasburg, 21 September 2021).

and I have been friends and colleagues for more than ten years, and he had been working on a project that involved fieldwork on the South African side of the river. I explained what had happened at Homsrivier and my conversation with Willem, suggesting that when Luregn came in November, we should go together to see him. 'Oh, I met Willem Basson last year!' Luregn remarked. 'Now that's a good story'.

Luregn Meets Basson

Throughout a good chunk of 2019, I was conducting archival and oral history fieldwork in Onseepkans, a small South African village on the Orange River which serves as a border post for car traffic to Namibia. This was part of a larger interdisciplinary project about landscape narratives along the Lower !Garib/ Orange River. Onseepkans is about one hundred kilometres upriver from Homsrivier Mission Station. Even though the village is quite isolated – most car traffic en route to Namibia uses the tarred road at Noordoewer – the people really know a lot about all sorts of places along the Orange River. In digging up records of irrigation schemes and missionary endeavours in the 1930s, I kept coming across the same family names of the African workers employed in the fields, which in those days mostly supported lucerne and crop vegetables. The oral history testimonies that I collected revealed that this was not just the case for Onseepkans, but for other mission stations and irrigation schemes along the river, including Pella, Witbank, Homsrivier, Goodhouse, Noordoewer and beyond. Some people moved up and down the river to offer their labour; others travelled up and down the river with their own livestock, earning a living outside the cash economy.

These family ties were not just from the old days; they continued in the twenty-first century. When a funeral takes place in the African communities of Onseepkans or Pella, for example, it is not just a village funeral, it is an Orange River funeral, and all areas are represented. Travelling along the South African side of the river, I constantly met people who traced their roots to Namibians fleeing the genocide of the Nama by the German colonial government, or to labour-related movements over the river, and many had ties to the Bondelswarts communal areas to the north. Some even referred to families who still lived along the Orange River on the north bank. This confused me, because I was under the impression that the Namibian side was just commercial farms owned by Whites, leaving little room for Black residents.

In 2020, I was sitting in the living room of a former politician in Karasburg, interviewing him about his life and the years he had worked on a diamond mine further downriver at Oranjemund. Suddenly, he changed the topic, mentioning that he had a friend who was staying on the Orange River, Willem, who had just learned that some landowners were about to open a lawsuit against him. 'You're a historian; you should talk to him. I think he's in town today'. Only minutes later, we drove to the petrol station, where Willem Basson III was with a few of his sons filling an old Nissan pickup. Our mutual friend suggested that Basson take me to see his kraals the next day. Basson agreed, though warning us that it was January, and it would be very hot.

We left Karasburg with his sons at 4 am for the long and bumpy trip. As the sun was creeping over the horizon, the pickup was climbing over the last mountain pass before descending to the Orange River near Homsrivier Mission Station. A guard employed by the property owners came to the road and demanded that Basson sign a form in order to pass through Homsrivier Farm. Willem was having none of this and shouted back at the guard that he wasn't trespassing and wouldn't sign any forms. Basson's son quickly opened the gate, and we sped through; they clearly did not want to stop for long at the church building where Willem had been baptised fifty years earlier. By the time we got to Basson's small homestead next to the Orange River, the thermometer read nearly 50° C; Willem was not kidding about the heat in these parts. After washing our faces in the river and cooling off a bit, Basson and his sons showed us around the area where he and his family had spent most of their upbringing, explaining that we were near to Girtis Farm, but technically on the state lands bordering the river. On the drive home in the evening, though, Basson took us to his family's graveyard, about 500 metres inland onto Girtis. He said a small prayer, and we continued on back to Karasburg.

When Bernie called me in 2021 to invite me to visit Basson's kraals, a part of me was hesitant, because November is almost as hot as January. But when he explained the ongoing hostility of Basson's neighbours, I realised that we needed to learn more about what was going on here. My Onseepkans work was coming to an end, so we decided to team up on this.

•••

Our curiosity about Willem Basson and his struggles against his 'conservationist' neighbours was initially understood by us as an epilogue to our respective PhD and post-doc projects. Indeed, many of our first interviews and communication with the individuals involved in this area were framed on that basis. However, the more we dug into the history of the African families, the surveyed farms and the foreign newcomer capitalists who have been accumulating land in this area, the more we realised that these questions were of equal or greater importance to what originally brought us to this arid corner of southern Namibia. So, we decided to put much of our work on hold to dedicate ourselves fully to this project. This was in part because the stories we uncovered fascinated us but also because our skills as historians were needed. The ongoing conflict between Basson and his conservationist neighbours is fundamentally a conflict over *history*.

Apart from Hartebeesmund and the Orange River itself, nearly the entirety of the lands south of Warmbad belong to a few wealthy individuals via corporate shareholdings. One company, Sandfontein Lodge & Nature Reserve (Pty) Ltd, owns four massive farms together covering more than 61,000 hectares in size. Marwilben Farming cc owns two large farms totalling more than 19,000 hectares. Tantalite Valley Estates (Pty) Ltd also owns two, measuring more than 16,500 hectares. All these farms are effectively controlled by one prominent English South African businessman, Sean Gilbertson, who is the CEO of Gemfields, the largest supplier of coloured gemstones in the world. Until very recently, he was also CEO of its subsidiaries, including Fabergé, which markets the stones via the famous Romanov-era brand. Another series of four farms, covering more than 45,000 hectares to the east, are either owned by or under custodianship agreement with the company Pelladrift (Edms) Bpk. Pelladrift is effectively controlled by the pioneer of Kenyan private nature conservation, Ian Craig alongside Peter van der Byl Morkel, the world-renowned Rhodesian-born rhinoceros veterinarian. Together, all these companies and properties have entered into an agreement to unify their conservation and business operations under the trade name ORKCA: the Orange River-Karoo Conservation Area (Pty) Ltd. It holds nearly 150,000 hectares of farmland and it is growing in size.[7]

As we explain in detail in this book, the Basson family – as a stand-in for all families along the Orange River – are being taken to court in Namibia by Gilbertson and Morkel (as representatives of their companies), who are demanding eviction of the river-folk and/or an interdict to prevent them from trekking their livestock from the Orange River to the north, east or west (moving southwards is already illegal, since it would mean crossing into South Africa).[8] This is

7 Information regarding land ownership has been obtained from records at the Windhoek Deeds Office. We explain in more detail in later chapters. From time to time, we use the term 'effective control' to signify that individuals with fewer than fifty per cent of company shares may exert disproportionate influence on company operations.

8 Archives of the Registrar of the Master of the High Court of the Republic of Namibia (hereafter, RoHC) File HC-MD-CIV-ACT-CON-2020/02424: Plaintiffs' Particulars of Claim – 25 June 2020. Please note that the plaintiffs often use the term 'ejection', rather than eviction. However, as will become clear in subsequent chapters, the political, economic and historic contexts render 'eviction' – in our judgement – a more apt description of the legal claim.

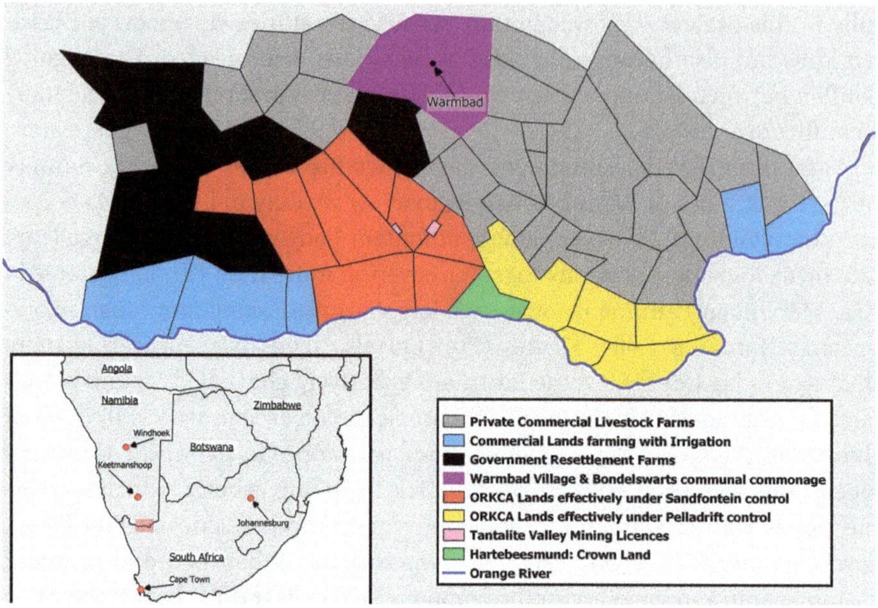

FIGURE 2 Present-day land tenure in southern Karasburg District
CARTOGRAPHY BY B.C. MOORE

not the first lawsuit faced by the Bassons, but it is the most unified and affects nearly one hundred kilometres of Orange River frontage, incorporating many state lands – which the companies have admitted in court that they do not own. For these companies, though, owning the river is less important than having it emptied of humans. Such legal proceedings, if successful, would mean the end of centuries of African existence along the lower Orange River.

In their defence, Willem Basson III and his lawyers (Windhoek's Legal Assistance Centre) have invoked the Roman-Dutch legal principle of *Vetustas* – sometimes called 'immemorial usage' – which refers to a right which has existed continuously since 'time immemorial', held against another person, property, or institution. Although this may seem an imprecise way to measure contemporary people's relationship to the past, there is a specificity to the legal science of it and it necessitates the skills that we historians possess. *Vetustas* does not need documentary proof in the form of title deeds; rather, it requires showing that the current generation has understood that a practice has existed for so long that neither they nor their forebears could point precisely to when it began. On the other hand, there also has to be no clear evidence that another state of affairs regularly existed. Given these two factors, *Vetustas* implies and assumes a lawful origin. Willem Basson's invocation of

Vetustas is based on the fact that he and his community near the Orange River have consistently traversed these farms to access other Nama communal areas since time immemorial, and they therefore have sought a public servitude to reflect this.[9]

One of the current authors (Moore) was brought on by the Legal Assistance Centre in 2022 to submit expert witness reports about the history of the Orange River region and the precolonial, colonial and postcolonial history of the Bondelswarts. In the historian's reasoning, these reports show that the river-folk's claim of *Vetustas* has merit.[10] This book reveals that despite losing most of their lands by colonial legal fiat, the river-folk continued to access their ancestral grounds and engage in subsistence farming, as well as wage labour for White farmers and miners on both sides of the river. Apartheid-era legislation made crossing these commercial farms more difficult but not completely impossible. This relationship with their ancestral lands began beyond the living memory of any individual person today or of previously-known generations, and it continues up to the present day, despite the actions of colonial, apartheid and capitalist settlers and/or 'conservationists'. The plaintiffs, Gilbertson and Morkel, have consistently denied these historical connections which the Bondelswarts claim to have to their ancestral lands, denying *Vetustas*, ultimately falling back on colonial land tenure as the natural state of affairs.

At the time of writing, the court case is still ongoing, so it is difficult to assess precisely the direction in which things will go. However, the Basson case is one of the first ancestral land cases in Namibian history, and it is the very first to invoke *Vetustas* as the basis for usufruct land use and/or tenure in resolving the legacies of colonialism and apartheid. Along similar lines, dealing with ancestral land claims on the grounds of *Vetustas* enables a better collaboration between historians and the court of law. For dispossessed humans, the question of ancestral land is ultimately a question about the past's relationship with the present. For the 'conservationist' enthralled with concepts of 'rewilding' land to a purported historical baseline, the same could be said. We historians must be present, doing our part to shape the knowledge necessary to build

9 RoHC File HC-MD-CIV-ACT-CON-2020/02424: Founding Affidavit, Willem Basson – 26 November 2020.

10 As elaborated in more detail in chapter 4, we use the term 'river-folk' to describe the heterogenous community of Bondelswart Nama with historical, familial and material connections to the Orange River. River-folk are a specific grouping of Bondelswarts.

a just society. This is not just building a 'useable past' for reasons of pride or national myth;[11] this is using our skills for something more material.

•••

This book seeks to accomplish three main tasks. First, we chart the history of the Orange River border region, stretching across the southern part of Namibia's Karasburg District and the northernmost part of South Africa's Northern Cape. Using close readings of archival records from Namibia, South Africa, Germany and beyond, we examine the precolonial, colonial and postcolonial history of these lands. Supplemented with oral history testimonies, we look into livestock farming, mining and irrigation, as well as where these farms fit into the regional apartheid-era economy and global market connections. Doing this enables us to demonstrate the long-term connections that the river-folk maintained with their ancestral lands, connections which have existed since time immemorial. This is not an easy task, and it required that we consult unconventional sorts of records, because the stories of Black Namibians and South Africans are not privileged in the colonial archive. Nevertheless, using varied sources from missionaries, the state, labour and mining inspectors, and even branding iron registers, we show that *Vetustas* is an apt description of the relationship between the Bassons/river-folk and the lands now falling under ORKCA.

Second, our study examines the complicated personal and financial networks that in the post-apartheid era led to the consolidation of these farm properties in so few hands. Through biographies of people, places and – *crucially* – companies, we show that post-apartheid land inequalities in southern Namibia depended not just on historically-conditioned processes going back to the colonial/apartheid years, but also upon specific contingent events. There was and is nothing inevitable about land being used in a certain way. This investigation was an equally-difficult task, as the chapters that follow show. It necessitated that we cross-reference the corporate filings of dozens of holding companies with land registry files, title deeds and court affidavits. Tracking the beneficial ownership of land in an era of limited liability companies (whether in Namibia, South Africa, British Virgin Islands, Panama or elsewhere) is a laborious process but one that in time bears fruit.

Third, although this study begins as a human history of land, connections to it and ownership of it, it ends as a story that includes private nature

11 See B. Jewsiewicki, 'African Historical Studies, Academic Knowledge as a "Usable Past", and Radical Scholarship', *African Studies Review*, 32, 3 (1989), pp. 1–76.

conservation and how it fits within this landscape. There is a reason why land ownership of these properties in post-apartheid Namibia is framed less along the lines of *using* the land for profit and more about being *custodians* of it. Throughout the chapters that follow, the reader is introduced to the history, political economy and ecology of nature conservation in Namibia, illustrating the reasons why conservation has gradually transitioned in the post-apartheid years towards being a *private*, for-profit endeavour. We examine the specific conservation initiatives that are taking place on the farmlands under investigation, and we place them in the context of the ongoing dispute over ancestral land rights in Namibia. Finally, we reflect upon what the case of the Orange River-Karoo Conservation Area (ORKCA) tells us about possible futures for nature conservation in Namibia and elsewhere. Ultimately, the purpose of this book is less to present an exposé of specific individuals and companies; rather, it is intended to delve into the broader politics of land tenure, nature conservation, and financial operations of each.

This book is written as a narrative, introducing the reader to the eccentric, peculiar and charismatic characters who walked within this southern Namibian landscape. From African sheep farmers to colonial settlers, to mining millionaires, to globe-trotting conservationists, all are given equal scrutiny. This has necessitated our stylistic decision to allow characters to appear, disappear and reappear chapter by chapter, enabling us to illustrate the unfolding trajectory of the lands under question. In addition, conflict over land in Namibia and beyond is not abstract; these conflicts concern specific landscapes and even specific surveyed farm parcels. For this reason, we are precise about not just the individuals about whom we write, but also the specific lands upon which they stood. Our fieldwork enabled us to crisscross this arid, mountainous region over and over again, and if we are to understand how African communities understood these unequal historical processes, we must situate these individuals in specific portions of specific landscapes. We provide ample maps, illustrations, and photographs to guide you along this journey with us to Namibia and beyond.

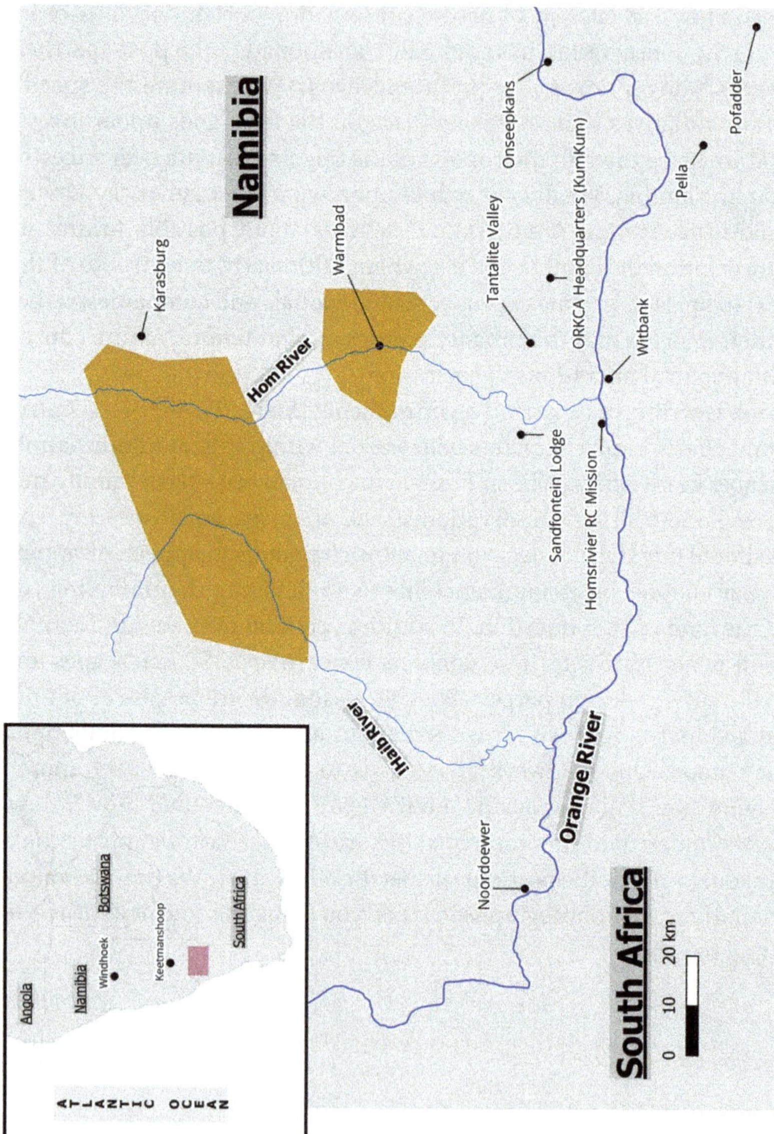

FIGURE 3 Map of selected locales along the lower Orange River. Bondelswarts communal areas are indicated in yellow
CARTOGRAPHY BY B.C. MOORE

Acknowledgements

This is a weird, unconventional book by the general standards of scholarship and journalism. When friends and colleagues would ask us what we were researching, our answers were different almost every time: sheep, coal brokering, Fabergé eggs, mesquite trees, Roman-Dutch law, rhinoceros diets, nuclear reactor claddings, heap leaching, offshore finance, apartheid segregation, voluntourism and more. When our friends and colleagues learned that our revolving door of temporary obsessions would all find their way into the same book, most were justified to feel confused. We hope that by the time you finish reading this book, the puzzle pieces come together a bit better and that the bigger reason for this project in the first place becomes clear.

This book would not have been possible without the tireless assistance and conversations with key individuals in Namibia and beyond. We would like to thank Mr. Willem Basson III and his children, siblings, and family for sharing their histories with us and opening up their personal lives to two curious foreigners. We are grateful for the assistance of Josef Rooi, Lazarus Kairabeb, and all those affiliated with the Bondelswarts Traditional Authority, for helping us understand the ways that the histories of a few dozen families near the Orange River were also the history of the Bondelswarts people. We were also closely assisted by Sr. Paule Elizabeth Atiogbe and all of her fellow nuns and priests at the Holy Trinity Catholic Church in Karasburg and the Cathedral of the Immaculate Conception in Pella, all of whom dragged us back to Sunday-school to understand the history of Catholicism in the region. We would also like to thank Catherine Cloete and her neighbours in Witbank, as well as Gerhard le Roux and the people in and around Onseepkans, who opened up the history of the South African side of the Orange River to us.

Researching this complicated topic required a large amount of technical assistance from key experts on many subjects. In Windhoek, we are grateful for the assistance from and collaboration with the team at the Legal Assistance Centre, particularly Chloe Brandt, Corinna van Wyk, Peter Watson, and Lotta Ambunda. We hope that this will not be the last time that we work together on a project like this. We thank the staff at the National Archives of Namibia, particularly Ndamian Hangula, Albertina Nekongo, Ralph Ntelamo, and Werner Hillebrecht for assisting us with locating key historical records. We appreciate the assistance from Martha Haludilu at the library and archives of the Ministry of Mines & Energy for help locating old, forgotten prospector's reports. Louise Niemöller of Pofadder was also kind in opening up her father's papers and photographs for us, which was incredibly valuable information. We are also grateful for the assistance of our legal teams in Windhoek and Amsterdam for their help and advice navigating treacherous waters.

Due to the unusual nature of the book and the rather sensitive topics covered, we made a deliberate decision to hold back on presenting most of our findings until publication. This stands in stark contrast to many academically-rigorous studies, which have done the rounds at several key conferences before any publishers are contacted in the first place. We did, however, choose to workshop some of our findings at two intimate weekend gatherings, which immensely improved how we approached this complicated subject. We would like to thank Jan-Bart Gewald, Duncan Money, Tycho van der Hoog, and all those who bent their ear towards us in Leiden in 2022. We are also grateful for the invitation from Ute Dieckmann to share our work at the Kloster Steinfeld near Cologne in 2024. The feedback from and conversations with all participants was crucial in making this book a success.

We would like to recognise the assistance of those who read and commented on portions of the work outside of professional conferences and workshops—offering their time, knowledge, and patience. Some also offered their expertise on more technical aspects outside our speciality. We appreciate the help from: Steven Archer, Caitlin Barker, Diana Rodríguez Cala, Emelie Danielsson, John Doyle-Raso, Matt Ecker, Graham Fox, Rob Gordon, Dag Henrichsen, Abdoulie Jabang, Aluteni Kamati, Will Lyon, Meredith McKittrick, James Merron, Michael Mhindu, Elsemi Olwage, Ann Pitcher, Stephanie Quinn, Lorena Rizzo, Diego Menestrey Schwieger, Paul Smith, Michael Stamm, Don Stevenson, Bayron van Wyk, Lydia Walker, and of course Silva Lieberherr for all of their support. We know that the final product benefitted from each of you.

We would also like to recognise all those interviewed in this book who shared information about their lives, their work, and their visions about this land and all who walk upon it. However, as the journalist Upton Sinclair wrote: 'it is difficult to get a man to understand something, when his salary depends on him not understanding it'. We are grateful that many of you took this project seriously, and we hope that the facts which we've presented on these pages are also examined and reflected upon with the same candour. We are pleased that only a small number of people decided to threaten us as a means to stop the information in this book reaching a wider audience. Happy reading.

We would like to recognise the assistance and cooperation of our various 'academic homes' around the world. The University of Namibia continues to put up with us after all these years, and for that we are extremely grateful. In Europe, the Centre for African Studies at the University of Basel hosted the Space in Time Project—led by Martha Akawa, Romie Nghitevelekwa, Maano Ramutsindela, and Giorgio Miescher—which was crucial in helping our narrow interests link to larger regional questions. We appreciate the assistance from all project participants, especially Sindi-Leigh McBride and Andrea Rosengarten, who were with us when we met Basson for the first time and also wrote about these farms elsewhere. We also thank Michael Bollig at the Global South Studies Centre of the University of Cologne and David Anderson of the Global History & Culture Centre at the University of Warwick. In North America, we would like to recognise the support of the History Departments at Alma College and at Michigan State University.

We are grateful for the help and assistance from our publishers, Brill: Preben Kaarsholm took a chance on a strange book; Franca de Kort and Nicolette van der Hoek consistently helped with the administrative side of a complicated publishing project; Sandy Shepherd did a great job transforming our words into meaningful prose; and Wai Min Kan assisted us with all aspects related to open access at Brill. Some of the research as well as the Open Access edition of this publication was supported by the Swiss National Science Foundation.

Finally, a project like this always becomes a part of the lives of those close to us as well. Bernie would like to thank his family and friends for constantly putting up with his escapes—of the body or of the mind—to southern Namibia and beyond. Luregn wishes to thank his family, friends, house-mates, and most of all Silva Lieberherr. Thank you all for your constant patience and support.

Figures and Tables

Figures

Tables

Abbreviations in Text

AALS	Affirmative Action Loan Scheme
BHP	Broken Hill Proprietary, Ltd
BRCP	Black Rhino Custodianship Programme
BVI	British Virgin Islands
CDF	Charbonnages de France
CMM	Congregation of the Missionaries of Mariannhill
DTA	Democratic Turnhalle Alliance
EPL	Exclusive Prospecting Licence
ESKOM	Electricity Supply Commission
GVE	Grootvee Eenhede (large stock units)
JCI	Johannesburg Consolidated Investments, Ltd
KVE	Kleinvee Eenhede (small stock units)
KWS	Kenya Wildlife Services
LPM	Landless Peoples' Movement
LTD	Division of Agricultural and Technical Services (SWA)
LWF	Laikipia Wildlife Forum
NAD	Namibian Dollar
NG	Nederduitse Gereformeerde Kerk
NKGCP	Northern Kenya Grasslands Carbon Project
NPP-435	Namibia Peace Plan Study and Contact Group
NRP	National Resettlement Programme
NRT	Northern Rangelands Trust
NTB	Division of Nature Conservation (SWA)
OCDP	Oana Community Development Project
OMI	Oblates of Mary Immaculate
ORKCA	Orange River-Karoo Conservation Area
OSFS	Oblates of St. Francis de Sales
PPF	Peace Parks Foundation
RSA	Republic of South Africa
SADF	South African Defence Forces
SANF	South African Nature Foundation
SATCO	South African Territories Company Ltd.
SOC	Soil Organic Carbon
SUAL	Sibirsko-Uralskaya Aluminium Company
SWA	South West Africa
SWALU	South West Africa Agricultural Union
SWAPO	South West Africa People's Organisation

tCO_2-e	Tonnes of Carbon Dioxide Equivalent
TFCA	Transfrontier Conservation Area
VCS	Verified Carbon Standard
WWF	World Wide Fund for Nature
ZAR	South African Rand

PART 1

Situating a Human Landscape

CHAPTER 1

Ostriches at the Altar

> 'I don't understand why they would kraal ostriches in the church. It was a terrible thing to do. That's a sacred place'.[1]

Near the confluence of Namibia's small ephemeral Hom River and the Orange River, there stands a small Roman Catholic mission station, called Homsrivier. Founded during the Great Depression by the Austrian branch of the Oblates of St Francis de Sales, it was to serve a dual purpose for the emerging Catholic community in southern Namibia and the Northern Cape of South Africa. First, using the waters from the mighty Orange River, furrows could be dug, and flood irrigation could be used to grow vegetables, grain and fodder to supply the churches and mission schools across the region. Second, the mission station could provide jobs and Catholic schooling for the several hundred Nama, OvaHerero and Coloured families living along the fringes of the lower Orange River. Most had already been baptised Catholic, though a one-hundred- kilometre mountainous trek to the mother station, Pella (on the South African side of the river), made attending Mass, services and school difficult for subsistence pastoralists in this parched corner of southern Africa. After all, mission stations were usually built where there were already souls to be attended to.

Not long after Namibia successfully shook off the shackles of formal apartheid and obtained its independence from South Africa in 1990, the church was being used by the new farm owner as a storage pen for ostriches rather than for Mass. Homsrivier Mission was closed and sold off in 1992 by the diocese to a foreign investor. Funds were needed to establish the post-apartheid Vicariate of Keetmanshoop, since Namibian churches would no longer fall under the South African hierarchy.

Katrina Pieters, whose statement begins this chapter, grew up between Homsrivier – where her father worked as an irrigation labourer – and Warmbad, where she attended school during the 1980s. She recollected how, each

1 Katrina Pieters, interview with Bernard C. Moore (Homsrivier Farm, 19 September 2021).

 | DOI:10.1163/9789004712454_002

FIGURE 4 Homsrivier Roman Catholic Mission station
PHOTO: B.C. MOORE, 2021

FIGURE 5 Thousands of chipped recesses mark the floor of Homsrivier's Church of the Visitation. They were made by ostrich beaks pecking the floor in the late 1990s
PHOTO: B.C. MOORE, 2021

Friday, her father's friend Lukas Swartbooi would leave Homsrivier in a truck filled with vegetables, lucerne, lemons, watermelon, spanspek (cantaloupe) and dates. He would drive north across Homsrivier Farm along the tiny dusty mountain road #284, cross the Orange River mountains and stop briefly at

Sandfontein Farm to sell some food and fodder to the workers of Mr Du Toit's massive karakul sheep farm. He would then continue along farm road #217, leaving Sandfontein and entering Norechab Farm, where he had to carefully avoid sharp rocks as he crossed the dry beds of the streams leading into the Hom River. Once past Norechab, it was easy driving to Warmbad (with a population of about 1,000), where Swartbooi would deliver food to the Catholic church and hostel there. He would take the rest to the church in Karasburg. The drive was about 150 kilometres, and rough terrain meant that it could take a whole day to traverse, but it represented the network of longstanding connections between Warmbad – the political capital of the Bondelswart Nama nation – and the Orange River, the fertile lifeblood of this arid, mountainous region.

Traces of narrow footpaths [*trekpaaitjies*] line the landscape between Warmbad and the Orange River, and they give credence to the continuity of pastoralism and mobility in southern Namibia. For the pastoralists in this region, sheep, goats and cattle are a walking larder and, following the seasonal grazing availability, they have to be moved on the hoof between Warmbad and nearby communal areas and the Orange River, where water and grazing are sufficient, especially during the early summer months of November and December. These *trekpaaitjies* formed the basis of Namibia's network of small two track [*tweespoor*] farm roads, many of which – such as #217 and #284 – were proclaimed public roads in the mid-1950s, during the height of apartheid-era farm subsidies.[2]

Many *trekpaaitjies* were constructed within or alongside ephemeral riverbeds; these would enable livestock to avoid climbing over steep passes and also ensure sufficient grazing and browsing during transhumance between locales. The Hom River was the most important of these for the Bondelswarts throughout the precolonial and early colonial era, because it runs in a more-or-less direct line from the Orange River to Homsrivier, Sandfontein and Norechab, through to Warmbad and other Nama communal areas like ǂGâbes. Furthermore, the Hom River has several small natural springs along its bed, ensuring that livestock that travel along it reach Warmbad in a healthy condition.[3] Warmbad, as its name implies, represents the largest of these natural hot springs, hence its centrality in Bondelswarts history.

Although this kind of transhumance is far from rare today, it is a shadow of the former practice. During the 18th and 19th centuries, travellers, traders and scientists encountered Nama, Herero and San communities living along the

2 Legal Assistance Centre Archives (hereafter, LAC) M/B: *Offisiële Koerant van Suidwes-Afrika* (15 February 1955).

3 Surveyor General Windhoek, Warmbad, 2818 (1:250,000), 1990.

Orange River and these visitors regularly traded with the residents, picking up provisions and continuing their journeys from the Cape to Namibia. Orange River folk are described in surprisingly rich detail in the travel diaries of J.T. Rhenius (1724), C.F. Brink (1761–1762), H.J. Wikar (1779), G. de Vylder (1873–1875), and many others.[4] Brink himself describes encountering a Bondelswart family's kraal just near the current location of Homsrivier Mission Station, in October 1761. He then trekked along the Hom River for several days until reaching Warmbad. The springs in the river sustained his horses and trek animals.[5]

It is not climate or geography that have made Bondelswart Nama movements between the Orange River and communal areas to the north well-nigh impossible; rather, it is capitalism, colonialism, apartheid and their legacies. For settler colonialists, Namibia (called South West Africa before independence) would not remain a distant frontier for long. By the end of the 19th century, the Bondelswarts were caught between the informal colonial ambitions of the British authorities in Cape Town and Springbok, the increasingly militarised German colonial regime in Windhoek, and exploitative and fraudulent businessmen representing land speculation syndicates. Many of the Nama nations in southern Namibia at that time were connected with the British – such as Bondelswart Kaptein Willem Christiaan, who was on the payroll of the magistrate at Springbok.[6]

Despite extensive discussion about whether or not the British should simply annex Namibia,[7] they decided to roll back their colonial ambitions, allowing the Germans to colonise the territory, apart from the port of Walvis Bay. These terms were finally laid out in the Anglo-German Treaty of 1890 (known colloquially as the Heligoland-Zanzibar Treaty), which demarcated the boundary between the Cape Colony and German South West Africa[8] – the Orange River. But unlike virtually all treaties and legislation preceding it or emerging after

4 On the politics and history of ethnicity in this region during the 18th and 19th centuries, consult A. Rosengarten, 'Entangled Networks: Ethnicity, Mobility, and Exchange in the Lower !Garib/Orange River Region in the Late-18th Century', in L. Lenggenhager et. al. (eds.), *The Lower !Garib/Orange River: Pasts & Presents of a Southern African Border Region* (Bielefeld, Transcript Verlag, 2023), pp. 27-50..

5 *The Journals of Brink and Rhenius*, ed. E.E. Mossop (Cape Town, Van Riebeeck Society, 1947), pp. 31–33.

6 National Archives of the United Kingdom, Kew (hereafter, NAUK) Colonial Office (hereafter, CO) 879/21: J.T. Eustace, Resident Magistrate at Namaqualand to Hon. John X Merriman – 26 December 1883.

7 See E.L.P. Stals, *Môrewind oor die Karasberge: 'n Kultuurhistoriese Verkenning van die Karassreek van die laat Negentiende Eeu* (Pretoria, Protea Boekhuis, 2009). W.J. de Kock, 'Ekstraterritorial Vraagstukke van die Kaapse Regering (1872–1885) met Besondere Verwysing na die Transgariep en Betsjoeanaland' (PhD thesis, Stellenbosch University, 1948).

8 Abkommen zwischen Deutschland und England von 1 Juli 1890 (Heligoland-Zanzibar Treaty), Article 3, in *Die Deutsche Kolonialgesetzgebung: Sammlung* [...], ed. K. von Gerstmeyer (1893).

it, the Anglo-German Treaty did not mark the central line or lowest point of the river (the thalweg) as the dividing line between the countries; rather, the boundary was the highest known flood lines on the north bank of the Orange River. Not only would South Africa own the Orange River waters, but it would also – per the treaty – have sovereignty over all lands that the lower Orange River ever touched.

For the river-folk who were grazing their animals along the Orange River, collecting honey and foods from the foliage, or hunting game using sophisticated traps and pitfalls, the 1890 treaty did not change the ecology or the cultural significance of their transhumance between the river and Warmbad; it changed the political nature of it. They were not just crossing terrain; they were now crossing an international boundary. The Orange River and the strip of land on either side of it were *legally* South African, and this remains so to this day. But a line on a map is only as significant as its enforcement on the ground. This book reveals various ways in which that boundary has over the decades been reinforced, challenged or simply ignored.

For pastoralists along the Orange River, the 1890 boundary demarcation has meant far less than what has taken place north of that boundary. With the turn of the 20th century, the German colonial government and affiliated land-speculation syndicates began to sell off farmland to White settlers. This began on a small, ad-hoc basis in precolonial years – which was even supported in isolated instances by Bondelswart leadership – but these processes gathered momentum after the 1903–1908 genocide of the Nama by the German military.[9] By the time of the First World War, significant portions of the Bondelswarts lands had been seized by the government and sold off. The process continued when South Africa took over Namibia in 1919. Even peripheral, unproductive mountainous farmlands were gazetted for sale and sold off to 'poor Whites' immigrating into Namibia from South Africa. The Nama, who had lived on these ancestral lands for generations, were expected to either become a tenant labour force or simply vanish from sight into the communal areas, reduced in size after the genocide. By the end of the 1940s, the *entirety* of southern Namibia was demarcated into private commercial farms and sold to White settlers, and while the demographics of farm ownership have been slightly less monochromatic in the post-apartheid years, a land tenure system of massive, private, surveyed and fenced farmland exists to this day.

Surveyed farmland makes transhumance difficult, though not impossible. Legal procedures exist to enable movement of livestock across private

9 Windhoek Deeds Office Archives (hereafter, WDO), 'Übersicht über die Farmverkäufe und Verpachtungen der Landgesellschaften in Deutsch-Südwestafrika' (File originally from Windhoek Magistrate's Office, 1913).

FIGURE 6 Extract of cadastral farm map of southern Karasburg District, with properties near the Orange River to the south. Note the farms Sandfontein, Homsrivier and Norechab, and Warmbad village
OFFICE OF THE SURVEYOR-GENERAL, WINDHOEK, 1972

farmland. After all, not all farms border public roads. The river-folk found their lives constrained during these decades, and their numbers certainly dropped from when Homsrivier Mission Station was founded, but they never disappeared completely. Many still live and farm along the Orange River – moving back and forth from Haakiesdoorn in the west to Beenbreek in the east; their knowledge of the river, its bounties and its perils have been passed from generation to generation.[10]

A thread which runs throughout the following chapters is the struggle of the Basson family to maintain their transhumant existence along the Orange River. Led by family head Willem Basson III, himself a *raadslid* (councillor) for the Bondelswarts Traditional Authority, the extended network of families (Basson, Botes, Roos, Agus, Simboya, Witbooi, Kordom, Van Rhyn, Kraai and others) keep subsistence numbers of goats, sheep, cattle and donkeys, which graze along the state lands bordering the Orange River.[11] For at least two centuries,

10 LAC BCF: Witness Statement of Mattheus Veldskoen – 2016.

11 LAC BCF: B.E. Simboya, Pella 'Verblyf teen die Oranje-Rivier' – 16 July 2004. As will be made clear throughout this text, for simplicity's sake we will refer to these families collectively as 'the Bassons'; not only are many of these individuals interrelated but all are facing the same issues as their spokesman, Willem Basson III.

these families and their forebears have lived along the river, using *trekpaaitjies* and eventually state roads to access the communal areas near Warmbad and farther north, where they sell livestock, attend school, see to community affairs and occasionally take on wage labour on farms and mines and in towns.

Unlike the flatter, alluvial Orange River lands farther upriver, near Upington, Keimoes and Kakamas, the lands along the lower Orange River, which form the boundary between Namibia and South Africa, are a sharp, rough landscape where the mountains abut right up to the river. For this reason, irrigation works are difficult to build along the Namibian side of the Orange River, and most of those that do exist, or have existed, are on small farm parcels where the mountains open up to land that can be contoured for irrigation. The productive portions of these surveyed and demarcated commercial farms were not traditionally the areas that bordered the river because these were hard to access, and only small portions were regularly used by White settlers for planting vegetables or fodder, or for grazing livestock.

Thus, throughout the 20th century, the Bassons and their forebears were able to eke out a sort of autarchic existence on the fringes of White-owned lands – albeit with difficulty. Whites turned a blind eye to 'squatting' – illegal under apartheid legislation – as long as they could occasionally tap into the Basson family's labour. 'Squatters' or not, the Catholic Church at Homsrivier continued to serve these river-folk along the Orange. The living situation of the Bassons was isolated enough that in 1971 the Warmbad police 'discovered' more than forty five residents staying near the river bordering state lands at Hartebeesmund Farm.[12] Although apartheid racism and governance structures made life difficult for the river-folk, they were able to live on the fringes of the state, disappearing when required and appearing when convenient. Like the highland societies in Southeast Asia described by social scientist James C. Scott, the Bassons had mastered 'the art of not being governed'.[13]

Whereas the Orange River was an international boundary under German colonialism, it was not under South African apartheid. When river waters were low, the Bassons could graze their animals on some of the large islands in the Orange River, like Marten and Krapohl islands, and they could easily cross to visit relatives in Witbank and Pella on the South African side. With the collapse of official apartheid and the coming of Namibia's independence from South Africa in 1990, democracy and economic openness have become a

12 TBK KUS B-615 File B26/2/1/3 (vol. 2): Landdros, Karasburg to Streekverteenwoordiger, Dept. KbRa 'Ongeoorloofde Verblyf op Staatseiendom' – 13 August 1970.

13 J.C. Scott, *The Art of Not Being Governed: An Anarchist History of Upland Southeast Asia* (New Haven, Yale, 2009).

FIGURE 7 Date production on Haakiesdoorn (in the foreground) gives way to lucerne under pivot irrigation at Ramansdrift. The Orange River is visible in the background
PHOTO: B.C. MOORE, 2021

double-edged sword for families like the Bassons. Once again, their story has become a transnational one. Independence brought the politics of passports, police and veterinary certificates. The art of not being governed became much harder, if not impossible.[14]

Namibia's independence has meant the removal of apartheid-era economic sanctions and improved transport networks; new economic opportunities have emerged within this remote corner of southern Namibia. Irrigation is no longer intended solely for the production of lucerne for livestock feed or vegetables for supermarkets and school hostels. New irrigation works along the lower Orange River – such as at Haakiesdoorn, Ramansdrift, Stolzenfels and Aussenkehr – focus on capital-intensive, export-oriented cash crops, particularly Medjool dates, table grapes and sultana raisins.[15] These crops are less reliant on land-levelling schemes for flood irrigation because they can survive on drip sprinklers, which can bring peripheral mountainous lands under production, increasing property values and the commercial viability of lands on

14 L. Lenggenhager and A. Rosengarten, 'Flood Lines and Borderlines: Livestock Farming and Evictions Resistance at the !Garib/Orange River in Southern Africa', *Environment & Society: Arcadia* (Summer 2020).

15 For more on these economic changes, consult B.C. Moore, 'Swimming Upstream: From "Poor Whites" to "Coloureds" along South Africa's Lower Orange River', in L. Lenggenhager et. al. (eds.) *The Lower !Garib/Orange River: Pasts and Presents of a Southern African Border Region* (Bielefeld, Transcript Verlag, 2023), pp. 119–144.

which the Bassons had survived. Although many of the new businesses have foreign investors – Israeli, Saudi, Australian, Serbian – most Black Namibians still identify these firms with the old apartheid regime, because the river-folk still have to deal day-to-day with White Namibian and South African foremen and managers.

Since the late 1980s, Namibia and South Africa have encouraged the expansion of cash-crop irrigation works along the lower Orange River, a process facilitated by the completion of the Gariep Dam in 1971 and ancillary water works in later years. The region's arid climate and perennial irrigation waters have made it attractive for producing varieties of grapes and dates that cannot thrive in the more humid and fertile parts of southern Africa. The European Union has willingly imported this produce because southern hemisphere crops are picked when local variants are not yet ripe. It is hard to overemphasise the significance of this transformation for some portions of the river. On the South African side, Klein Pella (now owned by multinational farming company Karsten Boerdery (Pty) Ltd) is the largest date plantation in the southern hemisphere. Slightly more modest, similar changes have occurred on the Namibian side. In 1985, Aussenkehr and Noordoewer had, respectively, 275 and 224 hectares of land under irrigation for lucerne, and virtually no dates or grapes were grown.[16] By comparison, in 2014, Noordoewer had more than 154 hectares of grapes under irrigation, and in 2018 Aussenkehr had 950 hectares of grapes under irrigation (with proposals to expand grapes and dates up to 4,800 hectares by 2030).[17] This growth replaced much of the lucerne production and claimed more land for irrigation.

As more land has come under irrigation since independence – especially to produce crops that are more valuable than lucerne or vegetables – denser settlements have meant more security and more fencing to protect properties, extended as far as pump infrastructure at the river. For the Bassons and other families who had followed traditional pastoral geographies, this new situation has meant reduced mobility, reduced grazing and the increased risk of stock theft and stock seizure by private security if the animals strayed from the river's edge. The traditional transhumance of the Bassons once reached as far as Haakiesdoorn, Noordoewer and even Aussenkehr to the west, and to

16 Aussenkjer Plase (Pty) Ltd, *The Potential for Agricultural Development of the Aussenkjer-Noordoewer Area* (Stellenbosch, 1985), pp. 28–31.

17 Vioolsdrift-Noordoewer Joint Irrigation Authority Scheme, Water Management Plan (Orange-Senqu River Commission, March 2014), p. 14. Orange River Vineyard Investments (Pty) Ltd et al., 'Aussenkehr Agricultural Developments: Environmental Assessment Scoping Report' (November 2018), p. 3.

FIGURE 8 Emerging date and citrus irrigation on Gaidip Farm, to the west of Homsrivier
PHOTO: B.C. MOORE, 2021

Velloorsdrift, Khaais and Beenbreek farms to the east. The grazing along the river becomes more abundant to the east, where islands in the river are large and dense with grass and bush. Willem Basson would often send a nephew or employee to herd a portion of his flock to this rich grazing during certain times of year, but by the mid-2000s, this was no longer feasible.

In 2002, the company Frontier Farming (Pty) Ltd purchased Khaais Farm (upriver from Homsrivier) with plans to establish date irrigation schemes surrounded by electric razor-wire fencing.[18] Gates were built across the traditional Bondelswarts *trekpaaitjies* and were locked and/or electrified. Accessing the river became increasingly difficult.[19] Conflicts reached breaking point in September 2009, when Frontier Farming's White foreman allegedly threatened to murder Basson's shepherd, Fransiskus Matroos, if he did not leave with the livestock immediately, despite the fact that Khaais Farm does not own the river frontage. Gesturing to the vast mountains behind him, the foreman said to Matroos: 'There is more than enough space for me to bury you here'.[20]

In the years since Namibian independence, the Bassons have been increasingly constrained, even *trapped*, inside a smaller and smaller area next to the Orange River. Crossing to the south bank is now illegal, and irrigation farmers

18 WDO File T6719/2002: Khaais no. 153, Deed of Transfer – 2 October 2002.

19 LAC BCF: Lesle Jansen (LAC) to Adv. N. Bassingthwaigte 're: Consultation with Bondelswarts Community Leaders' – 21 July 2010.

20 'Hier is genoeg plek waar ek jou kan begrawe.' LAC BCF: Willem Basson to Lesle Jansen, Legal Assistance Centre 'Khaais 143' – undated letter, likely August 2010. Quote is verbatim from reports to the Legal Assistance Centre.

FIGURE 9 Small goat kraal belonging to the Basson family, near Girtis Farm
PHOTO: B.C. MOORE, 2021

to the east and west have shrunk the pastoralists' traditional transhumance to just the heart of their ancestral lands: the river from Pelladrift in the east to Homsrivier in the west. Apart from small exceptions – like the mission's fields described above, or some isolated portions of Haakiesdoorn or Khaais – these lands are not suitable for irrigation. They are simply too mountainous and rugged. On the other hand, herding goats, sheep and cattle, diligently avoiding the poisonous melkbos (*Euphorbia gregaria*), as well as leopards and floods, is possible.

However, irrigation is not the only new development in post-apartheid southern Namibia and not the sole reason capitalists want exclusive access to the Orange River. Since independence, most of the farms within the Basson's ancestral corridor – stretching north along the Hom River nearly as far as Warmbad – have been purchased for a very different use: private, hyper-luxury nature conservation and tourism operations. Apart from occasional trophy-hunting, which is now being phased out on these properties, these new operations are a form of private, commercial land tenure that expressly does *not* wish to exploit the land for personal or communal gain, and which prides itself on this fact.

This is not a competition for the most profitable use of the lands – whether to irrigate or graze. Immediate profit means little to the companies and the individuals involved in purchasing this corridor of farmland, which measures an expanse exceeding 150,000 hectares. These newcomers have made their money elsewhere. However, it *is* a form of land tenure which demands *exclusive* use of the land, built on the aesthetic, philosophical and economic frameworks of what Dan Brockington and others have designated 'Fortress

Conservation'. It idealises 'pure' nature as an unspoilt wilderness that has never encountered human beings.[21] When facing the actual history of human interaction with the land, these kinds of 'conservationists' believe the solution is to *rewild* the landscape in pursuit of a mythical historical baseline where herds of rhinoceros, giraffe, zebra, gemsbok and hippopotamus lived side by side on these dry, mountainous lands upon which no humans – at least no humans who had not arrived by private aeroplane – could have or would have gazed. Those who do not fit into this vision of rural Africa – such as the Bassons and virtually all Black Namibians – must leave, either by choice or by force. For the Bassons and those Bondelswart Nama living in southern Namibia – where apartheid seems not to have released its grip – eviction, whether by apartheid police, post-apartheid irrigation businessmen or romantic conservationists, represents old wine in new bottles. As Willem Basson remarked to us, 'I still live under apartheid'.[22]

Evictions do not constitute merely the loss of livelihoods derived from the land, which would be reason enough to fight companies pursuing accumulation by dispossession. Evictions also constitute a loss of cultural grounding and historical reference points. As a fairly successful communal farmer, Willem Basson likely could hire grazing elsewhere or obtain additional rights in the other Bondelswarts communal areas. But the land represents more than just grazing. 'I was born near the river, at KumKum', he said. 'My parents and ancestors are buried near Girtis. If I give it up, I'll never see the graveyard again.' Whether for Basson fearing a loss of connection to his ancestors' graves or Katrina Peters witnessing her church become a storage pen for ostriches to be released as game for international trophy hunters, the loss of land through colonial or post-colonial processes constitutes a loss of both livelihood and ancestral grounding.

This book is about human connections to the lands that conservationists seek to rewild, and the winding historical trajectory that gradually centralised a massive area of farmland in an arid, mountainous corner of Africa into just a few hands. It is a story of global flows of ideas, individuals and investment capital that have conditioned and facilitated these historical events as well as contemporary and future trends on the horizon. This book is not a traditional critical history of nature conservation in Africa, in part because we are looking into processes that are in a state of flux, changing and developing as we

---

21 D. Brockington, *Fortress Conservation: The Preservation of the Mkomazi Game Reserve, Tanzania* (Oxford, James Currey, 2002).

22 Willem Basson III, interview with Bernard C. Moore and Luregn Lenggenhager (Karasburg, 11 November 2021).

FIGURE 10 Visiting family graves along the !Garib/Orange River
PHOTO: L. LENGGENHAGER, 2020

research and write about them. Ours is a moving target. Throughout the chapters that follow, we delve into the histories of the land, its animals, the individuals, the companies and the ideas which spread across the globe to coalesce in this far-off place.

The greater part of critical historical research on nature conservation in Africa considers case studies of national parks, as well as those individuals and groups living near to – or displaced by – the establishment and operations of these conservation areas.[23] These scholars and others rightly show how the national parks of the early twentieth century were intimately bound to

23 See Brockington, *Fortress Conservation*. J.B. Shetler, *Imagining Serengeti: A History of Landscape and Memory in Tanzania from the Earliest Times to the Present* (Athens, Ohio University Press, 2007). K.G. Nustad, *Creating Africas: Struggles over Nature, Conservation, and Land* (London, Hurst, 2015). J. Carruthers, *The Kruger National Park: A Social and Political History* (Pietermaritzburg, University of Natal Press, 1995). J. Dlamini, *Safari Nation: A Social History of the Kruger National Park* (Johannesburg, Jacana, 2020). U. Dieckmann, *Hai//om in the Etosha Region: A History of Colonial Settlement, Ethnicity, and Nature Conservation in Namibia* (Basel, Basler Afrika Bibliographien, 2007). Others have considered conservation in communal areas, as well as links between conservation and militarisation. See M. Bollig. *Shaping the African Savannah: From Capitalist Frontier to Arid Eden in Namibia* (Cambridge, Cambridge University Press, 2020).

the consolidation of the colonial project, both on the African continent and beyond it, in the minds of citizens of the European metropole. This process ultimately produced laws and policies regarding the protection of game and the provision of physical space for game, which were envisaged and lobbied for by international groups, such as the colonial agency Society for the Preservation of the Wild Fauna of the Empire (today known as Fauna & Flora International), often counter to the wishes or interests of local communities.[24] Furthermore, most of the goals and effects of this early colonial conservation legislation were to exclude local inhabitants – particularly Africans – from the material benefits of game utilisation and 'conservation'.

Scholars have examined an abundance of reasons – both implicit and explicit – why conservationists promote securing land exclusively for occupation by flora and fauna at the expense of human residents. These reasons and interpretations vary, from simple aesthetics[25] to socioeconomic and political factors, such as veterinary considerations,[26] military and counterinsurgency operations[27] or even the intent to conscript evicted residents to create an industrial reserve army to work as labourers for colonial capitalist enterprises.[28] It is not our intention to systematically interrogate each of these.

Notwithstanding, most scholars have framed these transformations as a form of *state capture*, whereby 1) the land as commons is transformed into state-managed national parks, and 2) game legislation reconfigures wild animals as a kind of *res nullius* – conceived of as ownerless – and converts them into state property, epitomised by the term 'royal game'. This historical trajectory was popularised and brought into the mainstream through the pioneering work of Scottish historian John M. MacKenzie, who characterised this period of the early twentieth century as a time of increasing *state custodianship* over game and game reserves.[29] Scholars could thereby approach

24 M. Cioc, *The Game of Conservation: International Treaties to Protect the World's Migratory Animals* (Athens, Ohio University Press, 2009).

25 R.P. Neumann, *Imposing Wilderness: Struggles over Livelihood and Nature Preservation in Africa* (Berkeley, University of California Press, 1998).

26 G. Miescher, *Namibia's Red Line: A History of a Veterinary and Settlement Border* (New York, Palgrave, 2012).

27 L. Lenggenhager, *Ruling Nature, Controlling People: Nature Conservation, Development, and War in Northeastern Namibia since the 1920s* (Basel, Basler Afrika Bibliographien, 2018).

28 A.B. Kelly, 'Conservation Practice as Primitive Accumulation', *Journal of Peasant Studies*, 38, 4 (2011), pp. 683–701. A.B. Kelly and N.L. Peluso, 'Frontiers of Commodification: State Lands and their Formalization', *Society & Natural Resources*, 28 (2015), pp. 473–495.

29 J.M. MacKenzie, *The Empire of Nature: Hunting, Conservation, and British Imperialism* (Manchester, Manchester University Press, 1988).

the history of the colonial state through the lens of environmental history, conjoining the consolidation of the colonial state and civil control over nature as interdependent.

Other, more practical, factors have facilitated the historian's fascination with national parks across the continent. A rich paper trail can be exhumed from within the colonial and postcolonial archive, covering historical events in relation to the operations of the parks. For example, in the National Archives of Namibia, the *Natuurbewaring* (Nature Conservation, NTB) collection holds thousands of files pertaining to various national parks established during the twentieth century and general environmental issues of interest to the state. Furthermore, regional and magisterial archival holdings contain additional information relating to parks and state-sponsored conservation initiatives, such as soil conservation schemes. A historian focusing on Namibia's famous parks, such as Etosha National Park or the ǀAi-ǀAis/Richtersveld Transfrontier Conservation Area, can find nearly all the historical data relating to these within the National Archives of Namibia. The state's custodianship over nature runs parallel to the state's custodianship over archival records about nature.

However, the framework of nature conservation under *state custodianship*, what historian Lance van Sittert calls the 'MacKenzian Orthodoxy', is only one side of the story. State-sponsored game reserves and national parks are not the only structures that have defined legal and economic relationships between people and the non-human world during the colonial and post-colonial era. Van Sittert critiques this framework, showing that within South Africa's Cape Colony (later, the Cape Province) there was a precedent for the privatisation of game for commercial use, and the impetus behind this legislation – conservation for commercialisation – did not come from urban cosmopolitan elites but rather from rural landowners.[30] Game law was formulated and reformulated not to remove game from market pressures into the custodianship of the state for preservation purposes, but rather to increase the process of privatisation and commercialisation, ultimately enabled by enclosure and fencing legislation in order to dictate ownership.[31]

As later chapters of this book explain in greater detail, Namibia went along with the Cape Colony in pushing for the privatisation of game, the privatisation of nature conservation, and the commercialisation of each. Although Namibia had national parks and other conservation areas under state custodianship, the vast majority of game – from the lowliest springbok to the grandest

30 L. van Sittert, 'Bringing in the Wild: The Commodification of Wild Animals in the Cape Colony/Province c. 1850–1950', *Journal of African History*, 46 (2005), pp. 269–291.

31 Ibid, p. 284.

rhinoceros – was kept on private, White-owned commercial farmland throughout the post-Second World War period up to today. This system in Namibia was not just legal but was also encouraged by the governmental Division of Nature Conservation, whose archival collection was referenced above. If historians of nature conservation in Namibia focus only on official, state-sanctioned endeavours, they get but a tiny snapshot of what eventually became a massive industry. As an isolated example from 1988, game utilisation (broadly defined) brought the Namibian state less than two per cent of that which the private sector reaped.[32]

After the moment of independence and the end of apartheid-era sanctions in 1990, the profitability and therefore attraction of private nature conservation – whether for trophy-hunting, photographic safaris or forms of ecotourism – would only grow, and it became an avenue for White farmers from apartheid-era families to shore up their land tenure in a new era of multiracial democracy. Private enterprise also underwrote the emergence of an international elite who sought to treat nature conservation as 'connoisseur conservation', as sociologist Justin Farrell terms it, whereby, for the ultra-wealthy, 'nature provides a special dispensation for purchases and practices that may otherwise be viewed as morally suspect, opulent, or greedy. Nature is priceless, but priceless experiences can be quite expensive'.[33] In this model, nature conservation businesses need not be profitable to be sustainable, for they are supplemented with funds reaped from other sectors.

None of the landscapes described in this book fall within, or border on, any national park in the Republic of Namibia, which means that we cannot easily fall back on the state's archival collections detailing conservation operations. Apart from the Orange River itself, *all* Namibian lands described here are surveyed, demarcated, privately owned commercial farmland, or else remnants of Bondelswarts communal areas boxed-in by White-owned farms surrounding them on all sides. To understand the history of this area in a holistic sense requires one to research the history of private lands, private individuals and, *crucially*, that of private companies.

This is a most difficult task, and it requires exhaustive excavation of governmental records from the National Archives of Namibia (NAN), the National Library of Namibia (NLN), the National Archives of South Africa (NASA), and the Western Cape Archives and Records Service (TBK). These records include

32 National Library of Namibia (hereafter, NLN) TXX 0742: Department of Agriculture and Nature Conservation: Annual Report, 1988, p. 8.

33 J. Farrell, *Billionaire Wilderness: The Ultra-Wealthy and the Remaking of the American West* (Princeton, Princeton University Press, 2020), pp. 100–101.

applications to purchase state-owned farmland, financial subsidies allotted to White farmers during the apartheid period, mining licences and returns, nature conservation and riparian policies, as well as labour affairs. While there are few collections that explicitly deal with the lands under question, it is possible for the historian to pull together threads from many varied collections, eventually weaving those threads into a composite historical garment.

Outside of national governmental collections, we obtained additional historical and contemporary documents from other public repositories and records centres, such as the archives of the Ministry of Mines and Energy (MME), the Registrar of the High Court of the Republic of Namibia (RoHC), the Windhoek Deeds Office Archives at the Ministry of Agriculture, Water and Land Reform (WDO), and the archives of the Business and Intellectual Property Authority of the Republic of Namibia (BIPA). Records from the MME vary from mine inspection notes, annual reports and exclusive prospecting licence applications to environmental impact assessments, dating back to the 1960s. RoHC records concern ongoing and past court cases, company liquidations and occasionally police investigations. WDO files primarily comprise title deeds of farms, certificates of transfer and other documents relating to land tenure and land sale. BIPA records include company filings, such as articles of association, annual filings and director and/or shareholder data, among others. Virtually none of these collections have been previously consulted by historians.

Finally, additional documents were obtained from private institutional collections, such as the Namibia Scientific Society (NSS), the Erfdeel Argief en Kultuursentrum (EAKS), the Legal Assistance Centre (LAC) in Windhoek, and collections belonging to private individuals in Namibia and South Africa. We also sourced corporate documents by way of stock-exchange filings, public annual reports and press releases, and others. We complemented our archival and documentary research with several dozen formal and semi-formal interviews, and informal conversations with key individuals involved with these lands, currently and in the past. Most of these interviews were conducted in person, others were done over the phone, and a few were via email communication.[34]

34 Some, such as the Bassons, repeatedly requested to be named and identified, because they recognise the need for specificity in understanding the region. Across the board, interviews were *not* conducted on the basis that anonymity would be guaranteed. Furthermore, anonymity, confidentiality and discretion are not necessarily relevant when research is in the public interest, as this project is. In addition, anonymity for individuals who hold prominent and/or elite positions in companies, government or society compromises the quality of research and the very point of anonymity in the first place. These practices are in line with codes of journalism and research ethics published within

Researching private nature conservation demands that we remember that the use of these properties for broadly defined 'conservation' purposes cannot be taken for granted and certainly was not inevitable. The following chapters show that the farms under consideration each have had many land use options open to them, from the most to the least 'sustainable', which include: communal subsistence stock-farming, capitalist commercial stock-farming, riverside irrigation, mining, hunting, photographic tourism, luxury conspicuous consumption or even straightforward land speculation. Often, several of these options occurred simultaneously.

In short, specifically targeted chapters, this book explores the background of one of the largest private conservation schemes currently in development on the African continent: the Orange River-Karoo Conservation Area (ORKCA), which mainly comprises lands belonging to two Namibia-registered companies: Sandfontein Lodge & Nature Reserve (Pty) Ltd and Pelladrift (Edms) Bpk. We explore land use in this area from a historical standpoint, examining ways in which the properties were used as ancestral transhumance lands by pastoralists like the Bassons, eventually changing under colonialism to become massive karakul sheep ranches, producing lambskin pelts that are sold to overseas fur merchants. We look at the long history of mining and prospecting in the region, observing how this remote corner of Namibia was intricately bound to global technological shifts in manufacturing and in the production of weapons of mass destruction. We examine the history of human interactions with wildlife in these areas, taking game-farming into account as well as subsistence and commercial hunting.

We also explore the foundations of post-apartheid nature conservation in southern Africa. We look at the networks of individuals and ideas that shaped the emergence of a broadly defined 'ecotourism' sector within the Namibian economy, and we consider the ethical and financial motivations of those involved. We examine, too, the changing ways in which nature conservation is

---

the Republic of Namibia, as well as by international bodies. See Namibia Press Agency, *NAMPA Editorial Guide and Style Book* (Windhoek, John Meinert, 2015), pp. 40–41. See also, New Zealand Media Council, *Statement of Principles*, paras. 8–9. Agence France Presse, *AFP Editorial Standards and Best Practices* (22 June 2016), p. 9. C. le Roux, 'Oral History Research Ethics: Should Anonymity and Confidentiality Issues be Dealt with on their Own Merit?' *Africa Education Review*, 12, 4 (2015), pp. 552–566. C.H. Ellersgaard et al., 'Say My Name? Anonymity or not in Elite Interviewing', *International Journal of Social Research Methodology*, 5 (2022), pp. 673–686. Quotations from interviews – or translation thereof – maintain the meaning of the interviewee, and precise phrasing is maintained as close as possible to the original. Copies of all archival documentation and notes from all interviews are in the possession of the authors, and they can be shared if required.

being financed, especially the role of the voluntary carbon offset market and the increasingly corporate-aligned manner in which conservationists approach funding ideology. Finally, we discuss the legal, ideological and practical ways in which dispossessed people – like the Bassons – are struggling to hold on to what lands they still have.

Ultimately, through deep biographies of people and place, we unmask the process of capital accumulation over the past two centuries and the gradual consolidation of land ownership near the Orange River into fewer and fewer hands. The narrative clearly reveals that there was nothing inevitable about these lands being used for conservation purposes. Only after more than a century of rural enclosure and colonial capitalism does conservation start to gain traction. This story does not begin as a history of nature conservation, rather it becomes one.

A thread running through this narrative is the ways in which the Bassons and other pastoralists and river-folk in southern Namibia managed to withstand the privatisation of their ancestral lands, whether to syndicates, missionary societies, private White farmers, mining companies or private conservationists. Only in facing the last have the Bassons struggled to maintain their residence along the river. Only fortress conservation allows no space for African pastoralists seeking to maintain a connection with their ancestral lands.

Acknowledging this does not mean romanticising the relationship between the Bondelswarts and White farmers in the area during colonial and apartheid periods: Willem Basson surely does not. Remembering the head of a prominent White farming family, Gerrit Luttig, he said, 'That man was an apartheid fanatic. If he saw us anywhere near Pelgrimsrust Farm, he would seize our animals and force us to pay *weifooie en skuldgeld* [grazing fees and penalties]'. Nevertheless, Whites regularly needed labourers, and if one farmer was not willing to have tenant residents or so-called 'squatters' on his property, another likely would. Hence the Bassons' gradual migration up and down the Orange River, carefully observing where they were welcome and where they were not.

For many people, White farmers in post-apartheid Namibia are *personae non gratae*, a legacy of the old regime quietly clinging to racist views and apartheid-era privileges. The low wages and atrocious working conditions of employees of White farmers are rarely looked upon with sympathetic eyes by those outside the White farming community itself. Landowners like Gerrit Luttig may have been quite wealthy men by the coming of Namibia's independence, but they lacked the *moral capital* to justify their treatment of Black Namibians, their relationship to the land and their position in society, especially when observed from cosmopolitan cities and the international community. In an era in which the majority of Black Namibians lack access to agricultural land while food shortages and hunger abound, and in which new political parties like the Landless People's Movement (LPM) win most constituencies south of Windhoek, the expansion of White land ownership is no longer fashionable nor accepted in post-apartheid Namibia. That is, unless you are a 'conservationist'.

In her ethnographic account of White property-owners in post-Fast Track Land Reform Zimbabwe, Yuka Suzuki observes that through transforming their businesses from cattle and crop production into game-farming, safari tourism and trophy-hunting, Whites in Zimbabwe have exploited their alleged custodianship of nature to justify their place in society.[35] Land ownership by Whites is thereby discursively depoliticised as they reposition themselves away from being capitalist farmers towards being private environmentalists. Property rights move from one of verifiable colonial inheritance to one of perceived technocratic expertise. Under private conservation, White landowners in post-colonial and post-apartheid southern Africa regain the moral capital they had lost through their dispossession and exploitation of Black citizens. It would be a great controversy for a farmer to evict a Black family like the Bassons, but for conservationists to do so, clothing their actions in vague references alluding to the protection of endangered wildlife, it might attract little to no attention at all.

The ultra-wealthy capitalist 'conservationists' who have purchased these properties now exercise complete control. Their massive wealth allows them to transform the land as they please, with the law behind them, justifying their purchases, their land tenure and their ability to use the land as they wish. Finally, as 'conservationists', they now have the moral capital that enables them to remove more than 150,000 hectares from agricultural production and transform this area into a hyper-luxury 'conservation' operation. Compared to the power relations under White farmers, there has been a distinct shift under the regime of conservation. Historically, the scales rarely tipped in favour of people like the Bassons, but today they have tipped completely to their detriment. Whereas the river-folk families could once bargain with individual White farmers during the apartheid era, under the regime of private conservation – where ORKCA is directed by influential global capitalists – there is little place for the ancestral rights of the Bassons today.

•••

35 Y. Suzuki, *The Nature of Whiteness: Race, Animals, and Nation in Zimbabwe* (Seattle, University of Washington Press, 2017). For a similar observation about Whites in Kenya, see G.R. Fox, 'The 2017 Shooting of Kuki Gallmann and the Politics of Conservation in Northern Kenya', *African Studies Review*, 61, 2 (2018), pp. 210–236.

When looking at the *longue durée* history of nature conservation in Namibia and Africa more broadly, the trends and the transformations we observe do not actually reflect the most efficient or effective means of protecting wildlife and nature. Rather, these trends reflect transformations in broader economic and political foundations. Under neoliberalism, we can see how the state has tentatively relinquished its place in the driver's seat of conservation practice. The private sector – both for-profit and charitable – has taken the reins. Although the government of the Republic of Namibia is not planning to sell off Etosha National Park or any of its nature conservation areas, policy since independence has facilitated the private sector to operate alongside the state and in spite of the state. Tourism concessions within national parks are granted to private companies, and some fees make their way to the state. Endangered species belonging to the government, like black rhinoceros, are allocated to private landowners on a custodianship basis, allowing the latter to profit from the game's residency and use.

In this book, through our sensitivity and attention to time, space, species and place, we historicise private nature conservation in Namibia. We show that while the state custodianship model is not dead, the state increasingly cedes ground to a corporate model or towards private initiatives by the ultra-rich for whom ownership of conservation land is more a form of 'connoisseur conservation' and conspicuous consumption. Furthermore, we show that despite the private nature of these presently operating and planned schemes, the history of the state is still present through the longevity of apartheid-era land-tenure laws and practices.

Nature conservation is in itself a claim about the past, the present and the future. Conservationists justify their presence and their actions based on a particular interpretation of the past ways in which humans have used and interacted with the natural environment. This interpretation is often formulated through simplistic binaries of degradation versus regeneration, and conservationists regard themselves as agents who provide a return to a particular kind of 'healthy' environment. From youths doing a 'gap year' volunteer project during which they rip up farm fencing, to multimillionaires buying up land for 'conservation initiatives', they all regard themselves as champions of regeneration, or rewilding, broadly-defined. When they declare that they believe in this binary and in the validity of their restorative efforts, we are expected to take what they say at face value. This binary and their participation are often framed in catastrophic terms, as though the work that they are doing is one of the last defences preventing a global rise of another one or two degrees Celsius.

For many individuals who are active in this sector, commitment is ideological, scientific and personal, which is part of what makes criticism of certain nature conservation initiatives so difficult to make. For its benefactors, employees and supporters, criticism of ORKCA's operations implies either that the critic does not believe in the scientific reality of climate change or else that that critic represents the interests of environmentally destructive fossil fuel and mining companies. (Ironically, these are some of the main funding bodies of ORKCA-aligned carbon credit operations). This short-sighted viewpoint has made dialogue about the human impacts of conservation schemes almost impossible, because they devolve into *ad hominem* attacks and feelings of personal rejection, as though the conservationists' 'hard work' is neither valued nor understood.

As we show in this book, when nature conservationists make claims about the past, present or future, we as historians must interrogate these claims, expand on their implications and challenge them when necessary. *Conservation* itself is not actually the problem here. Protecting nature and working to slow, reverse and ultimately cease the colossal damage that we humans have done to our planet is of utmost importance and a noble cause. The problem with private nature conservation – especially the Namibian cases described here – is that the individuals and institutions involved refuse to challenge or work outside of the two systems that created most of our present environmental chaos: capitalism and colonialism. By simultaneously ignoring, perpetuating and honouring these destructive systems, private nature conservation reduces itself to being less about *nature conservation* and more about being *private*.

Historians have a role to play in framing how we understand the present and the future, and not just how we view the past. In a postcolonial, post-apartheid country such as Namibia, the present is intrinsically linked to the past. Merely abolishing apartheid legislation does not automatically democratise the economy or society. We are not the first to argue and demonstrate how contemporary conservation schemes like ORKCA reinforce colonial or apartheid economic and social structures, and we will certainly not be the last. In the chapters and the stories that follow, we transport the reader to southern Namibia and beyond, illustrating the ways in which the past lives on in the present. As long as a system exists, which allows a small number of individuals to buy up massive tracts of land and unilaterally determine the exclusive use of whole landscapes, there is little possibility of social and ecological sustainability.

CHAPTER 2

Cessna Aeroplanes over the Hom River

On 29 December 2001, Sean Gilbertson and Finn Behnken departed Upington in two Cessna 172 Skyhawk aeroplanes, aiming for Keetmanshoop, Namibia.[1] The planes were registered with flight schools in South Africa and both men were keen amateur pilots. Gilbertson and Behnken were both twenty-nine years old and fast-rising within the field of mining finance, though both had actually studied mining engineering at the University of the Witwatersrand in the early 1990s. Their planes were loaded with cameras, supplies, friends, girlfriends, beer and brandy, and they carried the gate key to Gilbertson's recently acquired property – Norechab Farm in the far south of Namibia's Karasburg District. It would be the first of many properties he would buy in this region. From Keetmanshoop, the group refuelled their Cessnas and flew the remaining three hundred kilometres or so to an airstrip at Tantalite Valley mine, located between neighbouring farms Umeis and Kinderzitt. From there, the group took a short drive north-west to Norechab Farm.

Sean was born in South Africa, the son of mining magnate Brian Gilbertson, and his youth was impacted by his father's rise within apartheid South Africa's large, yet politically isolated, mining sector. Brian Gilbertson had graduated with a degree in maths and physics in the mid-1960s, at a time when apartheid-era military conscription was escalating. 'I was given the choice of going to boot camp on the South African veld, or to Paris to work on a missile system. It was a tough decision', Gilbertson said, 'but I chose the lights of Paris'.[2] The Council for Scientific and Industrial Research (CSIR) posted him with the French Institute for Rocket Research, where the two governments were engaged in the development of a surface-to-air missile system. France was one of the few countries still selling arms to apartheid South Africa.[3] In 1970, Brian was hired as a mining engineer for the Johannesburg Consolidated Investment Company (JCI Ltd), which was mining a large number of gold, uranium and platinum deposits in today's Gauteng province. JCI also owned significant shares in De

1 Unless mentioned otherwise, information relating to personal visits to Norechab is sourced from www.norechab.com (Accessed 16 May 2024).

2 Quoted in 'Brian Gilbertson: The Industry's Godfather', *Global Mining Observer*, 109 (2014).

3 P. Thompson and R. Macklin, *The Big Fella: The Rise and Rise of BHP-Billiton* (Sydney, William Heinemann, 2009), p. 233.

 | DOI:10.1163/9789004712454_003

Beers.[4] During Brian Gilbertson's employment with the company, JCI opened South Deep gold mine in Gauteng, which grew to become one of the world's largest gold mines.

In 1988, Brian Gilbertson was hired by the mining behemoth, Gencor Ltd, and was eventually promoted to the position of executive chairman. Gilbertson was a different kind of mining titan to most others in South African history. Though he certainly possessed the knowledge to focus on improving production and mining strategies, his legacy is that of a globally-minded dealmaker and negotiator. Gilbertson downsized Gencor and sold off or unbundled unproductive assets, and after South Africa's first democratic elections in 1994, Gencor acquired the aluminium assets of Royal Dutch Shell, as well those of its subsidiary, Billiton.[5]

Although Brian Gilbertson presided over one of South Africa's largest mining companies – listed on the Johannesburg, London, Paris, Frankfurt and Zürich stock exchanges – he apparently did not view Gencor as an efficient or attractive global company. Though Gencor did control a large number of profitable, attractive mines, it was also holding many unproductive, low-grade gold shafts, which had been profitable only under the conditions of apartheid-era labour relations.[6] These operations would not survive, for the cost of sinking these shafts deeper was unattractive to international investors looking to buy shares in a global conglomerate. In order to make the company more 'efficient' and investor-friendly, in 1997 Gilbertson split the company's assets: aluminium and non-precious metals were placed under Billiton Plc (now a British company), while gold, platinum and other precious metals remained under Gencor.[7] Gilbertson announced to the attendees of the grand launch of the new mining arrangement:

> If I want to expand bauxite production in Western Australia to feed a new smelter I want to build in Mozambique, or if I want to be a serious contender for the privatisation of Venezuela's aluminium operations, I need access to international capital markets ... We have no intention of acting as nurse maids to dying operations. We intend this to be a world-class business, and [we] will seek out opportunities.[8]

---

4 B. Sergeant, *Brett Kebble: The Inside Story* (Cape Town, Struik, 2006).

5 'Brian Gilbertson: The Industry's Godfather', *Global Mining Observer* 109 (2014).

6 See also, J. McCulloch, *South Africa's Gold Mines and the Politics of Silicosis* (New York, James Currey, 2012).

7 J. Walker, 'Gencor Pops Champagne on Achieving its Ambitious Dream', *Business Times* (22 June 1997).

8 Ibid.

For the large mining companies in South Africa, the benefits of opening up South Africa to the world may not necessarily have corresponded with the democratic or human-rights goals of the anti-apartheid movement. It may have meant taking the mining houses of the Rand and transforming their expertise, assets and capital into a global concern – with global investment potential. Gilbertson's dealmaking was fairly profitable. When Gencor was absorbed into Gold Fields Ltd in 1998, the latter became the second-largest gold producer globally.[9] After assisting in the merger of Billiton and the Australian giant Broken Hill Proprietary Ltd, Brian Gilbertson would eventually be appointed CEO of the unified BHP-Billiton in 2002. The conglomerate became one of the largest mining companies in the world.[10]

These years were a time when the public face of mining executives was changing, from providing an efficient means to remove minerals from the earth to one of global jet-setting, mergers, acquisitions and complex financial instruments. In her book, *Liquidated*, anthropologist Karen Ho posits that deregulation and economic transformations during the 1980s and 1990s brought about a 'shareholder value revolution' within the modern corporate world.[11] This represented a new conception of *efficiency*. It was no longer grounded in labour or industrial productivity but rather in the quantity and magnitude of 'deals' which would reap the largest short-term dividends for shareholders, even if a short-term rise in stock price did not correspond with long-term corporate growth. The 1990s witnessed the departure of the paternalistic 'welfare capitalism' that was common during the mid-century and the arrival of the notion of a corporation as 'synonymous with its shareholders'.[12] For Wall Street and mining finance alike, labour redundancies and the local effects of liquidation were no longer a concern for a company or its executives. Although Sean's father may have helped found a mining empire, it was not an empire bound any longer to any geographical space, nation-state, particular commodity or social or community obligations. Brian Gilbertson's deals brought profits and

9 Gold Fields, 'Our History, 1998–2003' https://www.goldfields.com/1998-2003.php

10 Brian Gilbertson would resign in 2003 after his dealmaking went a bit too far, and the BHP-Billiton board rejected his USD 100-billion proposed merger with Rio Tinto (see chapter 9). 'BHP Chief in Shock Resignation', *CNN* (5 January 2003). 'Brian Gilbertson: The Industry's Godfather', *Global Mining Observer* 109 (2014). Sean T. Gilbertson, personal communication with Bernard C. Moore & Luregn Lenggenhager (email, 28 October 2024).

11 K. Ho, *Liquidated: An Ethnography of Wall Street* (Durham, NC, Duke University Press, 2009).

12 Ibid, p. 209.

dividends, but they also mothballed countless peripheral mining operations deemed unattractive.

Sean was most likely aware that it could be his responsibility to continue the family mining legacy, and he enrolled in 1991 for a degree in Mining Engineering at the University of the Witwatersrand. He moved in circles of young people in similar circumstances, befriending fellow South African Finn Behnken and the Namibian Heye Daun, both of whom were enrolled in a similar programme. Perhaps more significantly, he began dating Sabine Severin, daughter of the Danish mining magnate, Steen Severin.[13] Severin and his wife, Franka, had recently acquired Tantalite Valley mine in southern Namibia, which was situated on Umeis and Kinderzitt farms. The mine was partly closed owing to the low price of tantalum oxide.

Sabine recalled that although the mine itself was inactive, her family would take holidays to Tantalite Valley from the late 1980s to the 1990s, by which time her father had also established a residence there. In the early days, they would stay in a caravan, tents or in the old mine housing. The Severins kept Yamaha motorbikes on the property, using them to explore, let loose and relax. 'It's just the stuff you dream of', she recollected, 'you know, having space, waking up and seeing a klipspringer grazing, and it's on your land!'[14] In 1992, Sabine brought her new boyfriend Sean to Tantalite Valley for a holiday. 'It was my first ever trip to Namibia', Sean recalled, 'and I was blown away – I loved it'.[15] During subsequent trips, they were joined by fellow students, Finn Behnken and Heye Daun, and like many who witness the incontrovertible beauty of southern Namibia – the present authors included – they fell in love with the landscape, and they sought any excuse to travel there as often as possible.

Sean Gilbertson and Sabine Severin stopped dating around 1994, when their studies were finished, but they stayed in touch, meeting up on trips to Tantalite Valley mine with their university mates over the following few years. Sean recollected that he and his friends had a dream of owning land in the area one day, but that day would not come for some time. Sean left Johannesburg in 1995 and took a job with Deutsche Bank in Frankfurt and thereafter with its newly acquired subsidiary, Morgan, Grenfell & Co. At both locations, he worked on project finance relating to power stations and infrastructure, gradually deepening his knowledge and connections in the world of mining finance more

13 Sean T. Gilbertson to Bernard C. Moore (7 January 2023): letter in authors' possession.

14 Sabine Andersen (née Severin), interview with Bernard C. Moore (telephone, 2 January 2023).

15 Sean T. Gilbertson to Bernard C. Moore (7 January 2023): letter in authors' possession.

broadly.[16] Sean seemed to be walking in his father's footsteps, and though he was no longer spending much time in mine shafts anymore he was entering the new age of computerised mining finance, commodity futures trading and all the dealmaking that goes along with them.

Sean moved to London on 8 July 1996, and about two years later he, Sabine and a few other university mates founded the company Global Coal Market Ltd (alias globalCOAL), which sought to build on internet trading to financialise, standardise and popularise the trading of thermal and metallurgical coal futures, as well as to facilitate the delivery of physical shipments.[17] Apart from collecting a significant amount of data on coal transactions around the world, globalCOAL set up a Standard Coal Trading Agreement (SCoTA) which specifies and defines the quality of coal produced at major hubs.[18] This data includes, inter alia, calorific value basis (in kcal/kg), ash content, sulphur content, Hardgrove grindability index and moisture content, geographically divided for each major thermal-coal distribution hubs, like Richard's Bay in South Africa and Puerto Bolivar in Colombia, as well as for metallurgical/coking-coal production in Newcastle, Australia.[19]

As a carbon-based substance, not all coal is the same. Depending on its intended use – even beyond the coking/thermal divide – different qualities are required if coal is to be used for electricity production. The effectiveness of coal cannot be measured merely on the basis of tonnage. In this way, globalCOAL facilitated the exchange of production data, sales data and projections such that the financialisation of the thermal coal sector could mirror that of other commodities, like those on the London Metals Exchange. In collaboration with his Wits University friend, Bevan R. Jones, Sean Gilbertson designed the Web and telephone interface to broker sales of coal consignments, as well as design a standardisation protocol and contract system.[20] In order to do

---

16 Gemfields Resources PLC, 'Letter to Shareholders from the Chairman of Gemfields', in *AIM Readmission Document* (13 May 2008). Sean T. Gilbertson, personal communication with Bernard C. Moore & Luregn Lenggenhager (email, 28 October 2024).

17 UK Company House Archives, Cardiff (hereafter, UKCHA) Company no. 03531881: 'Certificate of Incorporation of a Private Limited Company' – 6 March 1998. Note that in 1999, Global Coal Market Ltd officially changed its name to Spectron Global Coal Ltd after the energy broker Spectron acquired a 50% stake; see UKCHA 03531881: 'Certificate of Incorporation on Change of Name' – 11 August 1999.

18 See the Internet Archive Wayback Machine crawl (hereafter, IAWB) of 'globalCOAL, FAQs' – 8 October 2003. https://web.archive.org/web/20031008135008/http://www.globalcoal.com/general/faq.cfm

19 globalCOAL, 'Standard Coal Trading Agreement: Specifications and Definitions' – as of 9 October 2019. https://www.globalcoal.com/scota/specifications.cfm

20 See Bevan Robert Jones, LinkedIn profile (Accessed 1 February 2023).

this, Gilbertson negotiated through his family and business contacts a series of loan and investment agreements with major coal industry players, such as Anglo American, Billiton, Glencore and the California venture capital firm, Accel-KKR.

The online brokerage portion of the globalCOAL business was bought out by Gilbertson and Jones in July 2000, who transferred its assets to their special-purpose vehicle, Gigajoule Ltd.[21] They quickly sold off the rest of the business to a new company, Global Coal Ltd, in exchange for approximately GBP 1.2 million and – more importantly – a 3 per cent shareholding in the new company.[22] The software was delivered to Accel-KKR's offices in California; Sean used most of the cash to pay back the loans he received from the various global mining houses that had backed his enterprise, and he was soon appointed a director of the new globalCOAL company.[23] Gigajoule effectively became a holding company for the three per cent shareholding in globalCOAL,[24] and Sean eventually bought out Jones's shares.[25]

Sean would also serve as CEO of globalCOAL from 2001 to 2003, increasing his financial interest in the company through a purchase of an additional 60,000 shares by the end of 2001.[26] While his shareholding in globalCOAL paled in comparison to that held by BHP, Anglo American, E.ON Kraftwerke and Enel Group, Sean had established himself as a major player in a major industry – coal. And though his father's connections and name were clearly influential in getting the business off the ground, Sean led this endeavour himself with the new technologies of the twenty-first century.

During his rise within globalCOAL, Sean returned to Tantalite Valley in late 1999 to usher in the new millennium. Steen Severin had just built a large mansion on Umeis Farm, and many of Sabine's university friends were invited to ring in the year 2000 in one of the most isolated and gorgeous corners of the globe. Sean recalled that up to eighty people had accepted the invitation, and the group at Umeis was comprised of both locals and foreigners alike.[27] This

21 UKCHA #03531881: 'Financial Statements for the Year Ending 31 March 2000'.

22 UKCHA #4007764: Gigajoule Ltd & Global Coal Ltd, 'Sale of Business Agreement' – 18 April 2001.

23 UKCHA #4007764: Global Coal Ltd, 'Appointment of Director or Secretary: Sean T. Gilbertson' – 18 April 2001. He had received approximately GBP 1.1 million in loans from Accell-KKR, Anglo American, Billiton, Glencore, Rio Tinto and Global HubCo B.V. [BHP].

24 UKCHA #3936280: Gigajoule Ltd, 'Directors' Report and Financial Statements for the Period Ended 31 July 2001'.

25 UKCHA #3936280: Gigajoule Ltd, 'Annual Return' – 28 February 2012.

26 UKCHA #4007764: Global Coal Ltd, 'Report and Financial Statements, 31 December 2001'.

27 Sean T. Gilbertson to Bernard C. Moore (7 January 2023), letter in authors' possession.

occasion rejuvenated Sean's love for this remote expanse of Namibia, and it was around this time that he learned that a neighbouring property was up for sale. Perhaps he recognised that the dream to own land in this landscape could actually be realised.

Directly north-west of Tantalite Valley mine sits the mountainous Norechab Farm. If one compares it to farmland in more fertile parts of Namibia and South Africa, its 9,844 hectares would be considered massive, but when judged against neighbouring properties in southern Namibia, this size is a little below average. Although it was legally demarcated a 'farm', Norechab was never particularly productive for commercial sheep-farming, because the flats comprise less than half of the property, the greater part being mountains, koppies (hills) and dry riverbeds. Even so, since the Hom River passes through Norechab, over the decades the farm provided supplemental grazing to neighbouring properties during dry years. Sections were frequently grazed as pastoralists and shepherds moved their flocks from farm to farm, or farther afield to Warmbad village, where the animals could be sold or slaughtered. Sean was not interested in land for the purpose of farming, however. He wanted space, he wanted beauty, he wanted silence.

Norechab was being sold by business partners Burger W. Oelofsen, a medical doctor, and Willem F. Mans, who ran a trophy-hunting operation on nearby KumKum Farm, through the sale of their company, Marwilben Farming cc. In Namibia and South Africa, engaging in farm partnerships requires forming either a proprietary company (Pty) or a closed corporation (cc), which would actually own the farm, shares being divided among the investors. This practice became widespread after the passage of the Subdivision of Agricultural Land Act of 1970. Not only could farm parcels no longer be split into smaller sections, but direct individual shareholding of farmland was made illegal as well.[28] Although corporate ownership of farmland made business partnerships possible without subdividing the land itself, it also made tracking farm ownership a remarkably burdensome ordeal. In order to avoid transfer duties on the sales of farms and erven, shares in the company are sold rather than the farm itself being put up for sale. This arrangement allows the records held at the Windhoek Deeds Office to remain unmodified even though a farm may have actually changed hands several times. In those cases, the holding company itself was sold, assets included. In order to track land sales in southern Namibia, one must consult both Deeds Office records and those of the Business and

28 Subdivision of Agricultural Land Act 70 of 1970, *Government Gazette of the Republic of South Africa* (nos. 2867 and 2950). The act became law in Namibia in 1971.

Intellectual Property Authority of Namibia (BIPA), which contains corporate annual filings.

This system of corporate farm ownership is also the primary means by which foreigners purchase farmland in Namibia. With the passage of the Agricultural (Commercial) Land Reform Act of 1995, the Republic of Namibia banned sales of agriculturally zoned land to non-Namibians. It also banned Namibian citizens from acting as 'nominee owners', who would legally stand in for the buyers who actually purchased the land.[29] Corporate farm ownership in the form of cc or Pty companies enables foreigners to own 49 per cent of a property by holding 49 per cent of the shares in a Namibia-registered company. Although all owners must sign an affidavit stating that they are indeed Namibian citizens or representatives of a Namibian company, the corporate structure of farm ownership in Namibia renders it opaque for the public and government officials to ascertain who actually owns a property and whether that person is in fact a 'nominee owner'.[30]

As a dual British and South African citizen, Sean Gilbertson could not purchase Namibian farmland outright. He needed a Namibian business partner, and he subsequently teamed up with his old university acquaintance, Heye Daun, to buy out Mans and Oelofsen's shares, purchasing Marwilben Farming cc (Sean with 49 per cent and Heye with 51 per cent).[31] Like Sean, Heye Daun had graduated with a degree in mining engineering and worked in mining finance, in his case with Old Mutual Investment Group and Nedbank Capital.[32] Like Sean, he was an avid amateur pilot, and now, together with Sean, he owned Norechab Farm in southern Namibia.

With friends Finn Behnken (South African, also in mining finance, who would eventually own 24.5 per cent of Marwilben),[33] Mark Adams (a photographer) and Murli Bhamidipati (an Indian engineer involved in energy and commodity finance in Vienna and Bilbao), Sean founded the Norechab Sand & Tent Club. Once or twice a year for nearly a decade, the team – and their significant others – would take extended holidays in Namibia, often visiting traditional tourist sites such as the Fish River Canyon or Sossusvlei. Inevitably,

29 'Agricultural (Commercial) Land Reform Act 6 of 1995', *Government Gazette of the Republic of Namibia* (no. 1040 of 1995), sections 58–62.

30 Various Namibian attorneys and government officials are aware of, and occasionally investigate, purported efforts by individuals and companies to evade and/or avoid or circumvent land reform legislation. This has been an ongoing problem since the promulgation of the law in 1995.

31 Archives of the Business and Intellectual Property Authority of Namibia (hereafter, BIPA) File CC/1996/0628: Amended Founding Statement (CC-2) – 1 December 2000.

32 Today, he is CEO of the Namibian/Canadian gold mining company, Osino Resources.

33 BIPA File CC/1996/0628: Amended Founding Statement (CC-2) – 20 March 2007.

the trips would culminate in landing their Cessna aircraft onto the airstrip at Tantalite Valley and driving Toyota Hiluxes to Norechab.

In the early days, the Norechab Sand & Tent Club was in all likelihood an ephemeral escape from the stress of corporate mining finance, a sector in which nearly all club members were intimately engaged. The club members publicly portrayed Norechab as a place to camp, to ride motorbikes, to relax, drink and reconnect with old friends. They exchanged photos and stories on blog websites, even purchasing the domain name www.norechab.com for their website where they could muse on their journeys and plan future ones (see Figure 11). To Sean, Finn, Heye and their fellow club members, Norechab was an exotic, alluring place. They would comment on 'strange vegetation' common on the farm, such as the giant quiver tree (*Aloidendron pillansii*), or they would set up tripods to catch the setting sun or take long-exposure photos of the stars. None of these actions is surprising. Norechab – like the entire southern Karasburg District – is a striking and beautiful place.

Of course, like any boys' camping club, one of the favoured pastimes was evening drinking parties – 'sundowners' in sophisticated parlance. Daytime hiking trips inevitably led to finding new and unique drinking spots on the

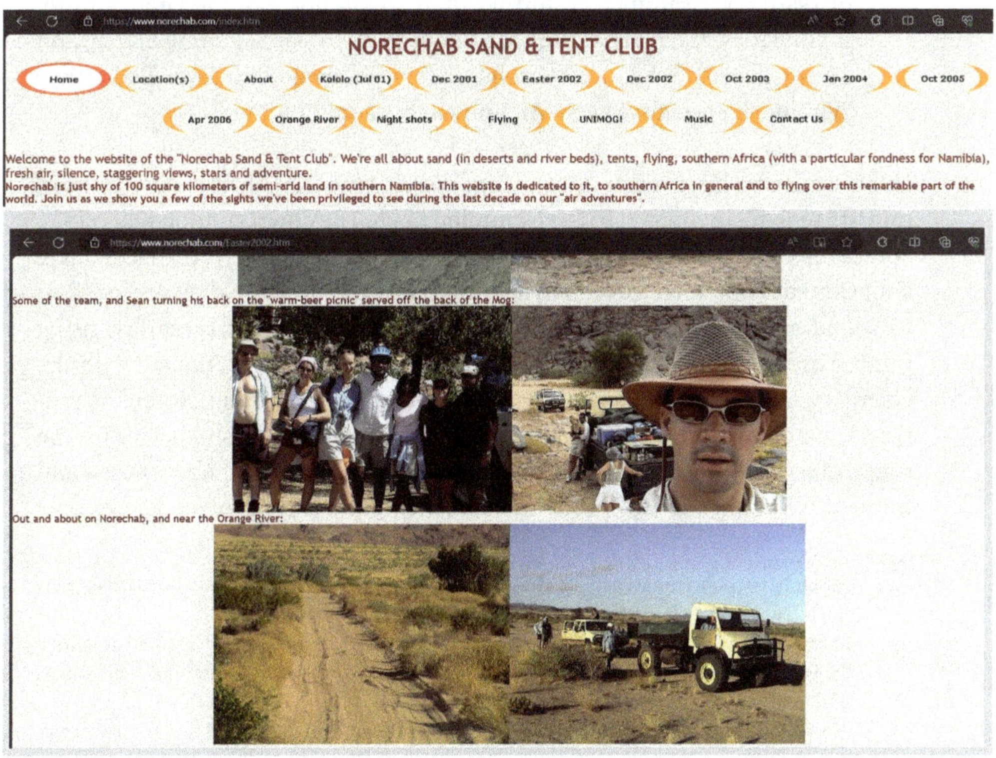

FIGURE 11 Excerpts from the Norechab Sand & Tent Club website

farm. 'Murli's Pub' was a small rock formation and occasional waterfall in the southernmost corner of the property, at the confluence of the ephemeral Hom River and a mountain stream from the Tantalite Valley mine. 'Grappa Hill' (listed on cadastral maps as Haanberg) was the favourite spot, however, named in homage to their much-adored Italian brandy, brought into Namibia by the case. Sean hoped one day to build a residence on Grappa Hill,[34] where they could look down upon the Hom River and, as he would later declare, 'listen to the silence of the wind'.[35] When a 'warm-beer picnic' was scheduled, one could simply lower the tailgate of Norechab's resident Mercedes Unimog and dig in. For these fast-rising executives within the field of mining finance, they had a regular holiday destination in the form of their huge farm in Namibia, and they were going to make the most of it.

On the club's website, these global jet-setters, confidently shaping a new era in mining – summarised Norechab's early purpose.

> Space is the ultimate luxury. No human beings in a 50 km radius. No electricity. No telephones. No clouds. Warm winds in your ears. Norechab covers 10 thousand hectares in a shape roughly 15 km by 7 km. It is located south of Warmbad, Namibia, no further than 20 km north of the Orange River. It's dry; it's stark; it's beautiful. It belongs to Heye Daun, Finn Behnken, and Sean Gilbertson. Silence: Absolute. Stars: Trillions. Dry River Beds: Two. Windhoek Specials (beers). An unbeatable combination.[36]

Gilbertson and his friends are not middle-class South African or European tourists seeking to rough it in the Namibian bush. They own London real estate, companies and other high-value assets. The members of the Norechab Sand & Tent Club did not need to sleep in a tent pitched on the sand; they could afford to spend a month in Namibia's most luxurious lodges. For these fast-rising mining executives, it was not *luxury* per se which was sought, but *space* in the clearly stated context of 'Space is the ultimate luxury'. For them, Norechab represented perhaps the only experience they could not simply purchase with the swipe of a credit card. Norechab is far; Norechab is dry; Norechab is stark – and most of all, *Norechab is theirs*. Norechab was their private place where they

34 IAWB: 'Norechab: Home Page' – 24 June 2002. https://web.archive.org/web/20020624033823/http://www.norechab.com

35 R. Swilling, 'Sandfontein Lodge: Far South of Namibia', Travel News Namibia (undated, likely 2012). https://www.travelnewsnamibia.com/news/stories/destinations/sandfontein-lodge-far-south-of-namibia/

36 'Norechab: About' – http://www.norechab.com/about.htm (Accessed 2 February 2023), italics ours.

could blast Sting albums on cassette on the Land Rover's stereo, where they could sit on Grappa Hill and drink brandy to their heart's content, away from people, away from everything.

When Sean Gilbertson and his mates purchased Norechab, they had little intention of purchasing more land. For the time being, they had enough space.[37] Pleasure projects rarely remain so benign, however. Friends drift apart; friends get married and move farther away. As the years went on, the Norechab Sand & Tent Club spent less and less time camping in Norechab's bush and more time in conventionally luxurious spots. The team's October 2003 journey transported eleven passengers spread across three Cessna planes; only two nights of the itinerary were spent at Norechab.[38] Swakopmund hotels and Sossusvlei lodges were the preferred accommodation. By April 2006, the club's annual tour was taken solely by Behnken and his partner.

Towards the end of Sean's tenure as CEO of globalCOAL, the general public was becoming increasingly aware of the dramatic future challenges caused by climate change, such that pressure on coal and other fossil fuel companies to change grew. It was impossible for these companies – and their representatives, like Sean – to hide from global warming, pollution and the long-term effects of fossil fuel emissions. At this time, Sean likely understood that the status quo with which he had grown comfortable was not sustainable, and steps would need to be taken to 'rein in' CO_2's impact. On the other hand, as a leading figure in the coal industry and heir of mining magnate Brian Gilbertson, Sean clearly moulded his individual concerns about climate change to fit within an overarching worldview of market efficiencies. In response to early discussions about carbon taxes, Sean wrote, in his capacity as CEO of globalCOAL, to the editor of *The Economist* in 2002:

> You suggest that the introduction of carbon taxes by governments is a plausible solution to the problem of CO_2. Governments should legislate and force producers to use, store, or be penalised for their CO_2, but transparent markets should determine prices. Producers should be free to choose how they pay for CO_2 creation, whether by storage, technological advancement, consumption (such as 'hiring' forests) or the purchase of emissions credits. This would give rise to a competitive industry that

37 Sean T. Gilbertson to Bernard C. Moore (7 January 2023), letter in authors' possession.

38 'Norechab: The Twin Rivers Trip (Oct 2003)', http://www.norechab.com/oct2003.htm (accessed 2 February 2023).

> determines accurate prices for the most efficient and cost-effective manner of reining in CO_2's environmental threat.[39]

Sean ultimately believed that problems involving climate change, CO_2 emissions and nature conservation could be solved through market forces and, as he implied, enlightened 'consumption'.

In his study of Teton County, Wyoming, the sociologist Justin Farrell argued that the ultrawealthy often engage in and use conservation practices (broadly and vaguely defined by them) as 'a tool to recapture something they've lost on their morally treacherous march toward economic triumph'.[40] If their wealth had been accumulated along an ecologically questionable pathway, they might seek to use their renewed relationship with nature to cleanse their conscience and greenwash their wallet. While Sean's initial purchase of Norechab had little to do with altruistic conservationism – or even conservation at all – this motivation would change over time. And though the Club's mantra, 'Space is the Ultimate Luxury', would never fade, the reason behind obtaining *space* and the justification for experiencing *luxury* would change over the coming years.

39 Sean Gilbertson, letter to the editor of *The Economist* (3 August 2002).

40 J. Farrell, *Billionaire Wilderness: The Ultra-Wealthy and the Remaking of the American West* (Princeton, 2020), p. 78.

PART 2

Terra Nullius?

CHAPTER 3

Of Syndicates and Surveyors

Sean Gilbertson and his Norechab Sand & Tent Club were not really entering lands which they had repeatedly described in language evocative of *terra nullius* [empty land]. After all, Norechab was technically a surveyed and demarcated commercial farm when they purchased it. Furthermore, since precolonial times, south-eastern Namibia's Karasburg/Warmbad District has been the ancestral lands of the Bondelswarts, whose captaincy is headquartered in Warmbad village. Like most Nama groups, the Bondelswarts have existed as transhumant pastoralists, moving with their flocks of sheep – and to a lesser extent, cattle – between sites of grazing and permanent water sources, such as along the Hom Riverbed from the Orange River towards Warmbad's hot springs.[1] During the nineteenth century, Warmbad was a busy, cosmopolitan place because it was one of the few secure water sources in this vast, dry area. The hot springs flow even in winter during drought years, and they proved to be a lifeline for all in this area. Warmbad was also the site of some of the first permanent missionaries and traders in southern Namibia, beginning with Pastor Josiah Nisbeth of the Wesleyan Missionary Society, in 1834; the Rhenish and Catholics soon followed – the latter becoming the most numerous by the twentieth century.[2]

Although the Bondelswarts and their lands were not annexed by the Cape government, they were certainly within its sphere of influence. Magistrate J.T. Eustace of Springbok regularly paid Bondelswart Kaptein Willem Christiaan in cash and ammunition in order to keep the peace along the northern border with the Cape Colony. This started in 1876 and lasted until 1890, shortly after the Bondelswarts signed a 'protection treaty' with the Imperial German government.[3] During his 1883 meeting with Christiaan, Eustace revealed the power politics present in the existence of the salary: the magistrate expected

1 B. Lau, *Southern and Central Namibia in Jonker Afrikaner's Time* (Windhoek, National Archives of Namibia, 1987).

2 Archives of the Namibia Scientific Society (hereafter, NSS): A. van der Colf, 'Stigting en vroeë geskiedenis' in *Historiese Oorsig van Karasburg* (1956). E.L.P. Stals, *Môrewind oor die Karasberge: 'n Kultuurhistoriese Verkenning van die Karasstreek van die laat Negentiende Eeu* (Pretoria, Protea Boekhuis, 2009).

3 H. Jäckel, *Die Landgesellschaften in den deutschen Schutzgebieten* (Jena, Gustav Fischer Verlag, 1909), pp. 70–71.

FIGURE 12 Grasses cover the hot springs flood zone in the otherwise arid Warmbad landscape
PHOTO: B.C. MOORE, 2015

FIGURE 13 Kharaskhoma Syndicate members (Willem Christiaan seated), early 1895, prior to reformulation as SATCO

that, if needed, the Bondelswarts would fight with the English/South Africans against the Germans.[4]

In 1889, before the Germans showed much interest in effectively colonising south-eastern Namibia, a representative of the Cape Town company Kharaskhoma Exploring and Prospecting Syndicate Ltd came to Warmbad. As the agent for the syndicate, Theophilus Hahn was tasked with securing mining, land and railroad concessions in the region.[5] Hahn was born to Protestant missionaries in Namibia, and he could speak Khoekhoegowab well. He was also an acquaintance of the local Rhenish missionary in Warmbad, Rev. Carl Wandres. Hahn took advantage of Kaptein Willem Christiaan's alleged weakness for Cape brandy and the Reverend's alleged weakness for cash bribes, and he thus managed to coerce the Kaptein to sign a land deal with Kharaskhoma.[6] In exchange for some share capital, seven per cent of all mining revenue and a share of land sale revenue, the Bondelswarts kaptein allegedly granted Kharaskhoma a massive concession area, covering nearly the entirety of contemporary Karasburg and Keetmanshoop districts.[7] The fraudulent concession zone measured more than 60,000 square kilometres.[8]

After the 1890 Heligoland-Zanzibar treaty with the British government, the Germans had de jure control over south-eastern Namibia. The Imperial government officially recognised the concession rights of the Kharaskhoma Syndicate in 1892, leading to Kharaskhoma's decision to restructure and reformulate itself into the South African Territories Company, Ltd. (SATCO). All shares belonging to the Bondelswarts were – with permission from the Imperial government – cancelled.[9] By 1900, when SATCO was finally taking steps to sell off some land within their concession zone, it became clear that they were no longer following initial agreements with the Bondelswarts. Instead of

4 NAUK CO 879/21: J.T. Eustace, Resident Magistrate at Namaqualand to Hon. John X Merriman – 26 December 1883.

5 Bundesarchiv Berlin-Lichterfelde (hereafter, BArch) R1001/1562: Memorandum and Articles of Association of the Kharaskhoma Exploring and Prospecting Syndicate, Ltd. (5 March 1889).

6 H.H. Hahn, 'Die Rol van Dr J. Theophilus Hahn in Suider-Afrika' (MA thesis, Stellenbosch, 1993), pp. 189–190.

7 BArch R1001/1564: Articles of Agreement between Willem Christiaan and the Kharaskhoma Syndicate – 10 October 1889. BArch R1001/1564: Kharaskhoma Freehold Land Concession (Bondelswarts Territory) – 8 April 1890.

8 Jäckel, *Die Landgesellschaften*, p. 72.

9 R.A. Voeltz, *German Colonialism and the South West Africa Company, 1884–1914* (Athens, Ohio University Press, 1988).

cooperating in choosing farms together for sale, SATCO targeted farms with existing Nama water sources.[10]

South-eastern Namibia was not a priority for settlement by the German government at that stage, nor did SATCO have sufficient capital to drill the boreholes necessary to make their farms desirable to sell. There was not yet a railway, and road networks were poor. By the time that Africans took up arms against the German colonisers in November 1903, only seven farms had actually been sold by the company, though another eighteen were under lease. Nevertheless, it was clear to the Bondelswarts that neither SATCO nor the German government had their best interests in mind. In Warmbad District, the most intense fighting took place in the second half of 1905 and the first half of 1906. German war graveyards lie near many permanent waterholes belonging to the Bondelswarts, such as at Warmbad, ǂGâbes and Churutabis. Some fighting even took place on drier plains and in the mountains, such as the 14 February 1906 battle of Norechab (won by the Bondelswarts).[11] It is estimated that during the war the German Schutztruppe killed about one third of the Nama population, though that number could be larger if we include forced exiles.[12] Many Nama fled to South Africa or to Botswana to escape a war long-regarded as genocide.

According to the 1906 'Treaty of Submission', the Bondelswarts were allowed to retain their two communal reserves in the district, though these were greatly reduced in size. They were forbidden from owning cattle, guns or ammunition.[13] Furthermore, their livestock had been reduced to such small numbers during the war that rebuilding herds of sheep or goats on such a small amount of land would be nearly impossible. That was not the concern of the German government, however, and it quickly took steps to sell off more land in Karasburg/Warmbad District. Although the majority of farms in the district had not been sold off by the time the Germans surrendered Namibia after the First

10 B.T.O. Mokopakgosi, 'German Colonialism in Microcosm: A Study of the Role of Concessionaire Companies in the Development of German Colonialism' (PhD thesis, SOAS University of London, 1988), p. 280.

11 For more on the genocidal war against the Nama, see W. Hillebrecht, 'The Nama and the War in the South', in *Genocide in German South West Africa: The Colonial War of 1904–1908 in Namibia and its Aftermath* (Monmouth, Merlin Press, 2004), pp. 143–158. See also, M. Biwa, '"Weaving the Past with Threads of Memory": Narratives and Commemorations of the Colonial War in Southern Namibia' (PhD Thesis, University of the Western Cape, 2012).

12 Marion Wallace's survey of the statistics about the genocide is very useful because it elaborates on some of the methods that various scholars have used to calculate mortality rates. See M. Wallace, *A History of Namibia: From the Beginning to 1990* (London, Hurst, 2011), pp. 177–178.

13 National Archives of Namibia (hereafter, NAN) Archives of the Secretary for the Protectorate (hereafter, ADM) 85 File 2163/2: Certified Translation: 'Treaty of Submission: Signed at Ukamas' – 23 December 1906.

World War, more significant progress in land expropriation had been made, especially after the completion of the railway line to Karasburg in 1909.[14]

After South Africa officially obtained the South West Africa 'Mandate' from the League of Nations in 1919, it moved fast to incorporate the territory into South Africa as a de facto fifth province. Although South African Prime Minister Jan Smuts might have been a key player in the Treaty of Versailles and at least a vocal proponent of the self-governance of peoples, he had little intention to bring this to Namibia. From the earliest days of the Mandate, he instructed his administrators in the territory to structure economic and legal frameworks to facilitate the incorporation of Namibia – with the broader assumption that South Africa would acquire mandates/colonies in other parts of the African continent.[15] For Karasburg/Warmbad District, this took shape first and foremost with the Land Settlement Programme of the 1920s.

There had already been a sizeable migration of mostly Afrikaans-speaking South African settlers to southern Namibia by the time the Germans lost the colony: in fact, the majority of settlers in the south during the German period were Afrikaans-speakers,[16] and this process continued unabated under South African rule. In 1920, the new South African government revised the 1912 Union Land Settlement Act to encompass Namibia as well. The Secretary for Lands in Windhoek issued a number of memoranda detailing the farms that were to be prioritised for surveying, sale and settlement. The new administration specifically targeted the southernmost district, Karasburg/Warmbad, and its neighbour to the north, Keetmanshoop.[17] Not only had this area been less systematically sold off during the German period but it also bordered the Orange River and South Africa itself, making it the perfect means to make Smuts' goal of annexing South West Africa a real possibility. As Figure 14 makes clear, in 1920, virtually all farmland in between the Bondelswarts communal areas and the Orange River was planned for formal surveying and sale. South-eastern

14 WDO: 'Übersicht über die Farmverkäufe und Verpachtungen der Landgesellschaften in Deutsch-Südwestafrika' (File originally from Windhoek Magistrate's Office, 1913). Bayerische Staats-Bibliothek (hereafter, BSB) Karte C1490015: 'Besitzstands-Karte von Deutsch-Südwestafrika' (1 October 1911). NAN Archives of the Lands Branch (hereafter, LAN) 241 File 539 (vol. 2): Joint Liquidator of SATCO to High Commissioner for the Union of South Africa, London 'Land in South West Africa' – 12 March 1936.

15 See the illuminating letter from J.C. Smuts to G.R. Hofmeyr (23 May 1923), in *Selections from the Smuts Papers* (*vol. 5*), ed. J. van der Poel (Cambridge, Cambridge University Press, 2007), pp. 175–177.

16 See R. Fitzner (ed), *Deutsches Kolonial-Handbuch, 1901* (Berlin, Hermann Paetel, 1901), pp. 180–187. Also, Jäckel, *Die Landgesellschaften*, p. 246.

17 See, among other reports: NAN ADM 23 File 86/16: Secretary for Lands to Secretary for the Protectorate – 24 March 1920.

Namibia was to be transformed into sheep farms for new White settlers from South Africa.

Surveying this amount of farmland takes significant time and resources, and the new SWA Administration certainly did not have much of the latter. Mr H. de Waal, stationed in Keetmanshoop, was the only surveyor for the Department of Lands responsible for the south of Namibia for the first several years of South African rule. For the first five years of farm inspections, De Waal rode on horseback – or drove by car when roads were permissible – travelling from farm to farm throughout Karasburg/Warmbad District. Even though most of these farms had not been sold off to settlers during the German period, many were occupied on a temporary basis by a farmer with a Crown land grazing licence. This gave De Waal a place to sleep, a plate of food, as well as information about the farm, the climate, the ecology and who was living nearby.

Two of De Waal's first farm inspections were, by chance, Norechab and Sandfontein, in April 1920. Norechab had not been occupied for a few years,

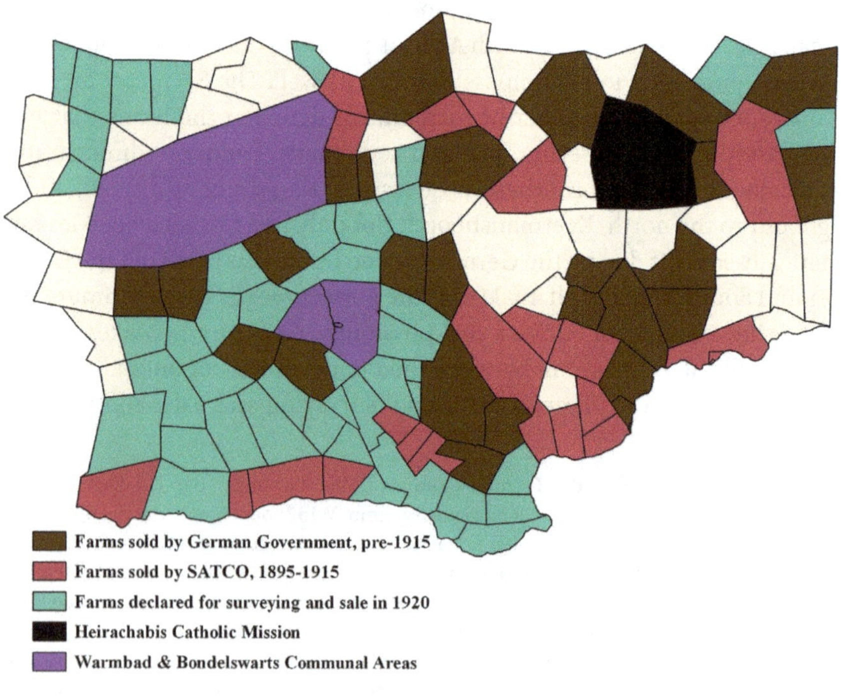

FIGURE 14 After South Africa took over Namibia in 1919, the new administration moved swiftly to survey for disposal nearly all remaining farmland in south-eastern Namibia
CARTOGRAPHY: B.C. MOORE

and although the lessee during the German period had dug a twenty one metre deep well, the water supply was not very strong. De Waal described the landscape as 'very mountainous with no vegetation' and reckoned that whereas goats could be kept on the farm 'in limited numbers', it was 'doubtful that any settler will ever find subsistence on this farm'.[18] Nevertheless, De Waal recommended that the farm still be gazetted for sale to a settler at the price of one shilling per hectare.[19]

De Waal's reports reflect a deeply pessimistic view of the fertility of southern Namibia's pastures. He was rarely impressed by the properties he inspected, yet he nevertheless went through the motions and allowed for nearly all to be gazetted for sale. One must keep in mind that surveyors' reports were shaped by the weather at the time of visiting, which almost without variation took place during the dry-season months of April through to September (when the temperature is cool enough to ride from farm to farm, days on end). Furthermore, the entirety of southern Namibia was affected by a strong drought during 1920.[20]

It is clear that the prospective settlers did not agree with De Waal's judgement about Norechab; eight separate applications for the farm were received by July 1921, most by recent incomers from South Africa.[21] J.P. Albertyn, the first occupant of Norechab under the Land Settlement Programme, was quickly able to increase his stock numbers to nearly sixty cattle and nine hundred small stock by 1923,[22] and in 1925 Albertyn constructed a successful perennial waterpoint near the Hom River.[23] Although the farm was clearly not Namibia's most fertile, it was performing more successfully than De Waal had anticipated.

Whereas De Waal lambasted Norechab, he was thrilled with the prospects of the massive 40,000-hectare expanse of Sandfontein, which he recommended splitting into two farms. He acknowledged that some parts of Sandfontein were 'very mountainous and practically useless', though he noted that there was a substantial section – as large as 20,000 hectares – where 'the veld

18 NAN LAN 333 File 703: Farm Inspection Report for Gazettal Purposes: Norechab – 9 April 1920.

19 NAN LAN 333 File 703: Extract from Minutes of the Land Board Meeting, Windhoek – 6 September 1920.

20 NAN Archives of the Magistrate at Keetmanshoop (hereafter, LKE) 3/1/19 File 17/16/20: Keetmanshoop District Annual Report, 1920.

21 NAN LAN 333 File 703: Land Board Resolution: Farm Norechab – 18 July 1921.

22 NAN LAN 333 File 703/2: Inspection Report: Norechab – 17 November 1923.

23 NAN LAN 333 File 703/2: Inspeksie-Rapport: Norechab no. 192 – 6 January 1928.

is good sweet-grass, with Tsamma [melons] and scrub bush, as well as supply of water'.[24] The eventual settler on Sandfontein, Sarel Jacobus Oberholzer, quickly constructed a very large kraal system and by 1926 he was grazing a large herd: 800 sheep, 400 goats, 210 cattle and some horses.[25]

By September 1922, De Waal finally reached the Orange River and could begin his inspection of the key farms of Pelgrimsrust, Pelladrift, Kambreek and KumKum. The main road from Warmbad to Onseepkans runs near Pelgrimsrust – the penultimate farm before the river – and De Waal was able to make this farm his base for inspecting the Orange River properties. At the time, the 10,000-hectare Pelgrimsrust was under monthly grazing licence to Dirk J.A. Brand, who had a small one room house where De Waal likely slept. Brand had been dedicating more effort to building infrastructure for his herds (780 goats and 40 cattle), including several stone kraals, two wells, dip tanks and windmills. The surveyor recommended that the farm be immediately advertised for sale, declaring that while the soil was 'stony and sandy', the farm was 'suitable for all classes of stock'.[26]

De Waal rode south from Pelgrimsrust, closer to the Orange River. With each kilometre, the mountains get steeper, and the landscape becomes, in the surveyor's words, 'very sharp' [*skerp*]. He reached Namibia's southernmost farm, Pelladrift, which at that time was divided into two individual holdings under lease to Calvyn Ludick – head of one of the old White *oewerboere* [shoreline farmer] families. Ludick's small homestead sat in a small valley in between the Skimmelberge and Spelonkberge mountain ranges, the latter of which steeply descends more than 1,500 metres down to the waters of the Orange River.[27] De Waal clearly had no idea how Ludick was making a living on this property, because he described the western half of the farm in harsh words: 'A very mountainous and rugged Orange River farm with poor veld on account of its stony nature. Very little can be done in the irrigation line along the Orange River'.[28]

24 NAN LAN 338 File 716: Farm Inspection Report for Gazettal Purposes: Sandfontein – 9 April 1920.

25 NAN LAN 338 File 716/1: Inspeksie-Rapport: Sandfontein no. 131 – 23 November 1926.

26 NAN LAN 659 File 1785: Farm Inspection Report for Gazettal Purposes: Pelgrimsrust no. 100 – 20 September 1922.

27 For a revealing description of the mountainous nature of Pelladrift, see NAN LAN 659 File 2135/1: M.J. Hayes, Pelgrimsrust to Hoofamptenaar, Afdeling van Lande, Windhoek – 2 March 1949. And note NAN LAN 2171 File GV.102: Inspeksie-Rapport: Plase Pelladrift & Meck – 24 December 1952.

28 NAN LAN 659 File 2135: Farm Inspection Report for Gazettal Purposes: Meck no. 103 (Pelladrift) – 19 September 1922.

FIGURE 15 View from the mountain ranges into the riverbed road of KumKum, 2019
PHOTO: KAROLINA MIKSLOVÁ, USED WITH PERMISSION

His view of the eastern portion was not much better:

> Orange River farm opposite Pella Mission Station in the Union [of South Africa]. Very mountainous and rugged, in some parts inaccessible. Outside veld, not of much value; sparse grass and scrubs. Along river, small patches could probably be brought under cultivation.[29]

De Waal made his way west, descending the Spelonkberge and crossing onto Kambreek Farm, a smaller parcel which he described in similar language to Pelladrift. It was leased by Calvyn Ludick's brother, Cornelius, who nevertheless wanted to purchase the farm, despite De Waal's disdain regarding its potential.[30] After remarking on the possibilities of running a small irrigation project near the Orange River waters of Kambreek, the surveyor progressed farther west, crossing into KumKum – in later years, the Ludick family farm. Although the mountains near the river were steep and inaccessible, De Waal

29 NAN LAN 659 File 1784: Farm Inspection Report for Gazettal Purposes: Pelladrift no. 102 – 19 September 1922.

30 NAN LAN 659 File 1783: Farm Inspection Report for Gazettal Purposes: Kambreek no. 104 – 19 September 1922.

was pleased with the grazing in KumKum's valleys on the northern portion of the farm.[31]

Ultimately, De Waal declared that the gazettal and sale of the farms that bordered the Orange River – Pelladrift, Kambreek and KumKum – should be delayed until the 'boundary question' was resolved. The 1890 Anglo-German Treaty had set the southern boundary of Namibia at the highest flood mark of the north bank of the Orange River, rather than at the middle of the stream. This meant that the farm boundaries did not actually extend to the river waters, and at this time South West African farms had no rights to irrigate because the Orange River and its waters belonged to South Africa.[32] This legal quandary meant that, during the 1920s, only Pelgrimsrust could be sold as part of the Land Settlement Programme.[33] The sale of the other farms was delayed until the 1930s, when conditions were made clear that the properties did not possess riparian rights. The South African Water Law was amended in the mid-1950s, which changed the situation slightly. It did not explicitly grant South West Africa riparian rights but rather removed any mention of these lands belonging to SWA 'as if such land were part of the Union'.[34]

Present throughout De Waal's surveying and inspections in southern Namibia was the near-religious belief that the purpose of these properties was to become private, territory-bound, family farms. This was reflected in his reports to the Land Board, where he reflected on the kind of 'settler' who could succeed on a given parcel, and in the Land Board's deliberations once settlement applications came in. The board would investigate the cash and livestock assets of each prospective settler, whether he was married, with children, and who his references were (NG Church ministers or local Afrikaner politicians were the best). Only occasionally would the surveyor or Land Board reflect on the local or inherited knowledge of the settler of given farming areas. One exception was Calvyn Ludick's application for KumKum, in which the Warmbad Magistrate stated: 'it is naturally very few people who can make a living

31 NAN LAN 710 File 2137: Farm Inspection Report for Gazettal Purposes: Kumkum no. 105 – 19 September 1922.

32 NAN LAN 28 File 45: Secretary for Lands, Pretoria to Secretary, SWA Lands Branch – 6 November 1920.

33 For KumKum, see NAN LAN 710 File 2137: Extract from Minutes of the SWA Land Board Meeting – 30 January 1924.

34 See NAN LAN 28 File 45: Director of Irrigation, Cape Town, to Secretary to the Prime Minister, Cape Town 'Use by South West Africa of the Orange River' – 17 May 1954.

in mountainous farms like Eselruh and KumKum. Mr Ludick is one of those people'.[35]

Despite the mixed feelings expressed in the land surveyor's reports, the SWA Administration went ahead with trying to settle most of the farms in the southern part of the district. There was political pressure to increase the White, South African share of the population as much as possible, and many of these settlers were those who could no longer obtain land on the South African side. Digging deeper into this question, De Waal's allegiance to the family farm model of the Land Settlement Programme clearly made it impossible for him to see that there were actually very few White families on these mountainous Orange River farms, and most were intermarried or related. Calvyn Ludick eventually took over KumKum, and he helped his son obtain land rights on the neighbouring Eselruh Farm. Dirk Brand, who farmed at Pelgrimsrust, married into the Ludick family, and they regularly moved their animals to whomever's farm had better grazing. The Hayes and Shaw families farmed cooperatively, moving their stock between Keimas, Pelgrimsrust, Kambreek and Pelladrift. Furthermore, like the Bassons, several of the White families – such as the Hayes, Herridge and Luttig families – had relatives on the South African side of the river, farming at Pella or Onseepkans, where they could obtain grazing.[36]

In many ways, De Waal's astonishment at the quantity of applicants for these purportedly barren farms revealed his ignorance about what farming in this area meant. In these early years, White farmers' actions did not match up with the nuclear family, homo economicus goals of the Land Settlement Programme. White farmers behaved like the Bondelswart Nama pastoralists who had been living there long before the Whites arrived. These White families *learned* from the Bondelswarts at the river, because the Bondelswarts had never really left these farms.[37]

Although the genocide by the Germans had stripped the Nama of much of their communal lands, the inaccessible nature of some of these farms meant that the Nama could hide from the state, technically residing as 'squatters'. De Waal himself noticed this. The assets list for Pelgrimsrust's official sale advertisement – based on the surveyor's report – declared that there were 'some

35 [*'Dit is natuurlik net sekere mense wat 'n bestaan kan maak op die bergagtige plase soos Eselruh en KumKum. Mnr Ludick is een van hierdie persone'.*] NAN LAN 1290 File 3951/2: Magistrate at Warmbad to Hoofamptenaar, Afdeling van Lande 'Applikasies: C.A. en R.P.C. Ludick' – 27 September 1938.

36 See National Archives of South Africa, Pretoria (hereafter, NASA) Archives of the Department of Social Welfare (hereafter, VWN) 1076 File SW458/10 (vol. 2): Rev. H.J. Thünemann, Vicar Apostolic, Pella to Secretary for Social Welfare, Pretoria 'Pella Mission Reserve' – 3 July 1946.

37 For a comparison on the South African side of the river, see R.J. Gordon, 'The Enigma of the Namaqualand Trekboer', in L. Lenggenhager et al., *The Lower !Garib/Orange River: Pasts and Presents of a Southern African Border Region* (Bielefeld, Transcript Verlag, 2023), pp. 83-98.

Hottentot [sic Nama] families on the holding'.[38] Indeed, a few years earlier, some reports emerged about Nama families residing on KumKum and refusing to take up wage labour for Whites.[39] In the same way that the ancestors of Willem Basson had moved their animals up and down the Orange River, following grazing and water, so did many of the Whites whom De Waal encountered. This would change after the Second World War, when large-scale subsidies enabled White farmers to improve the infrastructure on their holdings sufficiently that transhumance was not always necessary. However, at this stage it was the norm.

Beyond the racial demographic reasons for wishing to settle Karasburg/Warmbad District as quickly as possible after the Bondelswarts Uprising of May/June 1922 – in which the Nama waged a brief but unsuccessful war against the South Africans – there were also security concerns. The Bondelswarts had used the vacant and/or unowned Crown land surrounding their reserves to the south and south-west as a corridor along which fighters could cross from the Richtersveld on the South African banks of the Orange River. After all, among the grievances that had led to the uprising in the first place was the fact that the South Africans had not reallocated ancestral lands stolen by the Germans. In the aftermath of the drawn-out conflict, it was decided that the Land Settlement Programme should also create a buffer zone of White-owned and -occupied farms stretching from the south of the Bondelswarts Reserves down to the Orange River.[40] De Waal's survey of the Orange River farms in September 1922 was in part motivated by these goals.

Surveying farmland for sale, however, did not necessarily mean the immediate eviction of Nama residents. When De Waal advertised Pelgrimsrust with 'some Hottentot [sic Nama] families on the holding', he was describing a situation which was the norm across southern Namibia. Although the number of tenants and 'squatters' would gradually decrease over the years, it would take a very long time to eradicate the system.

Because 'squatters' were not farm owners, and since their labour was not governed by formal contracts, historians are often left with a dearth of documentation relating to their presence in the region. However, if we broaden our definition of evidence to include *both* writing on paper *and* writing on skin,

38 NAN LAN 659 File 1785: General Notice no. 8 of 24 April 1923.

39 NAN ADM 124 File 4672/5: Military Magistrate at Warmbad 'Explanations given by persons who signed the petition addressed to the Right Honourable General Botha on 1st October 1917' – undated, late December 1917.

40 NAN Archives of the Bondelswarts Commission of Inquiry (hereafter, A.486) vol. 1 file 1: Evidence of Maj. John Frederick Herbst, Secretary for SWA – Windhoek: 15–16 August 1922.

then new evidence comes into our purview.[41] In order to control the movement of cattle and maintain property rights, in 1923 the SWA Administration promulgated the Native Stock Brands Proclamation, which allocated a specific branding iron number for each Black stock-owner in Namibia.[42] Since the brands were held in the local magisterial or veterinary offices, stock-keepers had to present themselves before the officials to register for a cattle brand, who noted down their primary place of residence and the location where the livestock would be kept. In many cases, this was the first time that individuals' names entered the archival record. Branding iron registers are a very different sort of archive. Nevertheless, they give evidence of the historical connections of African people to these lands even after surveyors and settlers came.

Over the years that followed the passage of the law, the magistrates accumulated the names of African stock-keepers across Namibia. In 1929 they finally released the master list of branding irons for the whole territory.[43] A total of 419 individuals had registered for branding irons in Karasburg/Warmbad District alone, with only about one-third of them declaring their place of residence to be on the Bondelswarts Reserves. Most were registered as employees, tenants or 'squatters' on various White commercial farms in the district. Figure 16 reveals the Nama presence on White-owned farms, but it includes only those who officially reported stock ownership. Pelgrimsrust, Eendoorn, Gaidip and Oranjefall each had three 'native' irons registered. Sandfontein had ten and KumKum had two. Homsrivier had four, including those registered to Barend, Christiaan and Petrus Basson.[44] Despite this presence, the trend towards White dominance in the region was established, and the number of 'native' branding irons that were registered would gradually decrease from this date.

As shown in Figure 17, the SWA Land Settlement Programme of the 1920s and 1930s accomplished its envisioned goals. During this period, the number of farm parcels sold to Whites more than doubled from the German colonial era, and this was particularly stark in the less-densely settled southern districts, which had little German presence. This was not merely a triumph in getting

41 See L. van Sittert, 'Writing on Skin: The Entangled Embodied Histories of Black Labour and Livestock Registration in the Cape Colony, c. 1860–1909', *Kronos* 40 (2014), pp. 74–98.

42 NAN Archives of the South West Africa Secretariat: A-Series (hereafter, SWAA) 1439 File A.219/2: *Native Stock Brands Proclamation, 1923*. For more on the politics of cattle brands and veterinary control in Namibia and southern Africa, see K. Wedekind, *Impfe und herrsche: Veterinärmedizinisches Wissen und Herrschaft im kolonialen Namibia, 1887–1929* (Göttingen, Vandenhoeck & Ruprecht, 2021), pp. 342–360. G. Miescher, *Namibia's Red Line: A History of a Veterinary and Settlement Border* (New York, Palgrave, 2012).

43 'Naturelle Grootvee Brandmerke Register vir Suidwes-Afrika', in *Buitegewone Offisiële Koerant van Suidwes-Afrika* (No. 322 of 23 May 1929).

44 Christiaan and Petrus Basson were Willem Basson III's great-uncles.

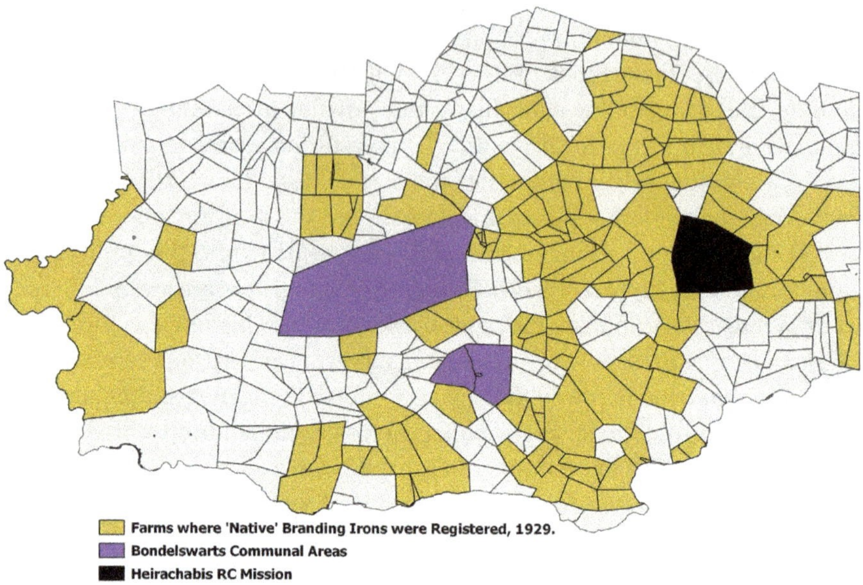

FIGURE 16 Farms and farm subdivisions where 'native' branding irons were registered during 1929, giving an indication of Nama tenant farmer or 'squatter' residence[45]
CARTOGRAPHY: B.C. MOORE

White farmers to sign long-term leases and title deeds; the number of livestock in this arid area increased dramatically, perhaps challenging De Waal's pessimistic view of the district. Even with the terrible drought that ruined many farmers in the late 1920s and early 1930s, commercial livestock farming continued to be viable in the southern Karasburg district. This was in part as a result of the rise of the karakul sheep-farming industry. Karakul sheep, originally from Central Asia, thrive in the arid, rocky areas of southern Namibia, partly because the value of the sheep comes not from wool or mutton, but rather lambskin pelts sold to the international fur fashion industry.[46]

Karakul performed better during the simultaneous great drought and Great Depression, such that after 1933 nearly all government-subsidised loans and advances to White farmers in the south were for the purchase of karakul breeding stock, which eventually brought in immense profits. For example, in the aftermath of the drought, Calvyn Ludick was keeping only 400 sheep and

45 'Naturelle Grootvee Brandmerke Register vir Suidwes-Afrika', in *Buitegewone Offisiële Koerant van Suidwes-Afrika* (No. 322 of 23 May 1929).

46 For more on the karakul industry, see B.C. Moore, 'Smuggled Sheep, Smuggled Shepherds: Farm Labour Transformations in Namibia and the Question of Southern Angola, 1933–1975', *Journal of Southern African Studies*, 47, 1 (2021), pp. 93–125. J. Swanepoel, 'In the Land of the Jackals: Postcolonial Aridity in Southern Namibia' (PhD Dissertation, University of the Free State, 2020).

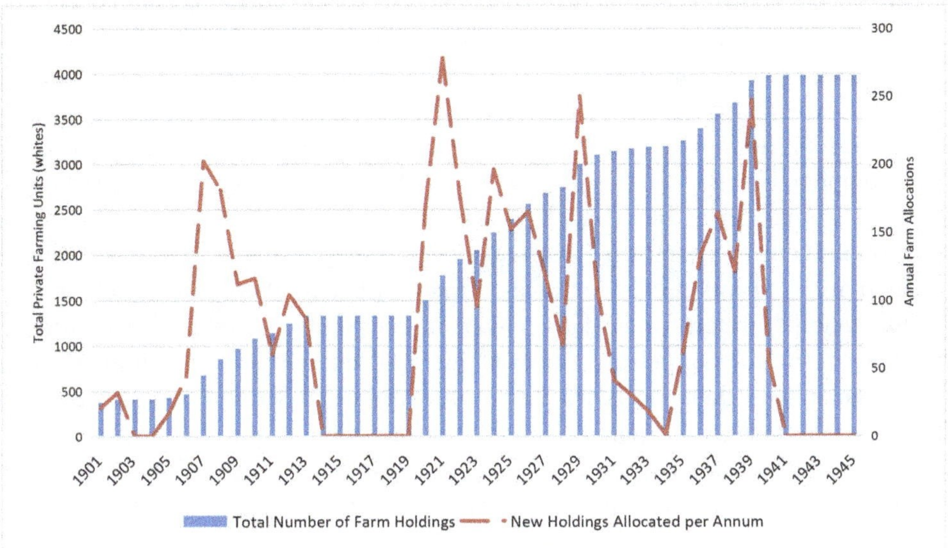

FIGURE 17 Growth of private farm ownership in Namibia, 1901–1945. De Waal's inspections occurred at a peak period[47]

200 goats (the number was lower during the drought),[48] but by 1937 he had increased his flock to 1,400 small stock, mostly karakuls, enabling him to invest pelt income into water infrastructure.[49] Even on the ultra-mountainous Peladrift, the eventual settler, R.J. Herridge, managed to earn between £500 and

47 Data from: Reichs-Kolonialamt, *Die deutschen Schutzgebiete in Afrika und der Südsee, 1909/1910* (Berlin, Königliche Hofbuchhandlung, 1911), pp. 39–40. Reichs-Kolonialamt, *Die deutschen Schutzgebiete in Afrika und der Südsee, 1910/1911* (Berlin, Königliche Hofbuchhandlung, 1912), p. 71. Reichs-Kolonialamt, *Die deutschen Schutzgebiete in Afrika und der Südsee, 1911/1912* (Berlin, Königliche Hofbuchhandlung, 1913), p. 76. Reichs-Kolonialamt, *Die deutschen Schutzgebiete in Afrika und der Südsee, 1912/1913* (Berlin, Königliche Hofbuchhandlung, 1914), p. 92. H.O. von Schöllenbach, *Die Besiedelung Deutsch-Südwestafrikas bis zum Weltkriege* (Berlin, Dietrich Reimer Verlag, 1926), pp. 122–123. NAN Archives of the General Rehabilitation Commission (hereafter, KGR) 1 File 1: First Report of the General Rehabilitation Enquiry Commission (March 1946). Union of South Africa, *Report of the Government of the Union of South Africa to the Council of the League of Nations Concerning the Administration of South West Africa* (vols. 1920–1938). Note that land settlement was halted during the genocide (1903–1904), martial law period (1914–19) and the Second World War (1940–45) and was severely curtailed during the Great Depression/Drought (1931–34). After the German period, these figures include farms allocated under the Land Settlement Programme (1920), the Crown Land Disposal Act (1920) and the special settlement programme for the Angola Boers (1929–1932).

48 NAN LAN 710 File 2137: Inspeksie-Rapport: KumKum no. 105 – 23 August 1934.

49 NAN LAN 710 File 2137: Farm Inspection Report for Gazettal Purposes: KumKum no. 105 – 15 April 1937.

£800 per year from the sale of karakul pelts, using the money to build a farm road, dig wells and improve other infrastructure. By 1956, Herridge had more than five hundred karakul sheep, three hundred goats, a three-roomed house, a truck and another automobile.[50]

Indeed, after the drought, the district-level flock was effectively rebuilt with a completely new breed of sheep: the karakul. As a smaller sheep whose value is not in meat prices, but rather lambskin pelts, karakul enables farmers in arid mountainous areas to bring in more revenue even when grazing is poor. The high price of karakul pelts in the fur markets of London, New York and Frankfurt enabled farmers to pay off their loans, invest in labour-saving infrastructure such as fences and pipelines, and even purchase lucerne fodder to supplement the grazing on their farms. By the Second World War, karakul accounted for more than seventy per cent of all small stock in Namibia as a whole, the figure being even higher for the south.[51][52] Whereas De Waal described the farms in somewhat pessimistic terms in the early 1920s, by the mid-1940s it was clear

TABLE 1 Selected statistics from 1947 Warmbad police survey of African farm residents[52]

Farm name	Adult family heads	Bondelswart nama	River-folk
Homsrivier	98	78	20
Sandfontein	6	2	0
Sandfontein West	21	12	0
Girtis	16	10	6
Gaidip	30	22	8
Pelgrimsrust	16	13	0
Oranjefall	6	6	0
KumKum/Eselruh	4	3	0
Ramansdrift	2	2	0
Haakiesdoorn	25	25	0

50 NAN LAN 2171 File GV.102: Opgawe van Vee en Gereedskap: R.J. Herridge, Pelladrift – 12 April 1956.

51 SWA Administration, *Report of the Long Term Agricultural Policy Commission* (Windhoek, 1949), Table VI.

52 NAN LKW 3/4/2 File 2/16/3: Station Commander, SAP Warmbad to Magistrate at Warmbad, 'Annual Report, 1947' – 18 December 1947. Depending on the farm, it is likely that many river-folk were also recorded under the heading Nama, given that they were Nama-speakers.

that the farms were supporting large numbers of sheep and farmers who had entered South West Africa as so-called 'poor Whites' were now highly specialised – and comparatively wealthy – fur farmers.

This increased agricultural production within southern Namibia during the mid-twentieth century led to additional documentation about the (omni)presence of river-folk and other Bondelswart Nama living on White commercial farms – whether as waged labourers or tenants. In 1947, the Warmbad police conducted a demographic survey of non-White residents on farms within its patrol area. The police officers visited all the farms south of Warmbad, collecting data and even separating the workers by ethnicity. The reports revealed that despite the survey and demarcation of Bondelswarts ancestral land into private farms, and their sale to Whites, Africans were likely still the majority of residents (see Table 1). The following chapter digs deeper into the history of the African presence along the Lower Orange River, also examining the role of Catholic mission stations in the human history of these riparian areas during the nineteenth and early twentieth centuries.

CHAPTER 4

River of God's People

When Bishop Jean M. Simon of the Oblates of St Francis de Sales arrived in 1892 in Pella – on the South African side of the Orange River – things could not have been more different from his native France, where he had previously been a schoolmaster in Lyon. Simon and his fellow priests took over an abandoned Lutheran mission farm, which the Protestants had departed several decades before, as a result of war and infighting along the Orange River. However, there was a large number of baptised Catholics and other Christians – Black, White and Coloured – living along the river and on neighbouring farms, such that the Oblates quickly found Mass full, and many Africans took up temporary residence near Pella to attend services.

Bishop Simon nevertheless could not understand why his parishioners would periodically pack up their belongings and depart for several months at a time. He reflected on this conundrum: 'Our Christians were nomads. They had an inborn need of change. So, they went away in the beginning of March with their children and their flocks, and I remained alone in Pella'.[1] The essentialism in his words aside, the Bishop was observing a very real phenomenon and dilemma: how does one design and maintain a Catholic mission station for a population which is often quite mobile?

The population along the Orange River that Bishop Simon had observed was fairly heterogenous. Although the Bondelswart Nama were at that time the most significant African polity in the area, 'Bondelswarts' as an ethnic identity and category is itself diverse. The group regularly took in non-Nama under their suzerainty, allowing them to become full members under the Kaptein. As this chapter will show, the Orange River's status as a lifeline for this arid region sustained an even more diverse group of people living along its banks, who eventually joined the church and became ardent Catholics. Many of these people today self-identify as Nama and answer to the Bondelswarts Traditional Authority, but their histories are far more complex. For simplicity's sake, throughout this book we use the term 'river-folk' to describe those Bondelswarts with close familial and material ties to the Orange River. If the Bondelswarts are a specific grouping of the Nama, so are the river-folk a specific heterogenous grouping of the Bondelswarts.

1 J.M. Simon OSFS, *Bishop for the Hottentots: African Memories, 1882–1909* (New York, Benziger Brothers, 1959), p. 38.

FIGURE 18 Cathedral of the Immaculate conception, Pella
PHOTO: B.C. MOORE, 2016

Pella was to become a centre for social relationships along the Orange River, but the Catholics were ultimately newcomers entering a region already rich with social and economic connections to regions farther afield – some of them peaceful, some anything but. More than 120 years before Bishop Simon came to the Orange River, a trader, surveyor and mapmaker named Carel Frederick Brink – employed by the Dutch Governor – had travelled along the Orange River during 1761. On the 18 September, he and his team arrived at the Orange River, likely in the area near Goodhouse. He described the richness of the fish stock in the river, supporting a remarkable number of hippopotamus.[2] Brink and his travellers trekked eastwards along the river for about two weeks to arrive at the confluence of the Hom River (ǁHomǀgâb in Khoekhoegowab)[3] and the Orange River (!Garib). They met some of the Bondelswarts (!Gamiǂnûn) who were staying there, and they tried to trade and barter with the people.[4] Over the following days, Brink's team was directed onto the best path to trek along the Hom River to Warmbad.

2 *The Journals of Brink and Rhenius*, ed. E.E. Mossop (Cape Town, Van Riebeeck Society, 1947), p. 25.

3 Joseph J. Rooi and Willem Basson, interview with Bernard C. Moore (ǂGâbes, 16 October 2022).

4 *Brink and Rhenius*, pp. 31–32.

The names for river-folk along the Orange vary from account to account.[5] Johannes Tobias Rhenius – a trader in the employ of the Dutch East India Company – described the river-folk he met in 1721 as 'Briquas'.[6] The traveller and naturalist, Robert Jacob Gordon, met these people not far from Onseepkans in 1779, and he described them with the ethnonym 'Einiqua', distinguishing different households by whether or not the men and boys were circumcised or had a testicle removed.[7] Meanwhile, the Swedish-born fugitive on the run from the VOC, Hendrik Jacob Wikar, fled to the Orange River in 1779 himself to escape the punitive efforts of the Company, and he lived alongside the 'Eynikkoa' for many months.[8] Wikar's account is likely the richest (it is often fantastical), dedicating dozens of pages to describing the lives, living habits, cultures and worldviews of these river-folk he encountered between Augrabies and Goodhouse.[9]

With the help of the translation of his guides from Goodhouse, Wikar described the lifestyle of those living along the river as quite secure and even prosperous to a degree.[10] He depicted the river-folk as bee-keepers, honey-gatherers, fishermen and horticulturalists.[11] Fish were described as abundant in the areas west of Augrabies Falls, and along the banks many Africans were cultivating small quantities of a kind of apricot, called 'royena'.[12] Wikar occa-

5 In general, our use of terms like Bondelswarts and Orange River, as opposed to !Gamiǂnûn or !Garib, respectively, reflects the terminology which our interviewees used in conversation with us. For historical context along these lines, see A. Rosengarten, 'Entangled Networks: Ethnicity, Mobility, and Exchange in the Lower !Garib/Orange River Region in the Late-18th Century', in L. Lenggenhager et. al. (eds.), *The Lower !Garib/Orange River: Pasts & Presents of a Southern African Border Region* (Bielefeld, Transcript Verlag, 2023), pp. 27-50.

6 *Brink and Rhenius*, p. 145.

7 Robert Jacob Gordon manuscripts, Rijksmuseum: Fourth Journey (MS107/3/1–2) – 28 September 1779.

8 *The Journal of Hendrik Jacob Wikar (1779)*, ed. E.E. Mossop (Cape Town, Van Riebeeck Society, 1935), p. 105.

9 Penn argues that most 'Einiqua' stayed upriver from Pelladrift, with the densest settlements around Augrabies. He holds that the river between Goodhouse and Augrabies was 'San' country, including a group of 'San' called the 'Nanningai' [mountain climbers] who lived opposite Pella on the north banks of the Orange River. N. Penn, *The Forgotten Frontier: Colonist and Khoisan on the Cape's Northern Frontier in the 18th Century* (Athens, Ohio University Press, 2005), pp. 158–162. However, his evidence for the claim that the land was exclusively San country is passages from Gordon and Wikar, which are quite inconclusive on this fact.

10 It is important to remember that many eighteenth and nineteenth century ethnonyms had little to do with cultural affinities or sociolinguistic elements, but rather production and socioeconomic strategies. They were often a product of the European imagination and less an accurate description of the populations encountered. See D. Henrichsen, *Herrschaft und Alltag im vorkolonialen Zentralnamibia: Das Herero- und Damaraland im 19. Jahrhundert* (Basel, Basler Afrika Bibliographien, 2011), pp. XVI–XXII.

11 *Wikar*, p. 57.

12 Ibid, pp. 59, 115.

sionally mentioned how the 'Eynikkoa' would use pit traps, known as 'Kaysi', to catch game that came to the river to graze and drink. Furthermore, these people were described as good shots with a bow and arrow, using toxic melkbos (euphorbia) sap to weaken the game and slow it, in order to make the kill.[13] It makes sense that where a perennial river passes through an arid and mountainous area like southern Namibia, people would survive alongside it, moving depending on the flow of the water, grazing their animals on the nearby pasture, and hunting wild animals for food.

The end of the eighteenth century would change things dramatically along the river. In March 1796, a group of newcomers came to the Orange River: the Afrikaner clan.[14] Led by Jager Afrikaner, this Khoisan group (sometimes called 'Nama-Oorlam') fled the Calvinia area after a serious disagreement with the prominent Dutch farmer Petrus Pienaar resulted in Jager's brother Titus killing Pienaar, who had employed the Afrikaner clan in hunting commando operations in the Northern Cape.[15] The clan as a whole had to flee farther north, outside the boundaries of the Cape Colony, to avoid prosecution and retribution by the Whites. Jager Afrikaner and his people settled for a while on Perde Eiland, an island on the Orange River outside of Kakamas.[16]

There, they developed a reputation for raiding the nearby Bondelswarts communities, taking advantage of their superior firepower to seize cattle and resources. At this time, the Bondelswarts, especially those living along the river waters, had few firearms. The Afrikaner clan lived there for several years until eventually taking up residence on the farms Jerusalem and Blyderverwacht, about fifteen kilometres north of the Orange River, from where they would gradually make their way north to Windhoek.[17] The clan eventually attacked the London Missionary Society stations at Pella and Warmbad in 1811, causing the missionaries to abandoning the sites.[18] For the Bondelswarts, especially the river-folk near the Orange, the withdrawal of the missionaries meant a drastic

13 Ibid, pp. 119, 179, 183.

14 Note that the Afrikaner group – descendants and followers of Jager Afrikaner – have nothing to do with the White, Dutch-descended Afrikaners (sometimes called Boers).

15 T. Dedering, *Hate the Old and Follow the New: Khoekhoe and Missionaries in early-Nineteenth Century Namibia* (Stuttgart, Franz Steiner Verlag) p. 54.

16 B. Pool, 'Die Geskiedenis van die Afrikaner-Oorlams in die Tyd van Jonker Afrikaner, 1790–1861' (PhD thesis, Stellenbosch, 1995), pp. 15–16.

17 For more on the Afrikaner clan's role in Namibian history, consult B. Lau, *Southern and Central Namibia in Jonker Afrikaner's Time* (Windhoek, National Archives of Namibia, 1987).

18 Dedering, *Hate the Old and Follow the New.*

reduction of security. Missionaries beget traders, and when missionaries flee, so do the traders.

•••

In 2021, we had the opportunity to sit down with Willem Basson III, a *raadslid* [councillor] for the Bondelswarts and a lifelong farmer along the Orange River and within the Bondelswarts communal areas. We met in a restaurant in Karasburg to speak about his life, and the history of his family, his people and the riverine lands where he keeps his livestock. We made arrangements to go to see his animals the following day. Basson spoke with us only in Afrikaans, and his frame of reference was rather local in scope. Yet, his awareness, knowledge and mastery of the history of the Orange River area and that of his family was immense. He explained to us a bit about the history of his people.

> You see, my people are Bondelswarts people, but we were not always Bondelswarts. When the Bondelswarts crossed the *Orange River* long ago, they met my people, and they called us 'Dama' [*ons was Dama genoem*]. Sometimes the books use the word 'Damara' to describe us back then.[19]

Basson told us that up until the end of the eighteenth century, the relationship between the Nama and the 'Dama' was healthy. Trade between the Orange River and Warmbad was symbiotic, and although conflict and violence were not unheard of, they were not common.

Things changed rapidly after Jager Afrikaner came to the river, though. The Afrikaner clan became much more violent after taking refuge on Perde Eiland, and this did not stop when they reached Jerusalem Farm, north of the Orange River. They would send raiding parties south to the Orange, where they would seize livestock and harvests. Willem Basson's great-grandfather's parents were killed by Jager Afrikaner's deputies, as were many of the 'Dama' or 'Damara' river-folk, and the Bondelswarts came to the river the next day to take in the orphans.

19 Willem Basson III, interview with Bernard C. Moore and Luregn Lenggenhager (Karasburg, 11 November 2021). Unless otherwise stated, information obtained about Basson's family came from this interview. Sometimes, government officials would refer to the river-folk and their relatives in Nama reserves to the north as 'Water Damaras'. See NAN LKE 3/1/52 File 2/10: Welfare Officer, Tses and Berseba Reserves to Native Commissioner Keetmanshoop 'Native Affairs Annual Report, 1937: Tses Native Reserve' – 1 January 1938.

His great-grandfather – Willem – was adopted by the Bondelswarts, entering the household of the Jossop family, and so his childhood name was Willem Jossop. As a teenager in the mid nineteenth century, he moved to Velloor Farm – just north of the Orange River near Onseepkans – and he became a tenant farmer on the property of a Dutchman named Jasper Basson. Jasper had apparently negotiated for usufruct land rights with the Bondelswart Kaptein at the time, and he became a fairly well-off farmer with a number of properties. He took in Willem and raised him as a son, and so the boy became remembered as Willem Basson I.

During our long conversation with him in Karasburg, Willem Basson III emphasised that even though his people were not originally Bondelswarts, they were fully-adopted as members of the Bondelswarts nation. According to him and others within the Bondelswarts Traditional Authority, group membership is tied less to a particular bloodline and heritage and rather more to whether the individuals and families live by the cultural, legal and social tenets of the Bondelswarts nation.[20] Nevertheless, Basson made clear that assimilation had its costs: his people lost their language – likely related to Otjiherero – and apart from a few given names[21] and surnames, such as Mohanda and Simboya, his people lost their naming structures. Most of his relatives today speak Afrikaans as a first language, with Khoekhoegowab as a close second. Even the original name of the river-folk clan has been lost in time. Basson refers to the river-folk historically as the 'Mabiro', though he acknowledges that this might not be the full moniker.[22] In short, Basson and his relatives were likely a distant branch of the Herero people, speaking a dialect of Otjiherero, and perhaps they were one of the groups described by Rhenius and Wikar as 'Briqua' or 'Eynikkoa', respectively, to distinguish them from the more numerous Nama in the area.

Once adopted into the Bondelswarts, the Basson family and their relatives continued their pastoral lives along the Orange River, though being absorbed by the Nama opened new doors for trade, movement and communal living farther from the river, such as in Warmbad and Karasburg to the north or Pofadder to the south. Indeed, in Klinghardt's ethnographic study of Pella Mission Station in the Northern Cape, a large portion of the residents were descended

20 Josef J. Rooi, interview with Bernard C. Moore (ǂGâbes, 28 October 2021).

21 Willem Basson III provided us with his Otjiherero given name, which he does not normally use in day-to-day life: *Kaveopeui*, which he translates as '*As hulle sou dood, van wat met die lewe*?' ['If they were to die, what would life be worth?']

22 Willem Basson Private Archive, ǂGâbes (hereafter, WBPA): W. Basson, Warmbad to President of the Republic of Namibia 'Petition to Reclaim Land Previously Occupied by the Mabiro Tribe along the Banks of the Orange River' – 4 June 2021.

from 'Damara/Herero' who came from the Namibian side of the river.[23] On both sides of the Orange, river-folk had developed a shared way of life and historical background of adoption and assimilation into the Bondelswarts, so it is unsurprising that many from these families intermarried and continued these pastoral strategies together. For example, in the 1870s, Willem Basson I left Velloor, eventually meeting Lena van Rhyn, a 'Dama' who was staying near Homsrivier and who had a similar story of adoption. They married, and between 1875 and 1895 they had four children, including Willem Basson III's grandfather, Samuel, born in 1892.

•••

Bishop Simon arrived at Pella in the aftermath of this violent period of southern African history, and he was able to observe at least a temporary calm after years of residents dealing with the violence of Jager Afrikaner. True, the Kharaskhoma Syndicate was busy trying to exact land concessions from Bondelswart Kaptein Willem Christiaan, but these transformations had not yet entered the daily lives of most people. The Bishop took a trip down the Orange River

FIGURE 19 Original Catholic church at Heirachabis. A new church was constructed in the 1970s, and the old one now functions as a kindergarten.
PHOTO: B.C. MOORE, 2022

23 G.P. Klinghardt, 'Missions and Social Identities in the Lower Orange River Basin, 1760–1998' (PhD thesis, University of Cape Town, 2005), pp. 14–15.

not long after he arrived in 1892, observing that the 'Damara' people seemed comfortable with their 'pastoral lives' along the river banks.[24] With funds from the missionary headquarters in France and the Oblates of St Francis de Sales branches in Austria, Bishop Simon set to work constructing a proper church at Pella, which could be a locus point for all Catholics in the area.

After nearly two years of firing bricks and constructing flying buttresses, the Cathedral of the Immaculate Conception was consecrated in 1895, the first Catholic church in the Northern Cape and southern Namibia; it remains in use today.[25] The following year, the Bishop purchased land at Heirachabis, to the east of Karasburg, which enabled the Vicariate of Great Namaqualand to expand over the river into Namibia and deeper into Bondelswarts territory.[26] Although the Catholics were not the only missionary society in this region, they were becoming increasingly the most influential and most present in the lives of the Bondelswarts.

By the end of 1904, the German military had more or less brought central Namibia under their control again, crushing the Herero rebellion with genocidal tactics. Now, General Lothar von Trotha and his lieutenants could focus their efforts on the guerrilla war of the Nama in the south.[27] Over the course of 1905, more than six hundred Nama and Herero refugees fled South West Africa over the Orange River, reaching Pella Mission Station to seek security and solace.[28] Many of these individuals became permanent settlers and residents of Pella or the Riemvasmaak area closer to Keimoes. At the end of October 1905, the flow of Nama refugees crossing into Pella increased dramatically. At this time, Nama rebels were under the leadership of Jacob Marengo.

Despite being born in the vicinity of Vaalgras (north-east of Keetmanshoop),[29] Marengo was a child of the Orange River, with close connections to the Riemvasmaak community east of Pella.[30] Over the previous month, Marengo had

24 Simon, *Bishop for the Hottentots*, p. 45.

25 Pella RC Mission Archives: R. Bientz OSFS, *Dorsland: Heldhaftige Stryd van Pionier-Sendelinge in Namakwaland* (self-published pamphlet, 1982), p. 8.

26 Simon, *Bishop for the Hottentots*, p. 113.

27 See, W. Hillebrecht, 'The Nama and the War in the South', in *Genocide in German South West Africa: The Colonial War of 1904–1908 in Namibia and its Aftermath* (Monmouth, Merlin Press, 2004), pp. 143–158.

28 Simon, *Bishop for the Hottentots*, pp. 201–202. They also treated some German soldiers who were injured in the Battle of Hartebeesmund.

29 J. Masson, *Jakob Marengo: An Early Resistance Hero in Namibia* (Windhoek, Out of Africa, 2001), p. 2.

30 Marengo was orphaned at a young age, and his adoptive father, Carolus Damara, raised him between Pella and Okiep/Springbok. Bishop Jean Simon offered an obituary to Carolus in his diaries, revealing that the extended family had connections between Riem-

FIGURE 20 Nama refugees on the South African side of the Orange River, opposite Ramansdrift, 1905

fled from his fortifications in the Karas Mountains and moved near the Orange River, setting up a small camp in the mountains on Hartebeesmund Farm. On 25 October 1905, the German military finally caught up with Marengo and the 400 or so armed soldiers under his command. However, the Germans sustained heavy casualties in the fighting.[31] Bishop Simon did not discriminate between who was taken in by his nuns for medical treatment, and both Nama and German wounded in the Battle of Hartebeesmund recuperated at the Catholic mission.[32] Marengo and his deputies (including the famous Abraham Morris) regrouped at Homsrivier,[33] but by this point in the war the Germans had more

vasmaak, Pella, Tses and other areas along the Orange River. J.M. Simon, OSFS, *African Memories: Part Two, 1909–1932* (Keimoes, RC Diocese of Keimoes-Upington, 2006), pp. 161–162. Fr. Zaby, missionary amongst the Riemvasmakers, wrote extensively about these river-folk, their connections to the Bondelswarts, and their displacement during the war. See, A. Zaby OMI, 'Die Riemvasmakers: Geschichte und Kultur einer Gemeinschaft', *Namibiana*, 4, 1 (1982).

31 BArch R1001/2137 (bd. 5): Oberstleutnant von Semmern to Generalleutnant von Trotha – 29 October 1905.

32 Simon, *Bishop for the Hottentots*, pp. 200–203.

33 BArch R1001/2137 (bd. 5): Generalleutnant von Trotha to Generalstab, Berlin – 3 November 1905.

FIGURE 21 Jacob Marengo (centre) stands with some of his comrades
NATIONAL ARCHIVES OF NAMIBIA PHOTO NO. 18978

or less full control over most of the northbound roads to Warmbad. So the tactics of Marengo and the Bondelswarts became one of swift guerrilla attacks out of the mountains at dawn or dusk, often raiding the German horses and livestock, thus sustaining the rebellion for longer.[34] Reports from the German police station at Ramansdrift indicate that during January 1906 there were at least 400 armed Namas in the mountains along the Orange River between Vioolsdrift and Hartebeesmund.[35]

The Battle of Hartebeesmund on 25 October 1905 was something of an embarrassment for the German military, such that it was even discussed in the Reichstag and in foreign embassies.[36] Germany responded by increasing its troop presence in the far south of Namibia, to try to put down the Nama rebellion for good. On 14 February 1906, approximately two hundred Nama fighters emerged from the mountains on Sandfontein Farm and crossed into Norechab, where they attacked Lt. Eckert's detachment and seized horses and livestock. The firefight lasted for nine and a half hours, resulting in five German casualties and many more wounded.[37] The Bondelswarts had to retreat towards Homsrivier in the late afternoon after the German machine-gun battalion arrived to set up in the Hom River-bed near the southern boundary of Norechab.[38] The casualties of Nama fleeing southwards were heavy.

34 BArch R1001/2137 (bd. 5): Oberst. Dame to Generalstab, Berlin – 9 December 1905.
35 BArch R1001/2138 (bd. 6): Oberst. Dame to Generalstab, Berlin – 10 January 1906.
36 BArch R1001/2137 (bd. 5): Generalkonsulatsverweser, Cape Town to Auswärtiges Amt. – 6 December 1905.
37 BArch R1001/2138 (bd. 6): Oberst. Dame to Generalstab, Berlin – 16 February 1906.
38 This is roughly the location of 'Murli's pub' referred to in chapter 2.

In the following month, the Germans would finally break Marengo's defences on Hartebeesmund, requiring three battalions attacking from three separate approaches.[39] Marengo fled to KumKum and then to Ukamas near the road to Upington. On 4 May, he crossed into South Africa with some of his soldiers, but was pursued by the Germans, losing twenty three of his comrades in a skirmish near Van Rooisvlei. He was forced to give himself up to the South Africans a few days later. Marengo was brought to a Cape Town prison, where he was detained until June 1907 and then released on condition that he would not re-enter Namibia. When the South African police learned that he was planning to return, they intercepted and killed him at Eenzaamheid Farm, near the border, on 20 September 1907.[40]

Even though the Bondelswarts were forced to sign a 'Treaty of Submission' in late 1906 – technically ending the war – the genocide was not really over, nor had peace come. Most of the Bondelswarts livestock was taken by the Germans or perished during the war, and the treaty 'provided' them with only 1,500 goats 'for support of life'.[41] Their communal areas were reduced dramatically, and to leave the communal areas to seek grazing or work they had to carry a pass [*Dienstbuch*]. Not long after the treaty, the Germans placed heavy restrictions on, and in many cases banned, the sale of cattle and horses to non-Whites, further limiting whatever potential for pastoralism existed among the Nama.[42] Furthermore, both during and after the genocidal war, the Germans captured both Nama and Herero, rendering thousands as forced labourers. Africans were mandated to construct state infrastructure – like railways – or else were leased out to private industry, like the Woermann shipping company.[43]

It is not surprising, therefore, that hundreds of Nama, including significant numbers of Bondelswarts, fled over the Orange River to become near-permanent refugees in South Africa.[44] Individuals such as Abraham Morris – Jacob Marengo's deputy – lived in the area of Springbok, engaging in farming and occasionally working at the Okiep copper mine. The presence of Bondelswarts along the south banks of the Orange River unsettled the German

39 BArch R1001/2138 (bd. 6): Oberst. Dame to Generalstab, Berlin – 12 March 1906.

40 Masson, *Jakob Marengo*, pp. 41–43.

41 NAN ADM 85 File 2163/2: Bondelswarts Treaty of Submission: Signed at Ukamas (23 December 1906).

42 Hillebrecht, 'The Nama and the War in the South', p. 154.

43 C.W. Erichsen, 'Forced Labour in the Concentration Camp on Shark Island', in J. Zimmerer and J. Zeller (eds.), *Genocide in German South West Africa: The Colonial War (1904–1908) in Namibia and its Aftermath* (Monmouth, Merlin Press, 2003), pp. 84–99.

44 Some Bondelswarts would return to Namibia after the First World War, though many again fled south after the May 1922 uprising.

Administration, and officials wrote to the South African Minister of Native Affairs to request that the Bondelswarts be either moved further from the river or else returned to South West Africa.[45] While the South Africans refused to extradite Abraham Morris and his followers – likely knowing that the German promise of amnesty was an empty one – they nevertheless took steps to cancel all Nama grazing rights on farms that bordered the south bank of the Orange River, like Goodhouse, Abbasas and Witbank, and prepared the farms for sale to Whites.[46] The Bondelswarts who had previously held legal grazing licences were expected to either transform into tenant farmers – which many did – or be evicted to mission stations and 'native reserves'. Surveying more farmland was viewed as a way to address the problem of landless Whites and appease the Germans to the north.[47]

Facing assaults on their livelihoods from all sides, the Bondelswarts were placed in a very difficult position. Few felt that they could return to Namibia safely. Even if they were granted amnesty as promised, the German restrictions on livestock ownership, freedom of movement and rights to land meant that whatever life they could make near Warmbad would be a poverty-ridden one. Some chose to remain on farms once they were sold off to Whites. Johnny Damara, for example, held grazing rights on Abbasas and Goodhouse before the latter was sold to irrigation farmer and speculator Charles (Carl) Weidner. He managed to become the foreman for one of Weidner's other properties, Garganab.[48] Some chose to covertly remain on farms once these were sold to Whites; this was ordinarily the choice of the Bassons and those who moved to the mountains on the Namibian side of the river.

However, most chose to settle on church mission stations, particularly Pella and some of its future outstations on the Namibian side – like Heirachabis, and eventually Gaidip and Homsrivier. Klinghardt's ethnographic work in Pella – conducted during the late 1980s and early 1990s – revealed that the vast majority of Nama and 'Damara' resident families were originally refugees who had

45 NASA Archives of the Department of Native Affairs (hereafter, NTS) 7658 File 5/382: Imperial German Consul-General, Cape Town to Minister of Native Affairs, Pretoria – 15 June 1911.

46 NASA NTS 7658 File 5/382: Civil Commissioner, Namaqualand to Secretary for Lands, Cape Town, 'Cancellation of Exclusive Grazing Rights' – 3 August 1911.

47 For more on the surveying of the south bank of the Orange, see B.C. Moore, 'Swimming Upstream', in L. Lenggenhager et al. (eds.) *The Lower !Garib/Orange River: Pasts and Presents of a Southern African Border Region* (Bielefeld, Transcript Verlag, 2023), pp. 119–144.

48 Western Cape Archives and Records Service (Buro-KAB) Archives of the Department of Agricultural Credit and Land Tenure (hereafter, ACLT) 27 File 3588 (vol. 5): C. Weidner to Inspector of Lands, Upington 're: Garganab, Witbank, and Two Islands' – 31 January 1940.

crossed into South Africa during the genocide.[49] Pella was not so different from the Riemvasmaak Catholic community. Bishop Simon did not have to go out into the fields to find his flock and convince the people to be baptised as Catholics; the people came to him. The Bondelswarts sought refuge in the church as a means to escape the brutality of the German regime and the insecurity of maintaining a livelihood in the newly merged Union of South Africa.

•••

The movement of refugees into Pella in 1905–1906 and the slow yet steady trickle during the following years was met with enthusiasm by Bishop Simon and his clergy. Pella already had a garden to supply the small school, but over the following several years it was expanded dramatically into a fully-fledged irrigation farm along with nearly 48,000 hectares of village lands and grazing commonage. By 1913, Pella had 430 Black and Coloured residents, and seventy White residents: all were Catholics. The school was the mirror image of South African segregation – all seventy two children studied together. According to one observer, the irrigation works near the river allowed 'almost every description of fruit' to grow: dates, oranges, figs, lotus fruit, pomegranates, grapes,

FIGURE 22 The early moveable pump system at Pella, circa 1915
PHOTO: PELLA RC MISSION ARCHIVES

49 Klinghardt, 'Missions and Social Identities', pp. 15, 68.

vegetables, bananas, cereals and lucerne.[50] Most of the workers in the fields were the same refugees who had fled several years earlier. They earned proceeds from the sale of the crops, and housing, schooling, grazing and meals were offered gratis. Pella and the Roman Catholic church provided many Bondelswarts a livelihood that was simply impossible under the colonial governments on either side of the Orange River.

In 1912, the Union of South Africa inaugurated two new governmental units: the Department of Irrigation and the Department of Social Welfare. Both were tasked to alleviate what became known as the 'poor White question'. Changes in livestock farming economies and mining technology were leading to the increased urbanisation of the White population, causing fears that migration to towns would cause a White 'degeneration'. These new departments drew inspiration from a project initiated in 1898 by the Nederduitse Gereformeerde (NG) Kerk's charity division at Kakamas, on the Orange River. There, portions of the river frontage and islands had been cleared, flattened and subdivided into smallholdings for 'poor White' farmers, who were given training by the church to use river irrigation.[51]

By 1913, Kakamas was officially recognised by the South African government as a 'White labour colony' [*blanke arbeidskolonie*], and by 1921 more than three thousand 'poor Whites' were living there.[52] In the early days of the 'poor White question', many commentators felt that a state – church partnership was essential for the success of these 'close settlement' schemes – where urban influx, morality and alcohol could be controlled.[53] The NG Kerk was obviously preferred by the state and the Civil Commissioner for Springbok reflected that the efforts at Pella could be a model for non-White 'close settlements'.[54]

In the run-up to and during the Great Depression, the push for 'close settlement' irrigation schemes for 'poor Whites' grew dramatically, not just along the flat alluvial lands near Kakamas and Upington but also downstream where the steep mountains meet the Lower Orange River. One of them was

50 NASA VWN 1076 File SW458–10 (vol. 2): Civil Commissioner, Springbok to Secretary for Native Affairs, Pretoria 'Pella Mission Station' – 10 April 1913.

51 N. Roos, 'Work Colonies and South African Historiography', *Social History*, 36, 1 (2011), p. 58.

52 K. Visser and K. du Pisani, 'Die Vestiging van Skanskopeilande-Besproeiingsnedersetting as 'n Mylpaal in Selfwerksame Armoedeverligting', *Historia*, 57, 1 (2012), pp. 103–126.

53 J.W.N. Tempelhoff, 'Omgewingslagoffers of Armblankes? Water en Welvaartskepping in die Era van Hertzog (1924–1939)', *Generaal J.B.M. Hertzoggedenklesing XXXV* (21 September 2006).

54 NASA VWN 1076 File SW458–10 (vol. 2): Civil Commissioner, Springbok to Secretary for Native Affairs, Pretoria 'Pella Mission Station' – 10 April 1913.

Onseepkans, where an outstation of Pella mission already existed.[55] In 1928, the Department of Irrigation toured the north bank farms to identify irrigation potential. Charles Weidner of Goodhouse Farm was one of the main informants for the investigation, and he clearly hoped to sell the government some of his properties, like Haakiesdoorn on the Namibian side and Witbank and the Marten and Krapohl islands on the South African side.[56]

The inspectors ultimately recommended that Ramansdrift be converted into a poor-White irrigation settlement, and if Haakiesdoorn could be bought out by the government, then up to eighty three White families could be settled on plots of approximately six morgen (five hectares) in size. Similar recommendations were made for Homsrivier and Beenbreek; if they were bought out from their existing owners/tenants, then together they could become irrigation settlements for approximately seventy five White families.[57] Two separate proposals were made for Aussenkehr to become a poor-White settlement for Germans in SWA[58] or else Afrikaners from South Africa.[59] Ultimately, only Beenbreek and Vioolsdrift-Noordoewer became poor-White settlements, however, because the Orange River boundary issue meant that only through ministerial exemption could the north bank be granted riparian rights.[60]

Though the border/riparian question made it difficult for this type of government-sponsored poor-White schemes to develop, there were plans to set up several of the Orange River farms on the Namibian side for smaller-scale irrigation purposes. Charles Weidner of Goodhouse had bought Haakiesdoorn Farm and rented the western portion of Ramansdrift with the intention of expanding his operations to both sides of the river, only to be constantly

55 L. Lenggenhager, 'Onseepkans: Irrigation, Removals and Resistance in the Borderlands of Namibia and South Africa', in L. Lenggenhager et al. (eds.) *The Lower !Garib/Orange River: Pasts and Presents of a Southern African Border Region* (Bielefeld, Transcript Verlag, 2023), pp. 151–164.

56 NAN Archives of the SWA Water Affairs Branch (hereafter, WAT) 91 File WW.77/3: Inspeksie, Ramansdrift no. 135 – 16 December 1928. On Witbank and Gariepdale, see Moore, 'Swimming Upstream'.

57 NAN WAT 91 File WW.77/3: Inspeksie, Houmsrivier no. 133 – 16 December 1928. NAN WAT 91 File WW.77/3: Inspeksie, Beenbreek – 16 December 1928.

58 BArch R1001/1580 (bd. 3): Reichsstelle für das Auswanderungswesen to Auswärtige Amt, Berlin 'Farm Aussenkjer in Südwestafrika' – 25 November 1930.

59 NAN LAN 2173 File G.V.147: H.H. Wex, Heemstede, Netherlands to Administrator for SWA – undated, ~1936.

60 NAN WAT 91 File WW.77/3: Agreement Made and Entered into by His Honour the Administrator of SWA, on the one side, and Denys Reitz, in his Capacity as Minister of Lands of the Union of South Africa – 1933.

reminded by the SWA Administration that he had no right to do so.[61] Some of the other SATCO farms to the east – particularly Gaidip and Homsrivier – were quickly targeted by speculators who bought the properties with the intention of engaging in petty irrigation until the boundary question was solved. After exchanging hands several times, Gaidip was sold to Ludwig Weiss in June 1921,[62] who also acquired Homsrivier in August 1922.[63] Weiss was acting as a representative of the South West Irrigation Company Ltd, a South Africa-based syndicate presumably looking to mimic what Charles Weidner was successfully doing at Goodhouse Farm on the South African side of the Orange River.[64] He was clearly a rival of Weidner in many ways, and he constantly protested about every piece of irrigation that Weidner constructed on Goodhouse, alleging that the weirs and dams would flood his properties once the water level rose.[65]

They might have been business rivals, but both Ludwig Weiss and Charles Weidner were staunch Roman Catholics. After Bishop Simon and his priests set up a mission church at Warmbad in 1907, there was a desire to have an outstation closer to the Orange River to see to the needs of those working on farms in the area, and those living surreptitiously on the fringes of colonial society. Gaidip Farm was sold by SATCO in 1903 to J.P. Mostert, who was also leasing the neighbouring Homsrivier at the time, presumably with the goal of creating irrigation works.[66] In around 1908 he agreed to let the Catholic Church set up a temporary structure on his property, where visiting clergy could occasionally hold Holy Mass.[67] Once Ludwig Weiss purchased Gaidip in 1921, he expanded these privileges, demarcating two hectares where a school could also be built.[68] Father Martin Forster moved to Gaidip, and by 1930 there were more than

61 NAN LAN 28 File 45: Senior Officer, Lands Branch, Windhoek to C. Weidner – 22 October 1921.

62 WDO: Gaidip no. 146, Registrasie Afdeling 'V'.

63 WDO: Homsrivier no. 133, Registrasie Afdeling 'V'.

64 When it appeared that Weiss would not be granted riparian rights for large-scale operations, he offered to sell Homsrivier and Gaidip to the SWA Administration for the purposes of expanding the Bondelswarts communal areas. But because this was in the aftermath of the Bondelswarts rebellion – and went against all 'Native' policy promulgated in Windhoek – the government naturally declined the offer. NASA NTS 7658 File 5/382: Ludwig Weiss to Magistrate at Warmbad – 20 September 1922.

65 NAN LAN 28 File 45: L. Weiss to Administrator for SWA – 10 October 1923.

66 WDO: 'Übersicht über die Farmverkäufe und Verpachtungen der Landgesellschaften in Deutsch-Südwestafrika'.

67 A.P.J. Beris CMM, *Making the Desert Bloom* (*vol.* 2) (RC Diocese of Keetmanshoop, 2003), p. 119.

68 Ibid, p. 120. See also M. Forster, 'Der Bau einer neuen Schule in Gaidip', *Das Licht* (1930).

thirty Bondelswarts children attending his school.[69] Gaidip never had a permanent church structure, however.[70]

During the 1920s, there was an increasing push from within the Roman Catholic leadership in Namibia and South Africa for the church to acquire 'mission farms'. These farms would enable the church to be less dependent on the goodwill of property-owners (like Weiss or Weidner) and to build more permanent rural churches, schools and housing. Furthermore, the church would engage in agricultural production – either livestock or irrigation – to fund church operations and provide food for its schools and hostels.[71] While these farms would technically be the property of the church, they would also explicitly be a place of residence for African parishioners. In addition, because produce from the farm would be sold in town, the Catholic schools would be less dependent on government aid and could thus design a curriculum that would eventually go against Bantu Education laws, which restricted both the level of education and subjects offered to Black children.[72]

Whereas Charles Weidner managed to (barely) keep his operations going through the Great Depression, Ludwig Weiss could not. Along with SATCO itself, he declared bankruptcy and his assets were sold over several years. The growth of the Catholic Church in the region meant that a second bishop was appointed in 1931, based in Keetmanshoop, granting the South West Africa side of the river limited autonomy from Pella, which remained the archdiocese. Unlike Pella, which was overwhelmingly staffed with French Oblates, the Namibian side witnessed a flood of missionary priests from Vienna, under the leadership of Bishop Joseph Klemann. Under his authority, the church bought Homsrivier in 1932, with the explicit goal of converting it into an irrigation farm to supply schools in the region. Gaidip was also purchased by the church in 1936, granting the diocese a surprisingly large river frontage.

69 F. Wehrl, *Mission am Oranje: Geschichte der Oblaten-Mission der Vikariate Keimoes und Keetmanshoop nach Briefen, Tagebüchern und Visitationsberichten* (Eichstätt, Franz Sales Verlag, 1994), p. 412.

70 In 1929, Charles Weidner invited Bishop Simon to visit Goodhouse, offering to allow the Bishop to build a mission station on his farm. This never transpired, however, because the Bishop died not long after this meeting. J.M. Simon OSFS, *African Memories: Part Two, 1909–1932*, pp. 179–181. The Weidner family is remembered by many in the Catholic church as being generous patrons (see chapter 6). Sr. Paule Elizabeth Atiogbe OSFS, interview with Bernard C. Moore (Karasburg, 1 August 2016).

71 A.P.J. Beris CMM, *From Mission to Local Church: One Hundred Years of Mission by the Catholic Church in Namibia* (Windhoek, RC Archdiocese, 1996), p. 513.

72 Ibid, p. 527.

In August 1932, Father Johann Zettl arrived at Homsrivier Farm with four Catholic brothers, one sister and three African evangelists. They got to work building a church, constructing irrigation works and serving the needs of the Bondelswarts families living nearby.[73] The missionaries and evangelists were assigned to each of the seven families residing near the confluence of the Hom and Orange rivers, and a school was founded, with twenty four children attending lessons by 1933. Between the adults, the Brothers and some of the Bondelswarts, rough irrigation furrows were constructed to quickly cultivate several hectares of land to feed the school children.

After the massive Orange River floods of January 1934 subsided, Fr. Zettl and his followers located an appropriate spot for a small diversion weir and canal, about eight kilometres upriver from the mission station, and they started quarrying limestone to build the infrastructure. Despite his professed love for the church, Weidner protested the irrigation works on Homsrivier, writing to the government complaining that the north bank had no riparian rights.[74] His protests had less to do with Fr. Zettl and more to do with the fact that Weidner himself repeatedly had been denied rights to develop irrigation works on his own north bank property, Haakiesdoorn. Nevertheless, Bishop Klemann indeed received a letter stating that he should cease all construction, because he had no right to the river waters.[75] Bishop Klemann wrote back to the Secretary for Lands in Cape Town, framing his desire to use the waters within the language of church-backed poverty-alleviation schemes, of the Kakamas 'poor-White' sort:

> There may soon come the time when arrangements are made that the billions of gallons of the Orange River-water running useless in the ocean could be used for cultivation, i.e., thousands of poor people without work could make a living. We [the Catholic Mission] are prepared to do everything possible to realise this plan of helpful and social necessity.[76]

Perhaps because of the Bishop's charitable goals, Homsrivier Church was granted riparian rights via a 1934 amendment to the South African Irrigation Act, whereby he could apply for an extraction permit.[77] Together with fifty

73 Wehrl, *Mission am Oranje*, p. 413.

74 KAB ACLT 26 File 3588 (vol. 3): Weidner to Secretary for Lands, Cape Town – 9 April 1934.

75 KAB ACLT 26 File 3588 (vol. 3): Secretary for Lands, Cape Town to Bishop Klemann, RC Mission Keetmanshoop – 12 May 1934.

76 KAB ACLT 26 File 3588 (vol. 4): Bishop Klemann, RC Mission Keetmanshoop to Secretary for Lands, Cape Town – 16 June 1934.

77 Beris, *Making the Desert Bloom (vol. 2)*, pp. 126–127.

labourers sourced from among the Catholic river-folk, Fr. Zettl and his Brothers constructed a diversion weir and a canal, which brought fifteen hectares under cultivation by 1938, growing over the following few years to fifty hectares.

With the completion of the dam and canal works in 1938, Bishop Klemann set forth a new plan for Homsrivier. Clearly influenced by the push for small-holder irrigation settlements at Kakamas, Noordoewer, Onseepkans and elsewhere, Klemann wrote to his followers in Vienna, requesting funds to construct what he called 'Mariensiedlungen' [*Mary-settlements*]. He sought to demarcate 5–6-hectare plots within Homsrivier, which would be given to 'Native Catholic Families' where they could cultivate their own ground, providing some surplus to the mission station.[78] This would be a 'close settlement', enabling more effective proselytising and provision of health services and schooling. The Mariensiedlungen were never constructed because not enough funds were raised, but this was one of the first attempts in the region to provide the same kind of institutional poverty-relief for non-Whites that poor Whites were already enjoying. The proposed size of the Homsrivier plots was the same for poor-Whites in Noordoewer and Beenbreek. Nevertheless, though Catholic Bondelswarts did not get small plots at Homsrivier, many resided on the farm, worked in its fields, and attended its church and its school. Homsrivier, therefore, was less a Catholic form of Kakamas and more a miniature Pella on the north bank of the Orange River.

By the time that Homsrivier had been passed on to the Catholic Church in 1932 and the irrigation works were complete in 1938, it had become a centre of life along the river, not unlike Pella. Ana August (née Basson) was born at Homsrivier in 1936 and was baptised by Fr. Zettl during the early days of mission work there. 'We were a fully Catholic family since the early days', she reflected in her home at Warmbad. 'My parents were married at Gaidip in 1921, and they never wanted to live far from a church. My father was a goat farmer, and when he was away in the veld, my mother and I had to stay near the priests. Homs-Drift [Homsrivier] was our home for a long time.'[79]

Indeed, each Sunday – and especially on holy days – Homsrivier church was full. Catholic river-folk would make their way to the mission station to attend Mass and shop at the small store on the premises, which by the Second World War was turning a tidy profit – giving credence to the number of Bondelswarts living along the river as well as the farming potential of the area.[80] Homsrivier became a centre for religion and learning for those staying near the Orange

78 Pfarrer Joseph Klemann, 'Mariensiedlung am Oranjefluß', *Das Licht* 31 (1938), p. 6.

79 Ana August (née Basson), interview with Bernard C. Moore (Warmbad, 7 July 2015).

80 Beris, *Making the Desert Bloom* (*vol. 2*), p. 132.

River. According to documentary evidence, in 1946 Homsrivier church had 230 Africans attending Mass each week, and the school had twenty three learners enrolled.[81] In 1954, Fr. Steiner – the new mission priest for Homsrivier – expanded the irrigation works, adding sluice gates and a small diesel pump system.[82] This nearly doubled the area under cultivation, bringing new demand for workers, such that some even came from the South African side of the river, like Witbank and Pella.[83]

Regina Mutota (née Alberts) was born in Warmbad in the 1940s, and her father was a foreman on Sandfontein Farm during her school years. It was a huge farm stocking black karakul sheep, she recalled, and the owner (Andries Baard) stayed in South Africa, checking in only every two months or so. During holiday time, she and her classmates would go to stay with their parents on Sandfontein, and they would play in the hills with some of the other farmworkers' children. She reflected that they would hide in the old sconces in the mountains. 'We would find tins buried up there, filled with old bullets and gunpowder', she said. 'I thought that it was from the war with the Germans. Maybe the Nama had a hideout there.'[84] On Sundays, one of Sandfontein's trucks would drive the workers and families to Homsrivier to attend Mass – this was the highlight of Regina's week – she would see her friends and there would always be fresh fruit to eat. 'There were so many people at Homs-Drift: Warmbad people, Pella people and the Bassons, who lived in those reed houses covered with river clay. Everyone from the river came on Sundays.'

81 NAN LKW 3/4/2 File 2/16/3: Warmbad District Native Affairs Annual Report, 1946.

82 Beris, *Making the Desert Bloom* (*vol. 3*), pp. 18–19.

83 Katrina Pieters, interview with Bernard C. Moore (Homsrivier farm, 19 September 2021).

84 Regina Mutota (née Alberts), interview with Bernard C. Moore (Gibeon, 14 October 2022).

CHAPTER 5

Grass, Meat and Skin

For many river-folk like the Bassons, the process of surveying, demarcation and sale of farmland to Whites during the 1920s and 1930s transformed them into 'tenants' on a good day and 'squatters' on a bad one. These families had to either endure the shifting whims of White farmers regarding their presence or else move to one of the Roman Catholic mission stations, like Homsrivier or Pella. By the end of the Second World War, South Africa's colonial expansion into Namibia was nearly complete, and the situation switched to one of colonial consolidation. During these years, through the development of apartheid legislation and subsidies for Whites, the SWA Administration sought to stabilise and expand White agriculture in southern Namibia, particularly within the karakul sheep industry. For southern Namibia, apartheid was largely about finding ways to turn *grass* into *meat* and *skin*, and more fundamentally it was about ensuring White control over the lands upon which the sheep grazed. While the preferred meat was mutton and the preferred skins were karakul pelts for the global fur fashion market,[1] policies would also favour and support the development of the game-farming and hunting industry.

Eendoorn Farm lies a thirty-minute drive southeast of Warmbad. Today, it is one of the last commercial sheep farms in this area just north of the Orange River, almost surrounded by either mining operations or farms owned by 'conservation' businesses. Among the first Afrikaans settlers in Warmbad/Karasburg District was the Pretorius family, who arrived before the First World War. Eendoorn was originally a SATCO farm, surveyed for sale and settlement along with the nearby farms Haakiesdoorn, Homsrivier and Girtis.[2] The farm was originally leased to D.J.A. Brand in 1910, who, it appears, returned it to SATCO a few years later, after which point it was sold to the Pretorius family.[3] Brand eventually took over the neighbouring farm, Pelgrimsrust. It is hard to imagine

1 For more on the history of the karakul industry in Namibia, see B.C. Moore, 'Land, Labour, and Karakul in Namibia, 1910s–1960s' (PhD thesis, Michigan State University, 2025). Also see J. Swanepoel, 'In the Land of the Jackals: Postcolonial Aridity in Southern Namibia' (PhD thesis, University of the Free State, 2023).

2 NAN Zentralburo des kaiserlichen Gouvernements (hereafter, ZBU) 1973 File U.V.u.13: 'General Plan der Farmen der S.A.T. Ltd im Bezirk Keetmanshoop, Deutsch Südwest Afrika' (undated, likely December 1911).

3 WDO: 'Übersicht über die Farmverkäufe und Verpachtungen der Landgesellschaften in Deutsch-Südwestafrika' (1913).

 | DOI:10.1163/9789004712454_006

FIGURE 23 Landscape on Eendoorn Farm during the dry season
PHOTO: B.C. MOORE, 2021

Eendoorn being a very productive sheep farm. The grass cover is sparse and the Skimmelberge mountains take up a good deal of the hectarage. That said, for a century the Pretorius family have continued farming when many of their neighbours have sold out to miners or conservationists.

Hendrik Erasmus (Henry) Pretorius sat down at his kitchen table, and his wife brought us some coffee and rusks. 'This drought is going on too long', he lamented, 'but there was some rain up in Ariamsvlei, so we're hopeful that things will get better soon'. Henry was not always a farmer; he started working on the family farm only once his father had passed away. His brother, Nico, owns Lugeck Farm to the north of Eendoorn. Henry did his schooling in Warmbad during the 1950s and 1960s, staying in the state hostel in town and returning to the farm on the weekends and school holidays. 'Everyone here was a karakul farmer in those days', he said. 'Even farms which weren't so productive were being put to use because the pelts were so valuable back then. Karakul built everything you see here'.[4]

By the time Henry finished school, there were few farms left that the government could sell; more or less all were occupied by White settlers. For this reason, the SWA Administration started pushing White farmers to educate their children in more than just agriculture, because there would not be enough farms for the many male children born to Afrikaans families.[5] After his military service in Rundu, Henry went to a technikon in South Africa, where he studied to be a teacher, and eventually rose up the ranks of the SWA Administration, taking on teacher training duties during the 1980s. He even spent a year living in a caravan in Ovamboland – the apartheid-era homeland in

4 Hendrik Erasmus Pretorius, interview with Bernard C. Moore (Eendoorn farm, 16 May 2020).
5 SWA Administration, *Report of the Long Term Agricultural Policy Commission* (1950), p. 68.

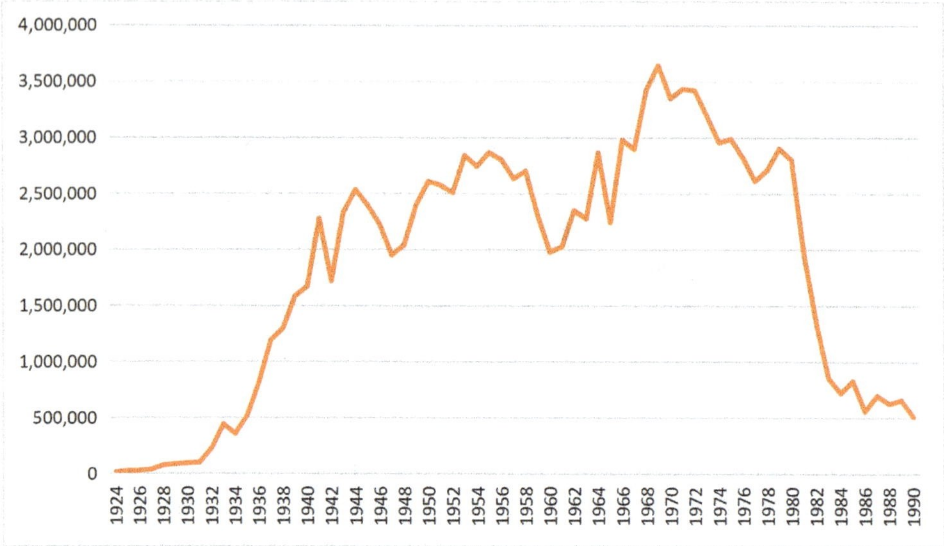

FIGURE 24 Karakul pelts exported from Namibia, 1924–1990[6]

north-central Namibia – 'advising' the homeland administration there on educational affairs.

Henry grew up at the height of the karakul industry in Namibia (see Figure 24). During the twenty five years after the Second World War, the karakul industry flourished, peaking in 1969, when more than 3.6 million pelts were exported, primarily to West Germany and the United Kingdom. Consumer demand for fur coats among the rising middle class in Europe and North America meant that there was a growing market for karakul pelts from Namibia: between twenty and thirty pelts are required to craft a woman's fur coat. Although Namibia was not exporting the gross quantity of pelts that the Soviet Union and Afghanistan were sending to the auctions in London and Frankfurt, SWA Persian lambskin pelts (known from 1966 by their trademark name 'Swakara') realised higher prices on average.[7] This was in part because Swakara production had a reputation of breeding pelts with a less-developed curl pattern, and therefore were on average lighter and easier to work with.

6 Data from L. Neubert, *The Karakul Industry: Policy Options for Independent Namibia* (Lusaka, United Nations Institute for Namibia, 1989). B. Bravenboer, *Karakul: Gift from the Arid Land, 1907–2007* (Windhoek, Karakul Board of Namibia, 2007).

7 KBN: Monthly Auction Reports, 1960–69, 1971–74, and *Swakara Nuus* magazine (Hudson's Bay & Annings, Ltd, 1973–1980). See Moore, 'Land, Labour, and Karakul in Namibia'.

FIGURE 25 Karakul pelt stretched on a drying frame, 1949
PHOTO: NAN PHOTO NO. 03693

The SWA Administration responded to the post-war boom in karakul by subsidising the erection of jackal-proof boundary fencing, camp fencing, pipeline and borehole construction, and the purchase of windmills.[8] By 1970, more than 40,000 kilometres of jackal-proof fencing and 95,000 kilometres of stock-proof camp fencing had been constructed, mostly in the southern sheep-farming

8 Moore, 'Smuggled Sheep, Smuggled Shepherds', pp. 116–119.

districts of Namibia. This enabled farmers to eliminate predators, reducing stock losses. Furthermore, rotational grazing strategies became more common, and farmers relied less on shepherds to move the animals throughout the farms. More scientific veld management enabled farmers to increase the stocking rate on their farms, supplementing the food for the additional karakul ewes with lucerne fodder grown at Noordoewer, Hardap Dam and Upington.

The jackal-proof fencing was subsidised by twenty five per cent by the SWA Administration, paid for in part by an export levy on the valuable karakul pelts en route to Western Europe. However, in order to claim the subsidies, farmers had to join a soil conservation association, contract an official ecological survey plan for their farm, and – crucially – evict all African tenant farmers resident on the property.[9] The SWA Administration sought to rationalise the farming sector in such a way that African-owned livestock was kept only within the so-called 'native reserves', effectively ensuring that all the livestock on White farms belonged to Whites alone. This provided new incentives for White farmers to evict families such as the Bassons, who were spread across the farms near the Orange River, including Eendoorn.

The White farmers who lived near major railway and transport corridors especially welcomed these labour policies, because they were able to switch about forty per cent of the workforce from local Nama labour to migrant contract workers from northern Namibia and Angola, who were paid less and did not bring their own livestock.[10] Under the new fencing regime, fewer shepherds looked after more sheep, with the average being 500 ewes per shepherd. The long-term infrastructural effects of the karakul boom were the gradual reduction of farm labour in southern Namibia and a shift in the geography of the labour-sending areas. Furthermore, with increased fencing and water infrastructure, farmers had better control over rotational grazing, predator management and monitoring of ewe pregnancies.

New technology improved production within the karakul sector, bringing more and more revenue into the pockets of White farmers. These were not 'poor-White' smallholders anymore but rather capitalist fur farmers. With that being said, technology does not remove farms from their ecological context. As Figure 27 makes clear, droughts still happen, and the period from the late 1950s to the early 1960s saw poor rainfall in southern Namibia. Nevertheless, these droughts coincided with large increases in pelt exports. We asked Henry Pretorius to tell us how his family overcame these struggles. He answered in a fairly matter-of-fact manner. 'The government at the time gave White Warmbad

9 NAN SWAA 1071 File A.138/25 (v. 2): *Soil Conservation Ordinance, 1952*.

10 See Moore, 'Smuggled Sheep, Smuggled Shepherds'.

FIGURE 26 Shepherd with karakul flocks
PHOTO: KBN ARCHIVES

farmers emergency grazing in the northwest of the country, near Khorixas', he said, referring to properties bought up to form the Damaraland homeland. 'We had to pay a small rent to graze our animals there, but the government discounted the cost of the train transport, as well as lucerne to feed them during the journey'.[11] According to Pretorius, this emergency grazing [*noodweiding*] arrangement lasted from 1962 to 1968.

Pretorius's story reveals some of the ways in which the apartheid-era system of individual family ownership of fenced, demarcated farm parcels did not really function this way at all. Although technological improvements enabled a degree of rotational grazing, allowing the stocking rate of the farms to increase, variations in rainfall meant that it was still necessary to trek to other locations during droughts.[12] This system kept the White farming sector stable,

11 Hendrik Erasmus Pretorius, interview with Bernard C. Moore (Eendoorn farm, 16 May 2020).

12 This included the Namib Desert *Sperrgebiet*, as well as the Bondelswarts Reserve after 1972. See NAN Archives of the Division of Agricultural Credit and Land Tenure (hereafter,

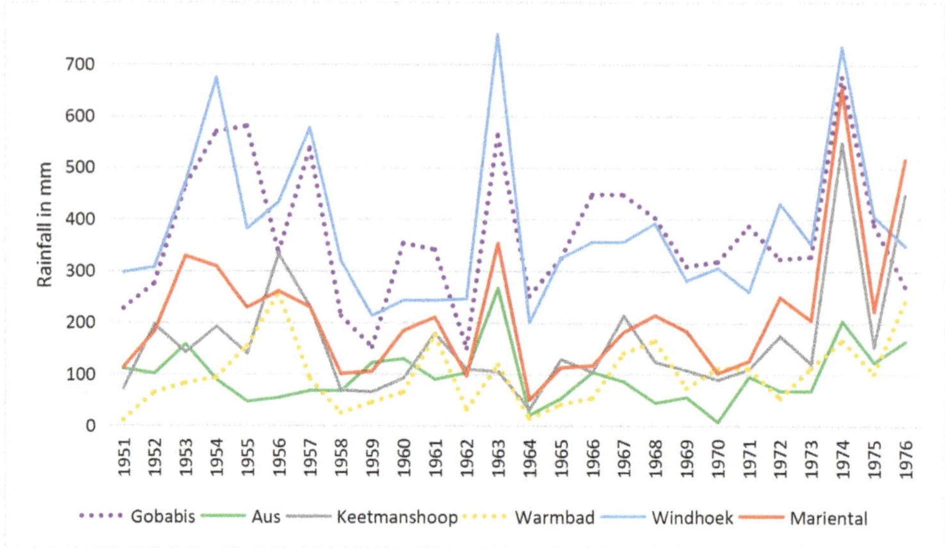

FIGURE 27 Selected Namibian rainfall patterns (1951–1976)[13]

but it was not sufficient if plans to increase karakul flocks were to continue. A new system was needed.

The drought troubles led the SWA Administration to institute a subsidised *Veeverminderingskema* [Livestock Reduction Scheme] in 1970, which would pay White commercial farmers to voluntarily cut the gross numbers of livestock on their farms by approximately one third. The farmers could then choose instead to upgrade to more valuable stock or more desert-adapted breeds. The scheme was a crucial way in which farmers, supported by the government, sought to make farming more adaptable to drought. Those who participated for the full five year term and reduced their stock numbers by at least one third received an annual cash payment per animal culled/sold. The scheme operated on a sliding scale, such that for the first 600 small stock culled, a farmer earned ZAR

LKG) 56 File 9/24/1/2: Landdros, Lüderitz to Hoof, Departement van Landboukrediet en Grondbesit 'Aansoek: Noodweiding op Staatsgrond' – 22 May 1972.

13 Data from NAN Amptelike Publikasies (hereafter, AP) 4/1/11: Departement van Vervoer: Weerburo Verslag vir die Jaar 1951; Jaarverslag 1952; Jaarverslag 1953; Jaarverslag 1954. NAN AP 4/1/12: Weerkundige Jaarverslag 1955. NAN AP 4/4/3/1: Weerkundige Jaarverslag 1954; Jaarverslag 1958; Jaarverslag 1970. NAN AP 4/4/3/3: Weerkundige Jaarverslag 1962; Jaarverslag 1975; Jaarverslag 1976. NAN AP 4/4/3/2: Weerkundige Jaarverslag 1971; Jaarverslag 1972; Jaarverslag 1973; Jaarverslag 1974. Georg O.J. Schmid (ed.) *S.W.A. Handboek* (Windhoek, South West Agency Co., 1969). H-M. Schindler (ed.) *S.W.A. Handboek* (Windhoek, South West Agency Co., 1972).

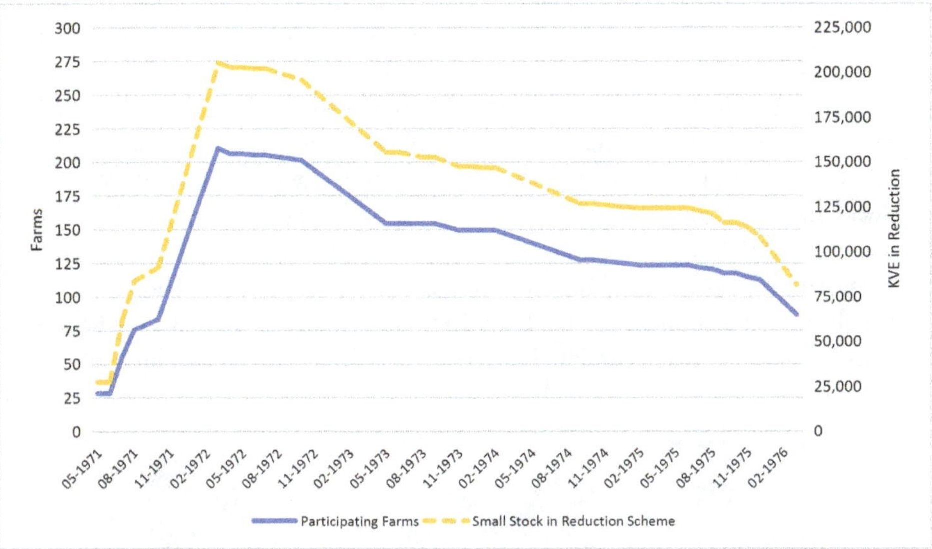

FIGURE 28 Livestock reduction scheme: Southern districts (1971–1976)[14]

3 per head annually; for the next 600, it was ZAR 2.50, and so on.[15] The scheme represented a push within the SWA Department of Agriculture away from production quantity and towards the quality of the breeding stock. Many farms in Karasburg/Warmbad District participated in the scheme from 1971, dramatically increasing over the following few years.[16] Through annual payments to farmers for killing off lesser-value livestock, the Administration sought to ameliorate the effects of drought.

We must remember, however, that the prices of Swakara pelts on the world market were still strong at this time, and for this reason farmers wanted to grow their participation in the karakul industry. The Karakul Breeders Association

14 Data from: NAN Archives of the SWA Agricultural Branch (hereafter, AGR) 533–535 File 68/6/30 (vols. 2–8): Landbou-Tegniese-Diens, SWA: 'Besonderhede van Deelnemers aan Veeverminderingskema' (1971–1976). Southern districts include Keetmanshoop, Maltahöhe, Lüderitz, Bethanie and Karasburg/Warmbad. Only about half of the participating farmers reduced their stock fully by one-third. Most, however, still made significant reductions.

15 NAN Archives of the Land and Agricultural Bank (hereafter, LAB) 17 File WM.70: Streeksveldbeampte, Landbou-Tegniese-Dienste, SWA 'Opsomming van Veeverminderingskema soos van Toepassing in Suidwes-Africa' (undated, likely April 1970).

16 NAN LAB 17 File WM.70: Land-en Landboubank van Suid-Afrika, 'SWA Deelnemers van die Veeverminderingskema' (1 January – 30 September 1971). NAN LAB 17 File WM.70: 'Deelnemers aan Veeverminderingskema soos op 20 February 1974'.

of Southern Africa (headquartered in Windhoek and Upington) funded a large amount of research to improve the quality and quantity of pelts produced. Various research and policy committees were formed with the goal of encouraging farmers to structure their breeding techniques to encourage particular hair and skin characteristics that were desirable for European furriers. Namibia could never compete with the USSR on the quantity of pelts produced, but Swakara had a higher reputation than Soviet pelts and the quality needed to be maintained.[17] Furthermore, throughout the 1970s, the karakul industry funded research into the artificial insemination of ewes and the use of hormones to synchronise and standardise the breeding and birth cycles of the whole flock.[18] This strategy allowed karakul farmers to use kraal systems, which enable

FIGURE 29 Swakara exhibition at the Keetmanshoop showgrounds, c.1975
PHOTO: KBN ARCHIVES

17 Results published in Archives of the Karakul Board of Namibia (hereafter, KBN): P.H. Hugo, 'Follikulêre Ondersoek na die Verskillende Patroonvormende Eienskappe van Karakoellammers' (MSc thesis, University of the Free State, 1982), pp. 42–46.

18 KBN: P.J. le Roux, 'Oestrus synchronisation in karakul ewes' (DSc thesis, University of Pretoria, 1974).

intensive lucerne fodder use during drought periods, and become less dependent on pasture grazing.[19]

In 1974, the rains improved across Namibia, such that during 1975–1976 at least 977 participants ceased participation in the Livestock Reduction Scheme across the territory – either voluntarily or because their five year tenure was ending.[20] For those in southern Namibia, however, the rainfall gains of 1974–1977 were ephemeral, and by 1978 drought again took hold, stronger than before.[21] This caused many farmers to rejoin the Livestock Reduction Scheme, and between 1978 and 1982, the southern districts culled more than half of the cattle and a third of the sheep.[22] During the drought, the apartheid government tried to ease the pressures on drought-stricken sheep-farming districts in the south by providing a ZAR 2 subsidy per pelt exported.[23] This was in part to offset the additional costs that sheep farmers had to spend on lucerne, and it was also a response to the global decline in pelt prices from the late 1970s, which was causing sheep farmers to increasingly slaughter their ewes for mutton.

In 1982, the SWA Administration hired a team of South African consultants to study the commercial farming sector systematically to identify structural problems and propose possible solutions. With regard to the south of Namibia, the team identified that even after significant progress by the Livestock Reduction Scheme, the average stocking rate in 1982 was still fairly high, yet the desired profits were not attainable. The commission estimated that assuming the karakul pelt price remained at its low rate at the time, fewer than one quarter of Namibian sheep farms would ever reach the levels of profitability they had once attained, even if pastures were fully stocked post-drought and unproductive neighbouring parcels were bought out to increase economies of scale for mutton production.[24] The consultants saw the problem not as a crisis

19 R.G. Niemöller Private Archive, Pofadder (hereafter, GNPA): L. Mundell, 'Carry on Kraaling', *Farmer's Weekly* (25 July 1973). KBN: A.J.A. van Niekerk, ''n Studie van Teeltprobleme by Karakoelskape met Spesiale Verwysing na die Niemöllerkudde' (DSc thesis, University of the Free State, 1972), pp. 142–146.

20 NAN AGR 535 File 68/6/30 (vol. 8): Direkteur van Bodembeskerming, Windhoek to Sekretaris, Landbou-Tegniese-Dienste, Windhoek 'Veeverminderingskema' – 29 December 1975.

21 Warmbad received 165 mm and 242 mm in 1974 and 1976, respectively; in 1980 and 1981, it recorded only 70 mm and 69 mm. Data courtesy of Namibia Meteorological Service, Windhoek.

22 Erfdeel Argief en Kultuursentrum, Windhoek (hereafter, EAKS), JWF Pretorius Versameling: S.A. Farm Consultants (Pty) Ltd, *Proposals for the Restoration of Profitability in of Farming in SWA/Namibia* (1983), p. 109.

23 Ibid, p. 17.

24 Ibid, p. 139.

of agricultural production but a crisis in the *perceived* effects of real decline. The decline in profitability was real, but it affected a highly subsidised and privileged class seeking to live at the economic zenith for White commercial farming within Namibia's rural areas. For the commissioners, this was less an existential Malthusian crisis of rangeland viability; rather, the situation reflected the declining relative size of farmers' incomes from a very high baseline. The commission was quite favourable towards the notion of properties being bought out by non-Whites, who 'for a variety of reasons, are prepared to accept a way of life resulting in a lower standard of living than apparently is acceptable to most White farmers at the present time'.[25] Unsurprisingly, the National Party buried the report. Presumably, they were unwilling to confront the unrealistic financial expectations of their rural constituencies. It was easier to blame economic changes on the weather and the climate.

•••

Not all farmers in those days were like Henry Pretorius, soldiering on through drought by switching how they farmed. Some saw the writing on the wall and switched *what* they farmed. While he is perhaps most famous as one of independent Namibia's first ministers of agriculture, Anton von Wietersheim started his career as a karakul farmer. He grew up on Gras-Süd Farm, in the area near Kalkrand village – about halfway between Karasburg and Windhoek. The farm had come from his mother's side: the Gusinde family.[26] His grandfather, Erich Gusinde, was until his death in 1965 an Honorary Chairman of the Karakul Breeders' Association and a member of the SWA Executive Committee. The Gusinde stud rams were among the most sought-after breeding stock for karakul farmers in southern Namibia.[27] After the death of his father and the completion of his studies, Anton took over the family farm in 1976. He soon noticed that prices for pelts and rams were gradually declining along with the rainfall. Despite inheriting one of the most productive stud farms in the territory, he began to sell karakul, replacing them with dorper sheep intended for mutton production. By 1982, the karakul stud had all been sold off. From the point that he took over the farm, Von Wietersheim pursued a growing new industry: *game-farming*.

25 Ibid, p. 45.

26 Anton von Wietersheim, interview with Bernard C. Moore (Swakopmund, 26 October 2020).

27 KBN: 'Erich Gusinde [Obituary]', in *Karakulzuchtverein von Südwestafrika: Jahrbuch 1966*, p. 8.

As pointed out in chapters 1 and 12, much of colonial Africa followed the 'MacKenzian Orthodoxy' regarding game legislation: a system of state custodianship of wild animals which strictly regulated hunting.[28] Under this system, hunters applied for permits from the state to hunt game, regardless of where the game lived, and permits determined the species that were eligible to be hunted during a given season. Up until the end of the Second World War, game legislation in Namibia conformed fairly well to the 'MacKenzian Orthodoxy'. Under the Germans, game species were 'considered to be ownerless according to the provisions of the law'.[29] In the Police Zone (southern and central Namibia), only Whites were allowed to hunt – permit or not. All methods of hunting without a firearm were banned, and Black Namibians, such as the Bondelswarts, were not legally permitted firearm ownership. The German Administration in Windhoek and its regional magistrates throughout the White farming districts controlled the grant of permits to (White) hunters as well as the fines for those who violated the law. After the First World War, the South Africans kept the German game legislation in place, viewing it as more than adequate for the time being.[30] While altering some specific species classification, the 1921 Game Preservation Proclamation maintained the same understanding of state custodianship of game resources. The legislation instituted the categories of 'Big Game', 'Small Game' and 'Royal Game', which was the terminology used throughout much of the British colonies in eastern and southern Africa.[31]

The 1948–1949 SWA Game Protection Commission signalled a dramatic shift away from the 'MacKenzian Orthodoxy'. Chaired by H.F. Mudge (father of the Democratic Turnhalle Alliance politician, Dirk Mudge), the commission was tasked to investigate the requests by the newly founded and increasingly political Suidwes-Afrika Landbou Unie [Agricultural Union: SWALU], that White commercial farmers receive property rights [*eiendomsreg*] over the game within their fenced farm boundaries.[32] Over the previous few decades, farmers had been considered by government to be poor custodians of wildlife, and that giving them property rights over game would create a country of

28 See, J.M. MacKenzie, *The Empire of Nature*, and L. van Sittert, 'Bringing in the Wild'.

29 NAN ADM 11 File 40/2: Translation of Game Law: German South West Africa (1 March 1909).

30 NAN ADM 11 File A.40/15: Secretary for the Protectorate to Acting Secretary, Transvaal Game Protection Association – 9 December 1915.

31 Inter alia, Royal Game includes elephant, rhinoceros, giraffe; Big Game includes kudu, wildebeest, gemsbok; Small Game includes springbok, warthog, steenbok. NAN SWAA 1345 File A.205–3: Game Preservation Proclamation, 1921.

32 NAN AP 5/7/4: Verslag van die Suidwes-Afrika Wildbeskermingskommissie, 1948.

'biltong farmers', threatening the imperial hunting treaties that emanated from Europe.[33] The Game Protection Commission was ultimately riddled with conflicts of interest, because all the officials appointed to it were also Afrikaans-speaking members of SWALU itself. Even so, some had reservations about whether or not a stock-proof fence (four to six horizontal wires) could actually hold game within a certain geographical space so as to maintain property rights. The dissenters noted that 'property rights cannot be continuously transferred depending upon which side of the fence the game is sighted'.[34] Although some did view farmers as potential 'conscious protectors' of game ['*gewetens-beskermers*'], the 'biltong farmer' reputation was not yet extinguished.

However, the Commission agreed with most aspects of the SWALU proposal, and these recommendations were submitted to the SWA Executive Committee in 1949, with nearly all being passed into law with the 1951 Game Protection Ordinance.[35] The legislation decreed that 'any [White] owner or occupier of a farm shall *own all game other than protected game* while such game is lawfully upon such farm and while such farm is enclosed with a sufficient fence'.[36] The legislation eliminated the category of 'Royal Game', and game was simply classified as big/small ('huntable') game and 'protected game', the latter category including species like elephant and rhinoceros. Farmers, potentially, could now shoot 'huntable' game throughout the year on their own properties, having to report to the local magistrate only the number of big game killed, and they could hire out hunting rights to others to shoot on their land, for a fee.[37] This legislation did not grant property rights over game to Black Namibians, however. The reserves were technically the property of the SWA Administration, and game residing within the reserve boundaries still fell within state custodianship.

It is not a coincidence that this new game legislation coincided with the rise of the karakul pelt industry and the finalisation of the Land Settlement

---

33 NAN SWAA 1347 File A.205/1 (vol. 6): 'A Neglected Asset of SWA', translated article from *Allgemeine Zeitung* – 6 February 1926.

34 ['*Eiendomsreg kan nie voortdurend oorgedra word na gelang van die kant van 'n heining waar die wild sig bevind*']

35 Italics ours. Basler Afrika Bibliographien Archives (hereafter, BAB) Personenarchiv: Hubertus Graf zu Castell-Rüdenhausen (hereafter, PA.24) File VI.A.1: Ordonnansie op Wildbeskerming, 1951.

36 Ibid, sec. 6(6). A sufficient fence is deemed to be a four strand galvanised-wire cattle fence or a superior variety, such as jackal-proof. Concerning the legal concept of 'sufficient fence' and 'superior fence', see NAN SWAA 2206 File A.486/1 (vol. 2): Attorney-General, Windhoek to Secretary for SWA 'Fencing Proclamation, 1921: Jackal-Proof Fencing' – 20 April 1954.

37 BAB PA.24 VI.A.1: Ordonnansie op Wildbeskerming, 1951, sec. 8(2).

Programme. In the 1950s, the SWA Lands Branch settled the last substantial blocks of farmland in southern Namibia, those along the Fish River Canyon and along the Namib Desert.[38] A 1965 SWA commission into Nature Conservation affairs found that upon receiving leases or deeds to property along the Namib, farmers immediately decided to 'thin out' [*uitdun*] the game on their property to allocate grazing to the growing flocks of karakul.[39] Between 1955 and 1960, there was a substantial decrease in game numbers on farmland across SWA,[40] which accelerated further in 1961–1962, when the foot-and-mouth disease outbreak meant that many farmers increasingly believed that one could not farm stock and game simultaneously. A farmer had to apply for a hunting permit, but this was simply a formality now; game hunting on private land was not restricted by season.

Although the 1965 Nature Conservation Commission did recognise that farmers sometimes made too heavy use of their property rights over game, it ultimately argued that 'After all, game eats a farmer's grazing, drinks his water, tastes his licks and damages his fencing. If game had monetary value, like a cow or a sheep has, then it would automatically result in better protection'.[41] For the Commission and the Administration, favourable legislation was the motor to drive profitability. With the passing of the Game Parks, Nature Parks and Private Game Reserves Ordinance of 1958, a number of farmers immediately applied to register their properties (or group of properties) as private nature reserves. There were seventeen in operation by 1965.[42] The legislation was re-affirmed in 1967 in slightly modified form, making it clear that (White) farm owners were permitted to hunt game on their farms throughout the year, in any season.[43]

38 See Moore, 'Land, Labour, and Karakul in Namibia'.

39 NAN BB/0042: Verslag van die Kommissie van Ondersoek na Natuurbewaringsaangeleenthede in SWA (1965), p. 6.

40 One study saw decreases in kudu (by 16.1%), springbok (18.4%), gemsbok (9.2%), red hartebeest (31.3%) and zebra (17.7%). See NAN Archives of the Division of Nature Conservation (hereafter, NTB) 1/8 File N.13/2 (v. 2): J.S. van der Spuy, 'Die Huidige Status van Koedoe, Gemsbok, Springbok, Eland, Rooihartbees, Bergsebra en Volstruis op Plaasgebied in Suidwes-Afrika' – February 1961.

41 [*'Wild vreet immers 'n boer se weiding, suip sy water, vreet sy lekke en beskadig sy heinings. As wild 'n geldwaarde, soos 'n bees of 'n skaap het, behoort beter beskerming outomaties te volg'*]. NAN BB/0042: Verslag van die Kommissie van Ondersoek na Natuurbewaringsaangeleenthede in SWA (1965), p. 17.

42 Ibid, p. 42.

43 BAB PA.24 VI.A.1: Nature Conservation Ordinance, 1967.

As Bernabé de la Bat, the long-term director of the Division of Nature Conservation (NTB) wrote, 'we conserve nature for man, not for its own sake'.[44] In the context of the quest for maximum profitability within the agricultural sector – especially karakul sheep-farming – game legislation was transformed to enable the farmer to profit most from his property, either by culling game numbers to conserve grazing or by jointly farming with game for meat production or trophy-hunting. This was apartheid, after all, and apartheid in the farming areas of rural Namibia was primarily about turning *grass* into *meat* and *skin*, as well as who had the land rights and economic privilege to do so.

As the homeland system in Namibia expanded from the late 1960s, so did the budget of the Division of Nature Conservation, and the NTB was eventually merged with the Department of Agriculture. The state viewed nature conservation and agriculture as analogous: both involved *grass*, *meat* and *skin*. The NTB further earned revenue by venturing into the newly created homelands – particularly Damaraland, Kaoko and Caprivi – where their game capture team would (without homeland approval) remove thousands of wild animals for relocation to either national parks or onto private game reserves. This included both 'huntable game', like zebra, as well as protected species, like black rhinoceros, which were often sold to private game dealers for 'tens of thousands of rand', according to De la Bat.[45]

•••

'*Nagoes* started on my farm, Gras-Süd', said Anton von Wietersheim. 'It was where they did the first test operations in the 1970s'.[46] By the time that Anton took over his family farm, he had a good relationship with NTB and Bernabé de la Bat, and in the face of decreasing karakul pelt prices and decreasing rainfall, the NTB pushed for a fundamentally new relationship between farmers and game, called the night harvests [*Nagoes*]. Although White farmers themselves already had the right to hunt game for meat, skins, trophies, etc., night harvests were on a completely different scale. The goal was to cull up to one third of the entire game population in *a single night*. The goal was speed, efficiency and profitability. Selling game meat to Cape Town hotels was profitable (the price

44 NAN Archives of the SWA Administrator General (hereafter, ADG) 2/4: Bernabé de la Bat, 'An uneasy marriage' *African Wildlife* (vol. 27, likely 1979).

45 NAN ADG 2/4: Direkteur, Natuurbewaring en Toerisme 'Memorandum: Toedeling van die Regeringsfunksies ten opsigte van Natuurbewaring en Toerisme aan die Gesagowerhede op die Eerste en Tweede Regeringsvlakke' – 15 February 1980: Bylae A, p. 12.

46 Anton von Wietersheim, interview with Bernard C. Moore (Swakopmund, 26 October 2020).

FIGURE 30 NTB operations in 1980/81. The original caption reads 'In the biggest single game capturing operation ever launched in southern Africa, 1500 Burchell's zebra were caught and sold to farmers'.[47]

had increased almost fivefold during the 1970s): farmers were able to earn up to ZAR 28 net profit per head of springbok culled and ZAR 180 per head of kudu.[48] Of course, this was after the farmer paid the shooting teams.

Skietspanne ['shooting teams'] would arrive on the farm during the afternoon, and they would survey sites where they could corral game into corridors

47 NAN JX/0012: B.J.B. de la Bat, 'Etosha: 75 Years', *SWA Annual* (1982), p. 21. See also NLN TXX 0742: Departement van Landbou, Bosbou en Natuurbewaring: Jaarverslag, 1980, Bylae D, p. 2.

48 This is a very reassuring rate when the net profit per slaughter sheep had declined to less than ZAR 9 each. After the *skietspanne* costs (ZAR 10 per springbok), butcher costs (ZAR 3 per springbok) and transport costs (ZAR 0.10 per kg to Cape Town) were taken into account, the profit for the farmer per springbok was about triple that of sheep. The profit was higher for kudu as well. E. Joubert, P.A.J. Brand and G.P. Visagie, 'An appraisal of the utilisation of game on private land in South West Africa', *Madoqua: Tydskif vir Natuurbewarings – en Woestynnavorsing in SWA*, 13, 3 (1983), p. 205.

FIGURE 31 Skietspanne preparing for the night harvest
PHOTO: NATIONAL ARCHIVES OF NAMIBIA

to enable clean shots and kill large numbers. After all, a bullet that did not hit the head or neck would condemn the entire carcass; bullet fragments inevitably get into the meat. The hunting was collaborative, between the drivers, the marksmen and the spotlight operators. Night harvests were a particular arrangement, supervised by the NTB, who trained the hunters and granted them special licences to shoot at night. *Skietspanne* (often six to ten teams per night harvest) were highly skilled and well equipped, and in a single night, they could cull 500–600 springbok on a single farm.[49]

In order to manage the large quantity of game carcasses, a mobile abattoir was constructed on the farm, together with refrigerated trucks. Springbok carcasses need to be eviscerated and bled within thirty minutes, or else bacterial infection from the entrails occurs. Teams would routinely deliver hunted springbok to the abattoir, where the entrails would be rapidly cut out and the abdominal cavity sealed. The carcass would then be hung in the truck until the load was full or the *Nagoes* was over. The meat would then be shipped by road to Cape Town, 1,000–1,600 kilometres away.

49 Ibid, p. 206.

FIGURE 32 Springbok carcasses delivered to the mobile abattoir
PHOTO: NATIONAL ARCHIVES OF NAMIBIA

FIGURE 33 Refrigerated truck at a Nagoes operation
PHOTO: NATIONAL ARCHIVES OF NAMIBIA

Anton von Wietersheim accompanied the NTB to many of the *Nagoes* operations throughout southern and central Namibia, and his farm was where many of the hunters were trained and certified in early 1976.[50] In the early days, the precision of the hunters was the deciding factor because some of the early harvests had resulted in far too many condemned carcasses. It was only after the NTB began offering training courses the following year, specifically for *Nagoes*, that the system began to run smoothly. Furthermore, as the operations became more profitable, more *skietspanne* were registered, streamlining the process of culling game numbers quickly. NTB officials could survey a farm, and within a week or two, the night harvest could take place.

Von Wietersheim joined SWALU as their 'Game Farming Chairman', and he worked with Hartlief Slaughterhouse to open a large-scale game abattoir in Windhoek in 1980, which meant that the carcasses did not necessarily have to travel to Cape Town. The abattoir received European certification the following

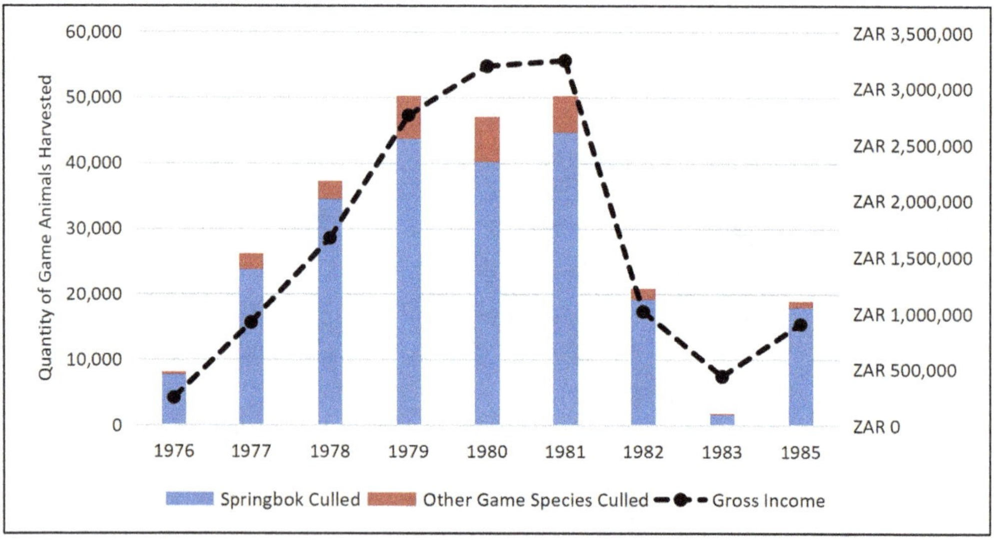

FIGURE 34 Night harvest returns (1976–1985)[51]

50 Anton von Wietersheim, interview with Bernard C. Moore (Swakopmund, 26 October 2020).

51 Data from: NLN TXX 0742: Dept. van Landbou, Bosbou and NTB: Jaarverslag, 1980. NLN TXX 0742: Dept. van Landbou and NTB: Jaarverslag, 1981. NLN TXX 0742: Dept. van Landbou 7 NTB: Jaarverslag, 1982. NLN TXX 0742: Dept. van Landbou, NTB en Veeartsenydiens: Jaarverslag, 1983. NLN TXX 0742: Dept. of Agriculture and Nature Conservation: Annual Report, 1985. Note that the 1984 Annual Report held no data about night harvests.

year, and venison exports to West Germany and France took off. Anton von Wietersheim saw the situation as Bernabé de la Bat did, in that the 'hunt as adventure' was maintained 'in combination of economic game use and the development of love and responsibility of the individual farmer, hunter and resident of SWA'.[52] The ultimate goal for Von Wietersheim was the long-term stability of farming, though others saw it as a much more short-term strategy: to complement the lost revenue from drought and the decline in the karakul industry.[53] As Figure 34 shows, most farmers who participated in night harvest operations likely saw eye-to-eye. Night harvest culling and profits inversely mirrored rainfall patterns. As the drought took hold, farmers culled their game to make up for lost revenue from karakul and other forms of livestock production. During the height of *Nagoes*, millions of Rands were paid out to White farmers.[54] There were also some farmers who saw night harvest operations as a 'get-out strategy': hunt everything on the farm and then sell out and move back to South Africa before a Black majority government could come to power. This view appears to have been more common in the Namib Desert farms of south-western Namibia during the 1980s.[55]

The farming economy in apartheid-era Namibia was primarily about *grass*, *meat* and *skin*. Even more fundamentally, it was about maintaining White control over those lands on which grass, meat and *skin was produced*. In the face of irregular rainfall and drought, as well as fickle prices for karakul pelts, White farmers like the Pretorius family changed how they farmed with sheep, culling unproductive stock and moving animals to reserve grazing far away. But by the late 1970s, it was not that easy to access emergency grazing on African communal lands anymore. As we elaborate upon further in chapter 7, apartheid planners understood that the political climate was changing. Night harvests were a solution for many like Von Wietersheim to reduce grazing pressure by reducing the

52 [*'In der Kombination von ökonomischer Wildnutzung und der Entwicklung von Liebe und Verantwortung durch den einzelnen Farmer, Jäger und Einwohner Südwestafrikas'*] A. von Wietersheim, 'Im Jagdrevier zwischen Kalahari und Namib (1984)', in *Vom Schutzgebiet bis Namibia 2000* (Windhoek, 2000), p. 404.

53 Berndt Rothkegel, interview with Bernard C. Moore (Windhoek, 23 June 2020).

54 The post-1978 files from the Division of Nature Conservation within the Administration for Whites have not yet been catalogued by the National Archives of Namibia, so we cannot yet disambiguate precisely which farms took part in *Nagoes* operations. Von Wietersheim remembers, however, that the farms in the area around Warmbad and Ariamsvlei participated, even though springbok numbers were a bit lower than farther north. He personally attended, as a member of a *skietspan*, one operation on Heirachabis farm, a portion of which was leased to a karakul stud-breeder.

55 Walter Theile, interview with Bernard C. Moore (Namtib farm, Lüderitz District, 25 September 2020).

number of non-domesticated ungulates on the properties. *Nagoes* operations revealed the ways in which farms in southern Namibia were remarkably adaptable and alive, despite economic downturn and drought. It is quite likely that the scale of these operations was unsustainable and determined more by the needs of the commercial farming lobby than genuine ecological assessment. But then again, they were not maintained for beyond the mid-1980s.

For Nama families in the Bondelswarts communal areas and for river-folk like the Bassons, the years between the Second World War and the coming of Namibian independence in 1990 were a period of White consolidation of the countryside. As sheep-farming became more profitable and capital-intensive, the Bassons' presence as tenants or 'squatters' became more precarious. This was further complicated once Whites obtained property rights over game and saw profits rocket with night harvests. Thus, commercial agriculture dominated the economy of the district. The next chapter explores the emergence of mining in the area, revealing other ways in which humans interacted with these mountainous lands near the Orange River.

CHAPTER 6

Namibia's Atomic Age

With increasing profits from the karakul industry – and later from hunting and game meat – White farmers in southern Namibia took advantage of the apartheid state subsidies in order to more fully utilise their properties. Higher capitalisation and higher profits meant a growing reluctance to allow for tenant farmers and 'squatters' on their lands. Alongside the growth in commercial agriculture came the rise of mining capital in southern Namibia and the Northern Cape. In the context of apartheid, it is not surprising that river-folk like the Bassons would not be able to seriously profit from this boom. Nevertheless, Willem Basson and others have surprisingly fond memories of the Weidner and Niemöller mining families during those years. To understand the increasingly restricted pastoralism by the river, we must examine the role of mining in the region, and the relationship between river-folk and these mining families.

For the Weidner family, everything would change after the Second World War. Charles Weidner, the family patriarch, would pass away on 12 February 1949, less than a month after his cherished irrigation farm Goodhouse was sold to the South African government.[1] Weidner's reputation was the only thing larger than his ego, and virtually all White travellers visiting the Orange River passed through Goodhouse, cementing legends about the man in well-known travelogues like Green's *To the River's End*, Birkby's *Thirstland Treks* and others.[2] For all the lifetime success that he had in land speculation and farming, it was nothing compared to what his son, Peter, would attain over the following thirty years.

Born in 1898, Peter Weidner grew up at Goodhouse on the Orange River.[3] Too young to be called up to serve for the South Africans in the First World War – though he assisted with feeding and boarding the SA Mounted Rifles during the SWA Campaign – Peter was sent to Grootfontein Agricultural College in the Karoo, one of the top agricultural schools in the country.[4] Presumably, his

1 KAB Archives of the Master of the Orphan Chamber (hereafter, MOOC) 6/9/15809 File 1076/49: Death Notice: Charles Weidner, signed at Springbok – 24 February 1949.

2 L.G. Green, *To the River's End* (1948), pp. 112–119. C. Birkby, *Thirstland Treks* (1936), pp. 25–37. For more on Charles (Carl) Weidner, see Moore, 'Swimming Upstream', and M. McKittrick, *Green Lands for White Men: Desert Dystopias and the Environmental Origins of Apartheid* (Chicago, University of Chicago Press, 2024).

3 For the rest of this book, the surname 'Weidner' will refer to Peter J. Weidner, rather than his father.

4 NAN ADM 38 File 349 (vol. 2): C. Weidner to Magistrate at Warmbad – 7 May 1917.

FIGURE 35 Peter J. Weidner at Tantalite Valley, 1967
PHOTO: P. HOLZ, 'THE PELLA GROUP STORY' (KARASBURG, PELLA GROUP, C.1969), P. 4

father sent him to master ostrich-breeding, which the family thought would supplement the irrigation works at Goodhouse. After completing his studies, Weidner remained with his father at Goodhouse until 1925, when he had the opportunity to start a dairy farm in the Transvaal. But the Great Depression doomed the business, resulting in him taking up a government position with the Ermelo municipality.[5] According to his memoirs, by 1942 Peter Weidner was faced with the option of joining the war effort as a middle-aged man or returning to the family home at Goodhouse and trying to help run his father's business; he chose the latter.

Weidner was not really interested in taking up irrigation farming again. Given his father's advanced age, Goodhouse was not as productive as it once was. Rather, he was intrigued by the public tenders offering up to GBP 2,000 per tonne of tungsten (wolfram), to support the Allied powers in the manufacture

5 P. Weidner, *The World Owes You Nothing* (Cape Town, self-published, 1969), pp. 7–15.

of weapons and equipment.[6] There were some outcroppings of scheelite ore (calcium tungstate) on Goodhouse and nearby, which Weidner and his father's workers mined with rudimentary tools. Because of the high prices offered during the war, they brought in enough profit to pay off any debts. With the eventual end of the Second World War, the price of scheelite collapsed, and Weidner needed something else to take its place. He invested some of the scheelite money into karakul sheep for a bit, but he was not really a farmer at heart. However, in late 1948, just before his father's death, Weidner would ride the Goodhouse pontoon across the Orange River with a geologist from a strange new agency, the South African Atomic Energy Board.

•••

Running across the Orange River in a south-easterly direction is a geological formation called the Pofadder Shear Zone (see Figure 36). In geology, a shear zone is a portion of the Earth's crust and mantle, which is deformed by comparatively small fault lines slipping past one another. This shifting facilitates lateral movement and allows igneous rock to rise closer to the surface and cool. As some of this magma and other molten rock starts to cool in confined quarters, it slows down and increases viscosity, enabling trace elements of rarer materials to come together and form much larger crystals within the igneous rock than would have normally occurred in volcanic processes. These kinds of stone are often called pegmatites.[7] Although pegmatites are related to granite, the specific ways in which the shear zone affects the viscosity of the rock enables much larger crystal precipitation than would be found in ordinary granite.[8]

Depending on the formation pattern and subterranean geological contents, the specific make-up of the crystals varies widely. Sometimes pegmatites present themselves primarily with quartz, feldspar and other common, non-valuable minerals. At other times, the crystallisation allows for extracting comparatively significant quantities of lepidolite, spodumene and alaskite (leucogranite): ores containing lithium, tantalum, columbite, beryl, niobium, caesium, thorium and uranium.[9] On the Namibian side of the Pofadder Shear

6 Ibid, p. 26.

7 C.W. Lambert, 'Granitic melt transport and emplacement along transcurrent shear zones: Case study of the Pofadder Shear Zone in South Africa and Namibia' (MSc thesis, Stellenbosch, 2013), pp. 19–21.

8 A.M. Evans, *An Introduction to Ore Geology* (2nd ed.) (Oxford, Blackwell, 1987), pp. 117–119.

9 D. Bradley and A. McCauley, *A Preliminary Deposit Model for Lithium-Cesium-Tantalum (LCT) Pegmatites* (Reston, US Geological Survey, 2013). Pegmatites are also sought out

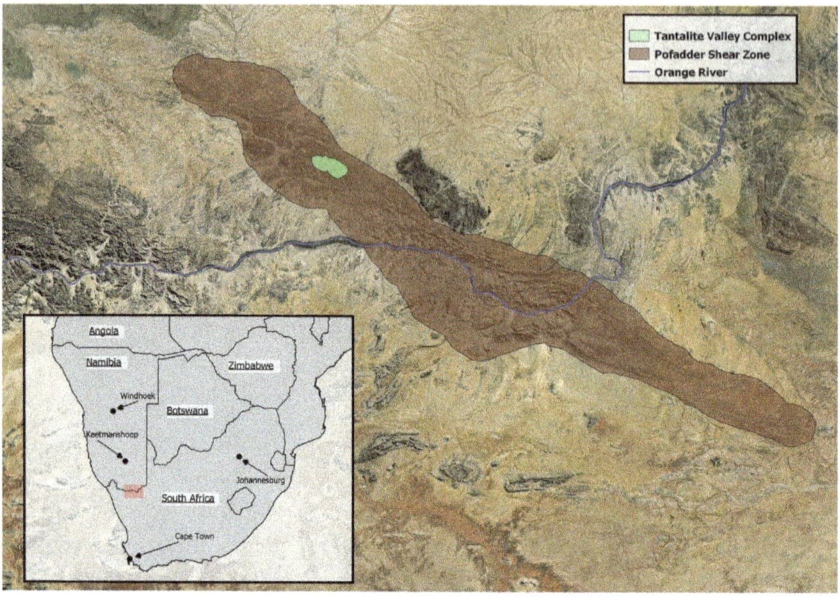

FIGURE 36 Tantalite Valley Complex in the Pofadder Shear Zone, crossing the Orange River
CARTOGRAPHY: B.C. MOORE

Zone lies a particularly rich pegmatite formation, today called Tantalite Valley, which has large amounts of lepidolite and spodumene.[10]

As early as the First World War, it was already known that certain kinds of leucogranite pegmatite deposits contained uranium or thorium, though, until the 1960s discovery of the massive Rössing deposits near Swakopmund, most of these deposits were thought to be too low grade.[11] For this reason, most of the uranium produced through the first three quarters of the twentieth century came as a byproduct or supplement of the gold and copper industries.[12]

---

for their gem-bearing qualities, such as aquamarine, tourmaline and other semiprecious stones, and occasionally more valuable stones like emerald and sapphire. See J.E. Shingley and A.R. Kampf, 'Gem-bearing pegmatites: A review', *Gems & Gemology* (Summer 1984), pp. 64–77. G. Giuliani et al., 'Emerald deposits: A review and enhanced classification', *Minerals*, 9, 2 (2019). G. Giuliani and L.A. Groat, 'Geology of corundum and emerald gem deposits: A review', *Gems & Gemology* (Winter 2019), pp. 464–489.

10 A.C. Moore, 'The petrography and regional setting of the Tantalite Valley Complex, South West Africa', *Transactions of the Geological Society of South Africa*, 78 (1975), pp. 235–249.

11 A.R. Newby-Fraser, *Chain Reaction: Twenty Years of Nuclear Research and Development in South Africa* (Pretoria, Atomic Energy Board, 1979), p. 82.

12 Ibid, p. 65.

Nevertheless, in 1940, under the auspices of the confidential United Kingdom 'Tube Alloys' atomic weapons programme, John Anderson of the British War Cabinet arranged a meeting with Field Marshall Jan Smuts, requesting that South Africa supply radium and uranium for Tube Alloys and the US nuclear programme, the Manhattan Project.[13] Smuts immediately took action, targeting specifically the Northern Cape and the Orange River region, and Namibia's Swakopmund district.

Between 1942 and 1944, a scientist named S.C. Wiid was sent by the South African Geological Survey to map out the pegmatite intrusions of southern Namibia and the Northern Cape, falling within what we today call the Pofadder Shear Zone. Using aerial photographs and boots-on-the-ground surveying covering 900 square kilometres, Wiid identified the presence of niobium, tantalum, bismuth, lithium and beryllium within the Tantalite Valley pegmatite complex, and passed on this data to the South African government.[14] Wiid's reports were not viewed as urgent, and uranium for the Manhattan Project was primarily sourced from the Belgian Congo. Nevertheless, after the passage of the Atomic Energy Act of 1948, increased funds were dedicated to prospecting and surveying for uranium sources, not just for military purposes, but also with the goal of developing atomic power within South Africa itself.

In 1947 Geological Survey had already contracted Rio Tinto (which was operating several mines in Zambia at the time) to drill boreholes and take core samples along the Pofadder Shear Zone in one kilometre intervals. This pegmatite data intrigued the Atomic Energy Board, who sent their own geologist the following year to examine the status of a spot called Signalberg, at the heart of the Tantalite Valley complex.[15] Around this time, Peter Weidner was running the pontoon transport across the Orange River from Goodhouse to Haakiesdoorn and 'roaming the mountains' looking for beryl crystals on the ground to sell.[16] The high price of beryllium during the mid-1940s was bringing in more money than sheep or irrigation farming, and Weidner was hoping to move full-time into prospecting and mining.

The geologist for the Atomic Energy Board happened to be crossing the pontoon on that particular day in late 1948, and was due to sleep at Goodhouse. Presumably, he knew old Charles Weidner from a long way back. Peter Weidner

13 Ibid, p. 21.

14 Archives of the Ministry of Mines and Energy of the Republic of Namibia (hereafter, MME) Report #560: J.W. von Backström, 'The Geology and Mineral Deposits of Tantalite Valley, Warmbad District, SWA' (Atomic Energy Board, April 1976).

15 Peter J. Cox, interview with Bernard C. Moore (telephone, 25 March 2023). Cox believes the geologist was a Dr Vermaak, though he cannot remember the given name.

16 Weidner, *The World Owes You Nothing*, p. 42.

was looking after his father in the heat of summer and was able to meet the geologist on the pontoon and share supper with him that night. Whether through collusion or simply letting facts slip after too much wine, the geologist allegedly informed Peter Weidner of fairly rich pegmatite intrusions on Umeis and Kinderzitt farms – north-east of Goodhouse – implying that Weidner should stake prospecting claims before he filed his paperwork in Cape Town in the coming weeks.[17] Although it was not as rich in uranium as the Atomic Energy Board had hoped, the area was abundant with other rare earth minerals relevant to the coming atomic age.

•••

The death of his father in February meant that Weidner would not reach Umeis and Kinderzitt farms until 14 September 1949. Prior to the discover of mineral reserves, the history of these farms echoed that of their neighbours. Kinderzitt had been leased out by the German Administration to a settler, who eventually returned it after the First World War.[18] Both farms were surveyed by H. de Waal in 1920, but were used just for grazing licences until 1938, when they were given out through the Land Settlement Programme. Like most of the farms in the area, Umeis and Kinderzitt were moderately successful sheep farms for most

FIGURE 37 Tantalite Valley's largest pegmatite vein, 'Homestead'
PHOTO COURTESY OF THE MINISTRY OF MINES AND ENERGY, WINDHOEK

17 Peter J. Cox, interview with Bernard C. Moore (telephone, 15 January 2023).
18 NAN LAN 239 File 535: Pachtvertrag, Kinderzitt: Ferdinand Otto Grimm.

of their existence. For example, on Umeis in 1955, F.A. Herridge successfully farmed with 570 sheep (mixed mutton and karakuls) and 450 goats, bringing in a respectable annual income from pelts and meat.[19]

When he arrived at Umeis, Weidner met Herridge, whom he likely already knew because the Weidner and Herridge families were long-time residents of the Orange River region. Herridge confirmed that a government geologist had found lithium in the mountain that stretched over the boundary between his farm and Kinderzitt.[20] Weidner pegged twelve mineral prospecting claims within the Tantalite Valley complex, declaring his primary interest in lepidolite and spodumene, and within four years he made an official application to combine the claims into a full mining licence,[21] which was approved in May 1955.[22] Having officially obtained a mining licence, in 1956 Weidner bought the small (approximately 450-hectare) Tantalite Valley subdivisions of Umeis and Kinderzitt. In 1969, he purchased the rest of Umeis and in 1971 the rest of Kinderzitt.[23] The mining licences, the farmland and the equipment were eventually registered under his company, Tantalite Valley Minerals (Pty) Ltd, a wholly owned subsidiary of Pella Refractory Ores (Pty) Ltd, which owned land and mines in South Africa.[24] Once Weidner realised that Tantalite Valley was a portion of a larger geological formation stretching across the Orange River – the Pofadder Shear Zone – he and his oldest son, Charlie, began prospecting in the area west of Pofadder in the Northern Cape, eventually getting a licence for what would be called Swartkoppies Mine.[25]

So, what were the Weidners actually selling? By September 1954, workers at Tantalite Valley – many of whom had lived by the Orange River for generations – had mined more than 6,700 kilograms of beryllium oxide, 100 kilograms of tantalum pentoxide, 19,200 kilograms of lithium oxide, and smaller quantities of niobium pentoxide, bismuth and microlite.[26] During the 1950s and 1960s, tantalum and columbium were considered 'minerals of strategic importance'

---

19 NAN LAN 240 File 535/3: Opgawe van Vee en Gereedskap: F.A. Herridge, Umeis no. 110 – 22 June 1955.

20 Weidner, *The World Owes You Nothing*, p. 50.

21 NAN Archives of the SWA Inspector of Mines and Works (IMW) 424b File P.T.4/3: P. Weidner to Inspector of Mines – 22 September 1953.

22 NAN IMW 424b File P.T.4/3: Offisiële Koerant (no. 1900 of 2 May 1955).

23 WDO: Registrasie Afdeling 'V': Umeis no. 110 and Kinderzitt no. 132.

24 NAN IMW 424b File P.T.4/3: P. Weidner to Registrar of Companies – 9 October 1953. NAN Company File F/53/211: Memorandum of Association: Pella Refractory Ores S.A. (Pty) Ltd – 27 January 1953.

25 Weidner, *The World Owes You Nothing*, p. 55.

26 NAN IMW 424b File P.T.4/3: Chairman's Report, Tantalite Valley Minerals (Pty) Ltd – 3 September 1954.

FIGURE 38 Peter Weidner (centre) and son Charlie inspect the pegmatite operations on the 'Homestead' outcropping
PHOTO: GNPA

by the governments of the United States, United Kingdom and many other western European countries. Although tantalum is better known today for its use in semiconductors, computers and mobile phones, after the Second World War it and beryllium were widely used for weapons manufacture. Mixing these minerals in carbides and alloys reduces corrosion, and their high temperature strength allows for their application for jet engines, gas turbines, airframes and ballistic missiles.[27] Furthermore, both tantalum and columbium (the latter mined in smaller quantities on Weidner's mines) are used in the manufacture of cladding for nuclear reactors.

Weidner saw himself as connecting southern Namibia to the world and having Tantalite Valley contribute to the atomic age. He wrote:

> You will appreciate that the world is crying for the atomic energy materials, both for military and commercial purposes. The atomic age is, as yet, in its infancy. Since 1946 when we were in dire financial difficulties, we have religiously clung to the "Theory" that we cannot go wrong if we peg for atomic energy minerals. The years that have passed have strengthened that belief. Providence has been kinder to us than to millions of

27 F.T. Sisco and E. Epremian (eds), *Columbium and Tantalum* (New York, John Wiley & Sons, 1963), pp. IX, 560–596.

> others. We hope that in return by our efforts, given good health and blessing from above, to put Tantalite Valley and the Southern Portion of South West Africa on the map.[28]

Tantalite Valley was profitable for Weidner over the years, it was not this that made him south-eastern Namibia's first multimillionaire. Swartkoppies sillimanite mine, west of Pofadder on the edges of Pella Mission Station would actually bring in most of Weidner's revenue over the years, though it was still tied to the industrial and economic changes of the post-Second World War era. It also showed that the Pofadder Shear Zone was not just about pegmatites.

Sillimanite minerals – including sillimanite proper, andalusite and kyanite – are an aluminium-silicate ore. When refined into mullite they can withstand extremely high temperatures (up to 1,840°C), which makes them industrially important as refractory ores for the operation of foundries and smelters.[29] For the first few decades after the Second World War, sillimanites were the dominant ingredient for the manufacture of alumina refractory bricks. In addition, the expansion of the steel industry – especially in Japan and West Germany – was dependent on a steady supply of sillimanite. The ore is not particularly rare, but finding it in exploitable quantities is more difficult. Two particular mines in the Northern Cape – Swartkoppies (belonging to Weidner) and Hotson (belonging to Rudolf Gerhard Niemöller) proved to be two of the largest sillimanite reserves in the world. Between the years 1954 (when Swartkoppies came into operation) and 1963, Northern Cape mines contributed annually between thirty per cent and sixty per cent of the global production of sillimanite.[30] Importantly, the particular corundum-sillimanite deposits at Swartkoppies contained the correct ratio of corundum and aluminium required by several major Japanese foundries.[31]

After Weidner staked his mining and prospecting claims at Swartkoppies in 1951, R.G. Niemöller discovered his own sillimanite deposits on Hotson Farm, and he received his mining licence in 1952.[32] While Weidner was in the process of building an empire, Niemöller was already the king of Pofadder. Gertjie

---

28 NAN IMW 424b File P.T.4/3: Chairman's Report, Tantalite Valley Minerals (Pty) Ltd – 3 September 1954.

29 E.R. Varley, *Sillimanite* (New York, Chemical Publishing Company, 1968).

30 Ibid, p. 30.

31 B.L. Gunsallus and F.P. Uswald, 'Kyanite and related minerals', *Minerals Yearbook, 1954* (United States Geological Survey, 1954), pp. 645–650. Peter J. Cox, interview with Bernard C. Moore (telephone, 15 January 2023).

32 D.H. de Jager and J.W. von Backström, *The Sillimanite Deposits in Namaqualand near Pofadder* (RSA Department of Mines, Geological Survey, 1961), p. 2. Some other sillimanite

FIGURE 39 Sillimanite mining on Hotson Farm, near Pofadder, owned by R.G. Niemöller (Edms) Bpk
PHOTO: GNPA

Niemöller was born in 1928 to German parents who had emigrated to South Africa, where they ran the Pofadder Hotel and owned a few other shops in town. He was the nephew of the famous German Lutheran minister, Martin Niemöller. During the Great Depression, the Niemöller family had acquired the nearby farms Klein Pella and Kambreek-Zandfontein. When the parents died, in the 1940s, the farms were divided among their sons, with Gertjie getting the Orange River south-bank frontage at Kambreek-Zandfontein.[33] Niemöller was not really a farmer at heart, but he was a businessman and an investor through and through. Observing a rise in the karakul pelt industry, he sought to participate in both sides of the trade. He and his workers levelled the land on Kambreek-Zandfontein and dug furrows to produce more than one hundred hectares of irrigated lucerne; he also imported prosopis trees to make additional fodder. Niemöller was also to amass the largest pure-bred karakul stud

deposits in the area were previously mined on a small scale by the Hayes family, whose family, like those of the Weidners and Herridges, straddles the Orange River.

33 Louise Niemöller, interview with Bernard C. Moore (Pofadder, 21 January 2022).

FIGURE 40 Trucks line up outside R.G. Niemöller's sillimanite mine, awaiting shipments to drive to the railhead at Kakamas
PHOTO: GNPA

flock in southern Africa by the 1970s, becoming one of the main suppliers of breeding rams to the South African industry.[34]

Unlike Weidner, who was in a constant state of near-bankruptcy in the early years, Niemöller had a diverse portfolio of commercial and agricultural activities when he obtained the rights over Hotson sillimanite mine. He also had some mining interests in pegmatites on the South African side,[35] but these were small compared to Tantalite Valley, so he poured his energies into Hotson. Also, unlike Weidner, who wrote philosophically about the rise of the atomic age, Niemöller was practical. Reminding himself of what sillimanite was for in the first place, he wrote in his notebook: 'You can't melt iron in an iron pot. The pot will melt too'.[36] Both Weidner and Niemöller had relationships with

34 As seen in the following chapter, he also engaged in date production. For more on Niemöller's farming, see Moore, 'Swimming Upstream'.

35 A. van der Westhuizen, 'Obituary: Hugh Jenner-Clarke', *Geobulletin* [Geological Society of South Africa] 60, 2 (June 2017), pp. 31–34.

36 [*'Jy kan nie yster in 'n ysterpot smelt nie, dan smelt die pot ook. Sillimaniet smelt by ± 1850 grade'*]. GNPA: Niemöller personal notebook on sillimanite mining.

FIGURE 41 Niemöller (right) with his friend Dr O. van Schalkwyk of Pofadder, in Japan to find sillimanite buyers, 1957
PHOTO: GNPA

South African exporters of ore to foundries in Sheffield and Essen, particularly Vereeniging Brick and Tile Co. in Johannesburg, who helped them get some of their first contracts.[37]

37 Weidner, *The World Owes You Nothing*, pp. 57–58.

Weidner and Niemöller were ultimately lifelong friends, but apart from the mutual desire for sillimanite prices to remain high, their business goals and strategies could not have been more different. Although Tantalite Valley and Swartkoppies were valuable and profitable, they were all Weidner had, and he wanted to maintain the longevity of the mines for as long as possible. For the younger Niemöller, mining was just the latest craze, and he planned to capitalise on it as quickly as possible in order to pour the profits into karakul or irrigation.[38] Niemöller mined extremely fast, such that newly interested Japanese buyers began to request shipments and deliveries of more than 1,000 tonnes per month, a production level that Weidner struggled to meet.[39] Niemöller had a fleet of trucks that would make regular deliveries of ore to the railhead at Kakamas, where the sillimanite would be transported to Cape Town harbour. He made regular trips to Japan to meet buyers face-to-face, eventually forming a long-term relationship with Nisso Iwai Corporation (now Sojitz), which lasted well into the 1970s when the demand for sillimanite began to decline.

•••

Both Niemöller and Weidner would become millionaires with their sillimanite mining, but by the 1960s, the strategic goals of the South African government

FIGURE 42 Buyers from Japan's Nisso Iwai Corp. fly in to visit Niemöller's sillimanite mines, 1970
PHOTO: GNPA

38 Louise Niemöller, interview with Bernard C. Moore (Pofadder, 6 April 2022).
39 Weidner, *The World Owes You Nothing*, p. 61.

would turn again towards Weidner's pegmatites, rejuvenating his interest in the atomic age. The changed variable was the 1964–65 discovery of a uranium-rich alaskite (leucogranite) pegmatite deposit east of Swakopmund, known today as Rössing. It had long been known known that pegmatites could hold uranium, thorium and other minerals necessary for the manufacture of nuclear fuel. However, the concentration of the Rössing mineralisation led not only to the development of the uranium mine there but also a new exploration programme across South West Africa with the new prospecting technologies of the 1960s and 1970s.[40] Rössing Uranium Ltd was jointly owned by Rio Tinto and subsidiaries (57.5 per cent), the South African state-owned Nywerheidsontwikkelingkorporasie (13.2 per cent), Total France (10 per cent), General Mining and Finance Co. (Gencor) (6.8 per cent), and other smaller investors. By the late 1970s, Rössing grew to be the single-largest open-pit uranium mine in the world, selling its uranium yellowcake to buyers in the United Kingdom, France, West Germany and Japan.[41] Uranium was also shipped from 1984 to Cape Town's Koeberg nuclear power station.

In 1968, the South African Atomic Energy Board appointed the geologist J.W. von Backström as chairman of the newly formed Pegmatite Steering Committee, which would survey pegmatite formations across the Northern Cape and southern Namibia, partnering with local mines if possible. As one of the foremost experts in pegmatite geology, Von Backström was well-positioned to take on this position, and he was already on personal terms with Peter Weidner and Gertjie Niemöller. The 1948 Atomic Energy Act had put in place a special mining policy for strategic minerals, such as uranium, and with the discovery of Rössing it was decided to apply this law across all pegmatites; thus, surveying needed to be performed across the territory.[42]

In 1969, after touring Niemöller's pegmatite prospecting claims (kimberlites) in the Northern Cape, Von Backström made his way to Tantalite Valley. Using newer sonar technologies, he discovered a series of at least six additional large pegmatite formations on Weidner's property, albeit outside the formal mining licence area. Taking core samples back to Cape Town for testing, he recommended that the Atomic Energy Board partner with Weidner to do

40 Newby-Fraser, *Chain Reaction*, p. 67.

41 United Nations Institute for Namibia, *Namibia: Perspectives for National Reconstruction and Development* (1986), pp. 303–304.

42 NSS: *Kommissie van Ondersoek na Beweerde Onreëlmatighede en Wanaanwending van Eiendom by Verteenwoordigende Owerhede in, en die Sentrale Owerheid van SWA: Verslag oor die Voorlopige Bevindinge en Aanbevelings van die Ondersoek na die Sentrale Owerheid se Bestaande Beheermatreëls oor die Mynbousektor van SWA* (Windhoek, March 1984), vol. 2, p. 333. (Hereafter, this report will be cited as *Thirion Commission: Grote Verslag*).

further exploration with diamond drilling rigs, combining the operations with their Northern Cape projects.[43] Von Backström believed that even if uranium was not located in Rössing-size quantities – which he suspected was highly unlikely – the South African Atomic Energy Council was in need of other additional rare earth minerals like tantalum, nickel, etc. for their plans to build the Koeberg power station at Cape Town.

South Africa's interest in pegmatites along the Pofadder Shear Zone occurred at a transition period within the apartheid regime's nuclear programme. Before the 1970s, research and mining in South Africa was primarily about civil uses of nuclear energy, such as the construction of Pelindaba – the nuclear research station west of Pretoria. The research reactor SAFARI-1 was inaugurated in 1965.[44] As the anti-apartheid armed struggle in South Africa and Namibia's War of Independence accelerated over the following decade, many securocrats in Pretoria confidentially agreed to switch policy, to focus on obtaining alternative nuclear fuel sources that would possibly allow for the building of a nuclear explosive device.[45] Viewed as a deterrent bomb, B.J. Vorster supported the investigation into such a weapon. After the Carnation Revolution, the dismantling of the colonial armed forces of Portugal (Pretoria's important ally), and the imminent independence of Angola and Mozambique, Minister of Defence P.W. Botha officially approved formal plans for a nuclear weapon in 1974. The secret remained so only for a few years – a nuclear test site near Upington in the Northern Cape was discovered by a Soviet satellite in 1977. By the mid-1980s, South Africa would develop six nuclear devices, which were destroyed only in 1989 under the direction of State President F.W. de Klerk.[46]

In the context of apartheid South Africa's refusal to sign the Nuclear Non-Proliferation Treaty, it is reasonable to assume that the Pegmatite Steering Committee was tasked not just to find Rössing-level deposits to sell uranium overseas but also to clandestinely enrich the country's own nuclear fuel. The results from Tantalite Valley, though, were disappointing. Uranium was virtually absent from the pegmatites on Weidner's property, and while there was some nickel within some of the drill cores along the shear zone (nickel is

43 NSS: Hoofinspekteur van Myne, Windhoek 'Tantalite Valley Minerals, Warmbad' – 30 November 1969. (Printed in *Thirion Commission: Grote Verslag*, bylae vol. 1).

44 Newby-Fraser, *Chain Reaction*, p. 50.

45 A-M. van Wyk, 'South Africa's Nuclear Programme and the Cold War', *History Compass*, 8/7 (2010), pp. 562–572.

46 Ibid, p. 568.

important for alloys that clad reactor cores) the quantity was not significant enough to merit mining yet. They would have to do further tests.[47]

Although the Atomic Energy Board was perhaps not interested, the regular presence of government geologists on his property granted Weidner credibility to negotiate with larger companies for either partnership of buyout. With Von Backström's information about nickel, Weidner negotiated a 1971 option agreement with Rio Tinto, granting them one year of exploration rights with the option to buy in 1972. Rio Tinto was not impressed, because they had never encountered any nickel mineralisation anywhere on Umeis that exceeded 0.31 per cent, well below economic grade for mining.[48] The Canadian nickel giants Falconbridge were informed of Rio Tinto trucks and drill rigs at Tantalite Valley, so they negotiated prospecting rights on neighbouring farms within the Pofadder Shear Zone, such as Arus, Keimasmund and Oranjefall. Like Rio Tinto, however, they came up empty-handed and ceased most of their drilling by the middle of 1972.[49]

By the following year, Weidner was despondent. He had made most of his money from sillimanite mining on Swartkoppies, but Tantalite Valley was his home. After spending a significant amount on drilling costs, Rio Tinto packed up and left, abandoning their purchase option. Uranium and nickel were simply not sufficiently present on his property, and the prices of most pegmatites – apart from the lithium-bearing lepidolite – had fallen in price dramatically. In an era of US – Soviet Strategic Arms Limitations Talks, the atomic age no longer looked as Weidner had imagined it.[50]

•••

While the decline in the value of pegmatite ore was a global shift in the production of materials of the atomic age, Weidner also saw the decline of Tantalite Valley as a transformation in local labour relations. Reflecting in 1973 on the prior two decades, he said:

47 MME File 557/EPL322: Tantalite Valley Minerals (Pty) Ltd: Exploration Reports and Statements – Book Year 1970–1971.

48 MME File 559/EPL322: Rio Tinto Exploration (Pty) Ltd 'Interim Report on the Exploration of Prospecting Grant M46/3/322' – 5 July 1972.

49 MME File 601/EPL359: Falconbridge Explorations Ltd 'Report on the Reconnaissance Investigation of the Pella Prospecting Grant no. M46/3/359' – May 1972.

50 MME File 559/EPL322: Peter Weidner, Tantalite Valley Minerals (Pty) Ltd 'Prospecting Grant no. M46/3/322: Exploration Report July 1972–June 1973' – 20 September 1973.

> One thing has become very clear and that is that wages are now too high to do any old time pick and shovel recovery. Prices for the concentrates are now much lower than they were in the 1950s whereas wages have now doubled. There are still approximately 30 Nama and Ovambos employed on the property and earning between R1.20 and R1.60 per day working a 5 day week. The men appear satisfied, and their work is satisfactory.[51]

Weidner believed that the success of Tantalite Valley lay not just in changing global consumption patterns of rare earth metals but also in mechanising and formalising the labour system in a way his mine had not done in the past.

For the first decade of Weidner's production at Tantalite Valley, he relied on a piecework labour arrangement, which he called 'tribute'. Local workers were employed on a casual, ad hoc basis and remuneration was calculated based on minerals recovered, rather than hours spent working.[52] According to Weidner, most of his labourers at the time were employed in this system – only a couple were full-time salaried employees, generally those who needed to drive trucks. The early days of Tantalite Valley were fairly unsophisticated – muscles and pickaxes – giving credence to the perceived richness of the pegmatites near the surface. For the first decade, Weidner did not believe in mechanical power: from the pickaxes to the Pleitz jig, all was manual. Partly this reflected his traditionalist vision of hard work, but it was also a reflection of the realities of labour relations under apartheid. Wages were low for adult men, and family labour could at times be tapped into as well. In an interview with the *Windhoek Advertiser*, Weidner proudly declared that some of the finer bismuth and tantalite gravel was sent to the Catholic mission hostels at Warmbad and Pella, 'where the children with their patient little fingers and keen eyesight diligently separate the yellow bismuth from the black tantalite'.[53]

The two shillings or so per day that these children earned after school were generally pooled as family income, because their parents were normally piecework labourers on Tantalite Valley itself. Like his father before him, Peter Weidner was a staunch Roman Catholic, and he sought to integrate the church into his mines and his mines into the church. To that end, the Weidner family was the single-largest private benefactor of the Catholic mission stations at Karasburg, Pella and Homsrivier. Peter and his wife Cecie constructed an amethyst and rose quartz shrine at Tantalite Valley, where travelling priests would

51 Ibid.

52 NAN IMW 424b File P.T.4/3: Chairman's Report, Tantalite Valley Minerals (Pty) Ltd – 3 September 1954.

53 'Tantalite valley is happy and its people work', *Windhoek Advertiser* (13 December 1957).

FIGURE 43 Family housing for pieceworkers on Tantalite Valley, late 1960s
PHOTO: WEIDNER, 1969

occasionally hold Mass or give Holy Communion.[54] He also eventually funded an orphanage at Pella Mission.[55]

Weidner was in many ways a typically patriarchal, Catholic businessman with close ties to the church and the local community, and he sought to put this into practice in the context of apartheid in Namibia. Under apartheid racist separate development policies, labourers for Namibia's mines were often sourced from northern Namibia on long-term contracts. Although Weidner did occasionally tap into this labour force, he understood the dire effects that the migrant labour system wreaked on his Catholic understanding of family life. For that reason, Weidner was one of the few mine-owners who from the very start allowed his workers to come to the mine with their wives and children, encouraging them, once the children reached sufficient age, to enrol them in one of the Catholic mission schools, sometimes at his expense.[56] At Swartkoppies in the 1960s, Weidner put into place a system whereby workers who had been at the mine for more than one year could apply for interest-free loans to

54 Sr. Paule Elizabeth Atiogbe OSFS, interview with Bernard C. Moore (Karasburg, 1 August 2016).

55 'These orphans have found a home', *Windhoek Advertiser* (13 December 1957).

56 See NAN Archives of the Magistrate at Karasburg (hereafter, LKW) 3/2/2 File N.3/13/2: Magistrate at Warmbad to Hoofbantoesakekommissaris, Windhoek 'Inspeksie van Tantalite Valley Minerals (Pty) Ltd' – 28 November 1977.

FIGURE 44 Tantalite Valley's football club in the late 1960s, which would regularly play other teams at Karasburg, Pella, Hotson mine and Haakiesdoorn
PHOTO: WEIDNER, 1969

construct family homes in Pella town. All materials – cement, windows, doors, wiring, etc. – could be purchased wholesale on the company account and shipped into Swartkoppies. This was much cheaper for the worker than having to buy the materials himself in Kakamas.[57] Weidner sought to promote family life not just to increase production through increased morale – or through additional child labour – but he also wanted to foster a culture of loyalty, to him and his operations.

Weidner believed that work, family and church were inseparably linked. And like his father before him, he was a paternalist who successfully combined his colonial/apartheid convictions regarding racialised stages of societal development with his responsibility for the local community on either side of the Orange River, whether Black, White or Coloured. He was a mining capitalist, but he was not an outsider – he and his family were intricately bound to the history and economy of this region. For this reason, he is remembered fondly today by some Bondelswarts, not just because of his work with the Catholic church but because he was a patron to whom people could turn for work when times were bad, and for whom Catholicism meant a bit more distance from apartheid-era race relations.[58] Like his father before him, Weidner clothed his paternalistic management style in the language of religion, family

57 Weidner, *The World Owes You Nothing*, p. 84.

58 Johannes Gertze, interview with Bernard C. Moore (Arus farm, Karasburg District, 16 October 2022).

FIGURE 45 Agricultural workers at Niemöller's irrigation works on Kambreek-Zandfontein in the 1950s
PHOTO: GNPA

and community. Although for many workers this was preferable to the terrible labour conditions on the other farms, Weidner had created a system based on *absolute loyalty* and deference to him. For those who stayed in the favour of the *Oubaas*, he was a welcome patron in times of need, but for those who crossed him, there was little chance of return.

Besides his self-professed distaste for the migrant labour system, there were simple practical reasons why Weidner could avoid it: there was an abundance of people living nearby. Weidner was able to tap into existing labour sources in the Bondelswarts communal areas and along the Orange River, something that other emerging mines in southern Namibia – like Rosh Pinah – could not do as easily. Indeed, the Bassons and other river-folk knew how to make use of Weidner's project. In 1971, when the SWA state ethnologist Kuno Budack visited the Basson family living at the Orange River near Hartebeesmund, he found that most of the men in the extended family had been employed from time to time as pieceworkers at Tantalite Valley, Swartkoppies or on Niemöller's irrigation works at Kambreek-Zandfontein.[59]

59 TBK KUS B-615 File B26/2/1/3 (vol. 2): K.F.R. Budack, Etnologiese Afdeling van BAO to Hoofbantoesakekommissaris van SWA 'Nama en Namasprekende Ovaherero op Hartebeesmund 108, Warmbad' – 11 March 1971.

Despite likely resulting in lower cash wages per day laboured, piecework was actually preferred by many river-folk in southern Namibia, because it allowed them selective contact and engagement with the apartheid economy. Facing decreasing grazing access as the White farmers sought to utilise the entirety of their properties for karakul production, people like the Bassons used Weidner's piecework system as a provisional means to offset these structural challenges and more temporary setbacks, like drought and predation.

River-folk like the Bassons were regular temporary employees of Niemöller, as well, who – although perhaps to a lesser extent than Weidner – understood his community role in a similar manner. Niemöller was an amateur photographer (eventually videographer), and he dedicated dozens of rolls of film to photographing his mining and agricultural workers, their families, their song and dance: in short, their lives. Many of these people are identified as belonging to the Basson, Witbooi, Simboya and other river-folk families.[60] Although this relationship was somewhat acceptable to these workers – or at least it was preferred to employment on the commercial sheep farms – it nevertheless meant a constant effort to remain in the good favour of the two largest employers along the river.

By the mid-1970s, the prices of Weidner's pegmatite ores were declining faster than his health; he was much older than Niemöller. But Weidner could not bring himself to completely stop production, so he decided to stockpile lepidolite ore, selectively selling it when the price of lithium rose.[61] Weidner sought to combine his Catholic worldview with his drive for profit as a mining capitalist. For him, Namibia's atomic age did not mean just his participation and that of his shareholders. It was, in his words, about 'putting the southern portion of South West Africa on the map'. The Bassons would never sit in the driver's seat in Weidner's atomic age, but as long as they played by his rules, there was a place for them within it.

•••

60 Hundreds of photos have been collected by Niemöller's daughter, Louise, who runs a small tribute page on Facebook, where discussions about the photographed individuals draw memories from residents – Black and White – from Pofadder and beyond. Her physical archives in Pofadder are even more expansive.

61 MME File 559/EPL322: Peter Weidner, Tantalite Valley Minerals (Pty) Ltd 'Prospecting Grant no. M46/3/322: Exploration Report July 1972–June 1973' – 20 September 1973.

FIGURE 46 Burial place of Peter and Cecie Weidner, Karasburg cemetery
PHOTO: B.C. MOORE, 2022

By the second half of the 1970s, the atomic age would come back to Tantalite Valley, but Peter Weidner was basically retired at this point. He and his wife Cecie had passed on most company operations to their son, Charlie, and the elderly couple retired to their home at Donata Street in Karasburg, eventually moving to a seaside villa in Scottburgh, Natal, which they called 'Peace Haven'. Peter and Cecie would pass away in 1984 and 1982, respectively. In some ways, Peter had succeeded in putting Tantalite Valley on the map and bringing in the atomic age, but this new iteration would look very different. The newcomers reflected a shift in global capitalism away from ostensibly well-meaning patrons towards faceless global companies; they just wanted pegmatites.

The San Francisco-based mining conglomerate, Utah International Inc., had little respect for the growing push within the United States for corporate divestment from apartheid South Africa. Utah was more than just a mining company; it was long a beneficiary of military contracts from the US government to build Minuteman missile silos, and its uranium mining in Wyoming supplied the American nuclear programmes.[62] From the 1950s, the company expanded throughout South America, Australia and eventually South Africa, where it operated subsidiary mining and prospecting companies. Utah's South African subsidiary, Southern Sphere Mining and Development (Pty) Ltd, was increasingly involved in buying up prospecting licences for strategic minerals like uranium, tantalum and lithium.

62 G.A. Sessions and S.D. Sessions, *Utah International: A Biography of a Business* (Ogden, Weber State University Press, 2002), pp. 247–248.

As a result of the 1973 oil crisis, many American conglomerates, such as General Electric (GE), sought to increase their focus on constructing nuclear reactors, viewed as a power source removed from OPEC [Organisation of Petroleum Exporting Countries] and geopolitical instabilities. For this reason, GE sought to acquire a mining division to supplement its manufacturing and metallurgical branches, especially its nuclear fuel division. This was GE's first major move outside the United States, so it decided to buy out Utah International's portfolio of overseas interests, leading to a merger of the two companies in 1976 in a USD 2.2 billion deal, the largest corporate merger in American history at that point.[63] GE was primarily interested in Utah's large Australian and South American assets, but buried within the paperwork was new ownership of Southern Sphere in South Africa.[64]

In 1979, GE declared a 'Search for Tantalite' project – a global strategy to increase its tantalum supply for its alloys and aircraft engine divisions. The company was already consuming almost 70,000 kilograms of pure tantalum annually. Although the most promising global tantalum deposit was likely Greenbushes in Australia, this would not be in full production until 1986, and GE therefore sought to obtain smaller mines across the world to supplement production until that point. One of its main targets was Tantalite Valley.[65] GE negotiated – via Southern Sphere – a one year prospecting agreement with Weidner in 1980, in order to mechanise tantalum production for aerospace products and electronic capacitors.[66] The drilling was successful enough that in July 1981 Southern Sphere bought out Weidner's shares in Tantalite Valley Minerals (Pty) Ltd for ZAR 850,000 – a bargain for the American behemoth.[67]

Tantalite Valley was no longer a small mine owned by a local capitalist but rather a small mine owned by a faceless international board of directors that stretched across the globe. When prices for tantalum and other pegmatite ores fell in the 1970s, Weidner kept some of his workforce employed, stockpiling what he could for the time being to pay bills and feed families. When the price of tantalum and other atomic age minerals collapsed in the mid-1980s amidst worries about the safety of nuclear power, GE/Southern Sphere ceased prospecting, dramatically cut back on production, and released the workforce. Tantalite Valley was no longer a local entity but one more intimately tied to

63 Ibid, pp. 284–286.

64 MME File 1234/EPL919: A Review of Southern Sphere Mining and Development Company (Pty) Ltd and the Company's Policy Towards Exploration in South West Africa, 1978.

65 MME File 1241/EPL919: Southern Sphere Mining and Development (Pty) Ltd 'Preliminary Feasibility Study on Tantalite Valley' – May 1981.

66 Peter J. Cox, interview with Bernard C. Moore (telephone, 15 January 2023).

67 *Thirion Commission: Grote Verslag*, vol. 2, p. 327.

the global vagaries of the minerals markets. This was ultimately a major shift for those who were living along the Orange River. Weidner and Niemöller were apartheid capitalists who deeply benefited from the system, but the paternalistic ways in which they ran their businesses meant that, in exchange for loyalty, the river-folk could find a path to tentative coexistence. The arrival of global mining capital to the region severed this relationship. The Bassons' roadmap had become outdated, and loyalty meant little to Southern Sphere.

CHAPTER 7

A River for Coloureds?

By the early 1960s, the South African government was increasingly recognising – particularly with regard to Namibia – that combatting the United Nations, international pressure to decolonise and the local rise of anti-apartheid movements required stronger intervention in the welfare of Africans. (Most state programmes during the previous forty five years had overwhelmingly benefited Whites.) The apartheid regime was long of the opinion that science, technology and planning were key to their racial policies of separate development. In the 1960s, this view hardened dramatically, and the regime used scientific planning to quieten the protests of the international community and local mobilisation, while maintaining the economic and political privileges of the White community. As it relates to southern Namibia, two concurrent state commissions investigated how to implement this.

First, the state targeted water. In the post-Second World War period, the lower Orange River was becoming much more economically important to the state. It served not just as a location to place 'poor-White' settlements, as during the Great Depression, but as a region of intensive livestock-farming, mining and irrigation. This region was no longer just the periphery of South Africa and Namibia. In response to the booming industry in karakul pelts and merino wool, between 1945 and 1962 an additional 18,500 hectares of lucerne fodder was planted along the Orange River between Buchuberg and Augrabies.[1] Indeed, the Orange River proved to be ideal for lucerne cultivation; the root system of *Medicago sativa* is deeper than most other leguminous plants, making the crop fairly resistant to drought.[2] Unlike many other fodder crops, lucerne performs well in dry, low-humidity climates under irrigation – the swift regeneration of stems and leaves following harvesting allowed many South African fodder farmers to get up to thirteen crops of lucerne hay in a single growing season.[3] Even the Soviet Union – which had more hectares planted with lucerne than any other country – was amazed at the productiveness of

1 J.L. Bolton, *Alfalfa: Botany, Cultivation, and Utilisation* (London, Leonard Hill, Ltd., 1962), p. 52.

2 J.B. Stevens, P.S. van Heerden, P. Reid, A. Liebenberg, E. Hagedorn and G. de Kock, *Technical Learner Guide: Irrigated Crop and Fodder Production (Part 8)* (Republic of South Africa, Department of Agriculture, Forestry and Fisheries [Water Research Commission], 2012), p. 4.

3 Ibid, p. 1.

FIGURE 47 R.G. Niemöller's irrigated fields on Kambreek-Zandfontein, opposite KumKum and Kambreek on the Namibian side, c.1980
PHOTO: GNPA

Orange River-irrigated lucerne. This yielded more than 19.5 metric tonnes per hectare, whereas the USSR's best irrigated harvests peaked at 5.7 and the USA reached only 6.9 metric tonnes.[4] In addition, lucerne has the ability to desalinate soil, making it a good primer crop for vegetables or fruit in the region.

The Orange River makes up approximately twenty two per cent of South Africa's surface-water resources, and the massive size of the catchment area means that without large-scale dam and hydrological infrastructure there are large differences in water levels between rainy and dry seasons.[5] If the river runs too low, irrigation is impossible; if the floods are too extreme, harvests are washed away. In order to stabilise the growing White livestock and irrigation sectors in the Northern Cape – and provide water to mines and municipalities – the Republic of South Africa (RSA) Department of Water Affairs launched a 1962–1963 investigation into large-scale infrastructural development along the Orange River, with the goal that – building dams, weirs and tunnels – up to 311,000 additional hectares of land could be brought under irrigation, including

4 A.I. Ivanov, *Alfalfa* (Russian Translations Series no. 63) (Rotterdam, A.A. Balkema, [1980] 1988), p. 81.

5 L. van Vuuren, *In the Footsteps of Giants: Exploring the History of South Africa's Large Dams* (Water Research Commission of South Africa, 2012), p. 195.

nearly 6,000 additional hectares downriver from Augrabies.[6] Some of the proposed development projects were massive in scale – such as the Gariep Dam (formerly H. Verwoerd Hydroelectric Dam) and the Fish-Sundays Tunnel – and construction costs would be higher than any prior project in South African history.

The water infrastructure schemes of the 1960s were intended not simply to provide work to poor Whites but also to utilise grand scientific and technological prowess to symbolise the compatibility of apartheid with the modern era.[7] Although the Gariep Dam and the Orange River Development Project clearly held symbolic prestige, they also regulated water flow along the Orange River to such an extent that larger-scale irrigation schemes could be attempted. No longer would irrigation schemes simply be of a 'poor-White', homestead nature. By 1970, the Gariep Dam started storing water, and by 1971 it was at sixty-five per cent capacity.[8]

In addition to engineering waterways, the administration sought to engineer African societies. In Namibia, this took shape with the Commission of Enquiry into South West African Affairs, known colloquially as the Odendaal Commission. Initiated in September 1962 and delivering its final report in January 1964, the commission investigated the means by which to provide 'accelerated development of the various non-white groups' in Namibia,[9] and like its South African predecessor, the Tomlinson Commission of 1954, the Odendaal Plan envisioned separate development in the form of ethnic homelands.[10] Simply put, apartheid-era planners envisioned homelands to be politically independent territories serving their ethnic-based citizenry. However, in reality homelands became puppet states, which did little to reduce dependence on White-owned industry and resulted in stripping the region's Black population of much of their remaining rights and citizenship. Although most of the Odendaal Report detailed issues of homeland construction (as well as electricity provision) in the northern homelands, particularly Owambo and Kavango, it did

6 RSA Secretary for Water Affairs, *Report on the Proposed Orange River Development Project* (1962–1963).

7 For more on this view of science, consult S.J. Sparks, 'Apartheid Modern: South Africa's oil from coal project and the history of a South African company town' (PhD thesis, University of Michigan, 2012).

8 L. van Vuuren, *In the Footsteps of Giants*, p. 202.

9 NASA Archives of the Secretary of the Treasury (hereafter, TES) 5808 File 33/841: Sekretaris van die Eerste Minister, Pretoria to Sekretaris van die Tesourie, Pretoria 'Kommissie van Ondersoek na Aangeleenthede van Suidwes-Afrika' – 22 September 1962.

10 See M. McCullers, 'Lines in the Sand: The global politics of local development in apartheid era Namibia, 1950–1980' (PhD thesis, Emory University, 2012).

contain a few sections detailing the ways in which southern Namibia would be transformed.

The Odendaal Report first advocated the alienation and eviction of the Bondelswarts Reserve and the purchase of 165 formerly White-owned farms in the vicinity of Gibeon, which would be given to the Bondelswarts and other evicted Nama to combine disparate reserves into a contiguous Namaland Homeland.[11] The Bondelswarts Reserve was to be subdivided for White-owned farmland. Second, the Report proposed the purchase of six farms along the Lower Orange River to be allocated to the SWA Coloured Council as an irrigation settlement: this included the farms Haakiesdoorn, Ramansdrift, Gaidip, Homsrivier, Girtis and Hartebeesmund.[12] This area was not to be considered a communal homeland, since there was no view that Coloureds constituted a self-governing ethnic entity. It was to be a *besproeiingsnedersetting* (irrigation settlement) based on the principle of privately operated *ekonomiese eenhede* (economic units).[13] The South African government and the SWA Liaison Committee included the Coloured settlement within the first five year plan of Odendaal implementation.[14] This was in part because both South West Africa and South Africa sought to combine the proposed Coloured settlement with existing 'Coloured reserves' on the South African side, including Pella and Witbank. This would also mean possibly including the Namibian farms KumKum, Kambreek and Pelladrift in the proposals.[15]

By the mid-1960s, the Nama in southern Namibia – including the river-folk on the banks of the Orange – had witnessed the economy changing. So-called 'poor Whites', who had settled in the region under the Land Settlement Programme, were no longer 'poor Whites' but highly capitalised karakul farmers, who increasingly evicted their tenant labourers. Mining capitalists like Weidner and Niemöller were gradually becoming local barons whose patronage was needed in order to maintain a semi-independent existence in the area.

11 NAN AP 4/1/13: Report of the Commission of Enquiry into South West African Affairs, 1962–1963 (1964), pp. 101–7 (hereafter, *Odendaal Report*). NAN Archives of the SWA Secretariat: AS-Series (hereafter, SWAS) 439 File AS.65/5/1: Reservation of State Lands for Natives Amendment Ordinance, 1974. See also, R. Kößler, *In Search of Survival and Dignity: Two Traditional Communities in Southern Namibia under South African Rule* (Windhoek, Gamsberg Macmillan, 2005).

12 *Odendaal Report*, p. 109.

13 TBK KUS 1/3 File 1/1/6 (vol. 1): Sekretaris van Kleurlingsake to Minister van Kleurlingsake 'Verslag van die Kommissie van Ondersoek na Aangeleenthede van SWA' – 4 March 1964.

14 NASA TES 5808 File 33/841: Memorandum: Decisions by the Government on the Recommendations of the Commission of Enquiry into South West African Affairs (1964), p. 5.

15 NAN Archives of the SWA Local Authorities Division (hereafter, PLA) 25 File 25/2/2 (vol. 1): Direkteur, Afdeling Waterwese to F.H. Odendaal, Voorsitter, Kommissie vir SWA – 12 June 1963. For more on the history of Coloured communities on the South African side of the Orange River, consult Moore, 'Swimming Upstream' and Lenggenhager, 'Onseepkans'.

Furthermore, the new apartheid regime sought to re-engineer the landscape, through building new infrastructure and the social engineering of population transfer and forced removals to the homelands. For many decades, the Bondelswarts and the river-folk had been able to keep a precarious balance of selective engagement with the colonial economy, but these new interventions further upset this balance.

•••

'We didn't have to move with the Gibeon Trek', said Monika Basson, relaxing on a sofa in her home in Warmbad. 'The Odendaal Plan was for the Bondelswarts who were living in the reserves here. The government at that time didn't know really about us down by the river. Only the Church did'. Monika Basson (Willem III's older sister) was born on KumKum in 1956, and she lived at the Orange River with her parents and brothers for her first twenty years. She started her schooling at Homsrivier Catholic Mission, eventually moving on to Warmbad mission school in 1966. 'A lot of my classmates were from Witbank, Pella, Goodhouse', Monika said, 'It was an Orange River school'.[16]

Monika was a just a child when the Odendaal Report was published, but as the years went on, she saw how much the apartheid homeland system divided her people. Evicting the Nama people from the Bondelswarts Reserve was central to apartheid in southern Namibia; it was in the government's interest for there not to be exclaves or 'black spots' (*swartkolle*) of homelands, because centralisation was easier to govern and control. The SWA Administration took steps to purchase White farms (twenty per cent above market rate) in the area near Gibeon to accommodate the Bondelswarts and the other Nama groups who would have to be evicted, and by the end of 1965, well over half of the properties had been purchased.[17] The Bondelswarts Reserve had not received very much rain during the early 1960s, and many Nama stock farmers looked to the so-called 'Odendaal farms' as a possible temporary relief to the drought.

One particular stock farmer and shopkeeper from !Haib village in the Bondelswarts Reserve, named Frank Basson (distantly related to the Orange River Bassons), wrote in 1966 to the Magistrate at Karasburg requesting that the state allow some stock farmers to receive emergency grazing (*noodweiding*) in the Gibeon area. Early the following year, most of the wealthier farmers – including

16 Monika Basson, interview with Bernard C. Moore and Luregn Lenggenhager (Warmbad, 9 November 2021).

17 NAN Archives of the Odendaal Commission Liaison Committee (hereafter, LUKS) 2 File 8: Odendaalplan: Vorderingsverslag tot 22 November 1965.

FIGURE 48 Monika Basson, photographed outside her home in Warmbad
PHOTO: B.C. MOORE, 20 21

Frank Basson – journeyed to Gibeon where the Bondelswarts received their first few 'Odendaal Farms'. The administration quickly realised that the ongoing drought was the most effective way to evict the reserves fully, and the 'Gibeon Trek', as Monika described it, became mandatory for all adults. Frank Basson was offered a paid government position as 'Acting Headman' for the Bondelswarts in Gibeon (a position he held until the late 1970s) if he assisted in convincing his fellow Nama to move.[18] Over the course of the second half of 1967, approximately 2,400 of the 2,600 Bondelswarts Reserve residents undertook the Gibeon Trek, resulting in a nearly depopulated area, which the Catholic missionary at ǂGâbes described as 'a ghost town'.[19]

Most of those who left, however, were under the impression that they would be permitted to return after five years once the pastures in their reserve had recovered. Apart from Frank Basson, few of those who trekked knew that the administration viewed this as a permanent solution. Those who remained in the reserves and refused to trek – a minority of the Bondelswarts, including

18 Martha Witbooi and Joseph J. Rooi, interview with Bernard C. Moore (ǂGâbes, 17 July 2019).

19 NASA Archives of the Department of Bantu Administration and Development (hereafter, BAO) 1/2393 File 188/1281/2: Hoofbantoesakekommissaris van SWA to Sekretaris van Bantoe Administrasie en Ontwikkeling 'Namatuisland' – 18 March 1968. A.P.J. Beris, *Making the Desert Bloom* (Vol. 3) (Keetmanshoop, RC Diocese, 2003), p. 110.

the Christiaan royal family – immediately were targeted by the police with violence and sabotage as part of the state's 'persuasion work' (*oorredingswerk*). Constables were given instructions to destroy all water infrastructure in the Bondelswarts Reserve, regardless of who owned the pumps, pipeline and wells. The Magistrate at Karasburg understood the actions as being performed not only to 'convince' the remaining Bondelswarts to trek to Gibeon but also to prevent anyone who had already made the trek from returning. He wrote: 'If there is no water, then the pastoralism will be constrained, and the remaining Nama will have to later trek under the force of the drought'.[20] Bondelswarts residents correctly accused the state of 'camouflaging [their] removal with emergency grazing' and alleged that in addition to destroying infrastructure, the Magistrate burned houses and forced residents to sell their livestock 'for the price of an apple and an egg'.[21] The fact that some 5–10 per cent of the Bondelswarts refused to trek ensured that their reserve was not fully evicted and sold off to Whites. In the late 1970s, the Bondelswarts would thankfully have their reserves reproclaimed, and some of the former trekkers were able to return. Most, however, did not have the funds to come home, and thus there are more than 2,000 Bondelswarts still living today in the Gibeon area.

The late Frank Basson has a complicated legacy among the Bondelswarts today. For those whose families come from Warmbad, ǂGâbes and other places in the reserve, he is often remembered as the embodiment of complicity with the apartheid regime. For those river-folk from the banks of the Orange, his legacy is more mixed. Monika Basson explained to us that although Frank Basson was distantly related to the Bassons at the river, he had lived in !Haib for most of his life. Yet, she said, for all his faults regarding the Gibeon Trek, one redeeming quality was that he deliberately concealed the fact that there were many Bondelswarts who had chosen to stay, living by the Orange River.[22] The river-folk were saved from the evictions and the Gibeon Trek not because the state had allowed them to stay but because, to all intents and purposes,

20 ['*Die rede vir hierdie aanbeveling is dat die moontlikheid bestaan dat bantoes* [sic] *weer kan terug trek en hulle om die pompe kan vestig. As daar nie water is nie sal hulle nie weer kan teruggaan nie … Indien daar nie water is nie sal dit die rondtrekkery ook beperk en sal die oorblewende bantoes* [sic] *later ook onder die druk van die droogte moet trek*']. NASA BAO 1/2393 File 188/1281/2: Landdros Karasburg to Hoofbantoesakekommissaris, Windhoek 'Verskuiwing van die Namas uit die Bondels- en Warmbad-Reservaat' – 22 February 1968.

21 NASA BAO 1/2393 File 188/1281/2: Inwoners van die Bondelswarts Reservaat to B.J. Vorster 'i/s die Uitsetting van die Bondelswarts' – 26 August 1972.

22 Monika Basson, interview with Bernard C. Moore and Luregn Lenggenhager (Warmbad, 9 November 2021).

FIGURE 49 Frank Basson's old shop in !Haib village, Bondelswarts Reserve
PHOTO: B.C. MOORE, 2021

they were invisible to the state.[23] Nevertheless, this invisibility further severed pastoral and transhumant connections between the Orange River and the Bondelswarts communal areas. The police were carefully monitoring all *intrekkers* who returned to the communal areas, and there was no reliable water in the reserve anymore.

• • •

Meanwhile, the implementation of the 'Coloured settlement' at the Orange River was continually delayed throughout the 1960s. Some of the challenges were technical. Since the Odendaal Plan envisioned an irrigation settlement, it was unclear precisely where this irrigation could or would be undertaken – given the landscape – and it was unclear what effects the ongoing construction of the Gariep Dam would have on the flow of the river and the propensity of floods. The Orange River Development Project (ORDP) made it clear to the SWA Administration that full priority would be given to the upriver schemes, and therefore it sought to complete them before considering any planning or soil surveys along the Lower Orange River.[24] The ORDP board declared that

23 The river-folk occasionally were made known to the police and the state, but the circulation of staff and the police's focus on the main Bondelswarts Reserve meant that the river-folk along the Orange were often out of sight and out of mind. See NAN LKW 3/2/1 File N.1/15/4: Stasiebevelvoerder, SAP Karasburg to Magistraat Karasburg 'Naturelle Bewoners op die Plaas Hartebeesmund' – 18 November 1952.

24 NAN PLA 25 File 25/2/2 (vol. 1): Sekretaris, Oranjerivier – Ontwikkelingsprojek to Sekretaris van SWA 'Oranjeriviernedersetting vir Kleurlinge' – 4 January 1965.

planning the settlement would not take place for several years,[25] and they decided to hold off buying Haakiesdoorn, Ramansdrift, Gaidip, Homsrivier, Girtis and Hartebeesmund.[26]

The Orange River settlement also floundered because most Coloured leaders in Namibia did not necessarily support it, and it was unclear from the very start how the state defined a 'Coloured person'. Coloured farmers were confused by the allocation of six Orange River farms in the Odendaal Report, given that most of them were livestock farmers working far from the river, such as at Keetmanshoop, Aroab and Aranos.[27] Most Coloureds wanted a livestock settlement [*veenedersetting*], which would solve many of their problems regarding grazing access and tenancy evictions. In 1964, after hearing the complaints of Coloured farmers in southern Namibia, *Kleurlingraad* [Coloured Council] Executive Committee member David Bezuidenhout travelled in a 4 × 4 vehicle to view the proposed Odendaal farms. He immediately did an about-face and advocated rejecting the proposal completely. He recalled: 'It was so inhospitable that a baboon would struggle to make a living in some places. I have no idea how goats, sheep or cattle could survive there. If this is all we could get, then "no thank you, we'd rather get by without".'[28] Despite this protest, the White National Party leader A.H. du Plessis insisted that the six farms were all the Coloureds could hope to get. The land could settle perhaps thirty families for stock-keeping or irrigation, and perhaps in time they could merge with the 'Coloured reserves' on the South African side of the Orange River.[29]

Of course, part of the issue here is that the legal definition of 'Coloured' was different on either side of the Orange River. Apartheid-era classification

25 NAN SWAS 888 File AS.127/1/3: Sekretaris, Adviserende Raad vir die Oranjerivier – Ontwikkelingsprojek to Sekretaris van SWA 'Oranjeriviernedersetting vir Kleurlinge' – 24 January 1968.

26 NAN SWAS 888 File AS.127/1/3: Uitvoerende Komitee Besluit: 'Oranjerivier-Nedersetting vir Kleurlinge: Onttreking van Sekere Plase aan die Afpenning van Kleims' – 27 March 1967.

27 TBK KUS 1/3 File 1/1/6 (vol. 1): Kleurlingraad van SWA, 'Verslag: Vergaderings met die Kleurlinggemeenskappe te Walvisbaai, Windhoek, Lüderitz, Keetmanshoop, Aroab en Aranos i.v.m die Odendaal Kommissie Verslag' – 22 February–14 March 1964.

28 [*'Die was so onherbergsaam dat 'n bobbejaan met 'n kierie sou moes sukkel om oor die weg te kom. Ek weet nie hoe bokke en beeste en skape daar sou kon bestaan het nie'*]. NAN AP 19/1/1 (vol. 4): Vierde Sessie van die Eerste Wetgewende Vergadering van die Kleurling Verteenwoordigende Owerheid (Khomasdal: 5 April 1982), p. 261.

29 TBK KUS 1/3 File 1/1/6 (vol. 1): Aanbevelings van die Subkomitee oor Aangeleenthede rakende Kleurlingsake in SWA: Die Moontlikheid van die Toepassing van die Wet op Landelike Kleurlinggebiede (no. 24 van 1963) op die Kleurlinge, Namas en die Rehoboth Basters van SWA – undated, likely November 1965.

categories in Namibia included the Nama, Baster and 'Boesman' [San/Bushmen]. After the 1950 South African Population Registration Act, these categories on that side were all lumped into the category 'Coloured'. One South African representative of Coloured Affairs remarked, regarding SWA Coloured identity:

> It is extremely difficult to differentiate a Baster from a Coloured, or even a Nama with Coloured facial features [*Kleurlinggelaatstrekke*] and who associates with Coloureds. The Coloured in SWA accepts them if they are family; one of the current *Raadslede*'s [council member's] mother is a Nama woman.[30]

Creating homogenous identities within heterogenous populations is normally an impossible task, but this was even more the case with the 'Coloured' category in Namibia, who included 'Cape Coloureds' who had emigrated to SWA from South Africa and Namibian Coloureds who were more likely had Damara or Nama grandparents.[31] The Cape Coloureds dominated the ranks of the SWA Kleurlingraad. However, most Namibian Coloureds in the southern districts were tied to Nama communal areas, and some were in fact Namas who had registered as Coloured in order to access better salaries or housing.[32]

The question of who could claim Coloured status did not just exist in the minds of councilmen and government functionaries; it was a significant issue that affected the livelihoods of those already living within this proposed Orange River Coloured settlement. In 1970, the administration discovered that a large number of Africans were already residing on Hartebeesmund, claiming Coloured status.[33] While the implementation of the Odendaal Plan would normally mean the eviction of these 'squatters', some in the administration thought that – if they were *Coloureds* – it would be more effective to

30 TBK KUS 1/3 File 1/1/6 (vol. 1): Afdeling Plaaslike Bestuur en Kleurlingsake 'Memorandum: Wetgewing: Gekose Kleurlingraad' – 11 November 1964.

31 M.J. Betts, 'Namibia's no man's land: Race, Space, and identity in the history of Windhoek Coloureds under South African Rule, 1915–1990' (PhD thesis, University of California, Los Angeles, 2009), p. 137.

32 See NAN Archives of the Department of Coloured, Rehoboth and Nama Affairs, Windhoek Office (hereafter, CRN) 106 File 31/1/2: Streekverteenwoordiger, KbRa to Sekretaris, KbRa 'Identiteitsdokumentasie en Bevolkingsregistrasie: Namas, Namaland' – 12 March 1973.

33 TBK KUS B–615 File B26/2/1/3 (vol. 2): Direkteur, Staatsnedersetting en Boerebystand to Streekverteenwoordiger, Dept. KbRa 'Plaas Hartebeesmund nr. 108, Distrik Warmbad' – 17 March 1970.

just offer them emergency grazing until the settlement had been finalised.[34] It was decided that if the 'squatters' were Nama, then all the residents would be forcibly evicted to Gibeon, but if they were indeed Coloureds, then they could remain.

Karasburg Police Constable Chris Coetzee was sent to investigate. Unsurprisingly, his vehicle could not handle the paths, and he had to walk nearly twenty seven kilometres from Homsrivier Mission Station to the settlement on Hartebeesmund, where he met a community of six households, forty five residents and about 2,000 livestock (mostly goats) between them. He described the community as 'mixed Nama and Herero', and he said that they had been living along the river for so long that the Hereros had all 'lost their language, and they now choose to go as Coloured'.[35] The administration was completely perplexed by the situation and chose to send the State Ethnologist, Kuno Budack, to investigate further. Budack was a fluent Nama (Khoekhoegowab) speaker and had written a 1965 MA thesis on Tses Reserve in Namaland.[36] Budack reported that the families included one Nama family and five families of 'Nama-speaking OvaHerero, mixed with Damara'. Most men would work from time to time on the farms and mines on either side of the river. Like Coetzee, he emphasised that these communities had been at the river for so long that they had 'developed a living situation accustomed to the Orange River valley', and like Coetzee, he reiterated that 'the people choose to be regarded as Coloureds'.[37]

Unlike Coetzee, however, Budack did not just count the livestock, he took notes of people's names and life histories. He spoke with Gert Witbooi, Abraham August, Matewis Kraai, Piet Botes and Willem Basson II. Budack's report was remarkably forgiving of the situation of these river-folk. It urged the Bantu Affairs Commissioner to let the people stay, not only because of their self-identification as Coloured but also because he felt that they would be

---

34 TBK KUS B-615 File B26/2/1/3 (vol. 2): Direkteur, Staatsnedersetting en Boerebystand to Streekverteenwoordiger, Dept. KbRa 'Plaas Hartebeesmund nr. 108' – 2 June 1970.

35 [*'en verkies om as kleurlinge deur te gaan'*]. TBK KUS B-615 File B26/2/1/3 (vol. 2): Landdros, Karasburg to Streekverteenwoordiger, Dept. KbRa 'Ongeoorloofde Verblyf op Staatseiendom' – 13 August 1970.

36 K.F.R. Budack, ''n Volkekundige Studie van die Tses Reservaat met Besondere Verwysing na die Geskiedenis en die Inter-Etniese Verhouding van die Bewoners' (MA thesis, University of Pretoria, 1965).

37 [*'Die mense self verkies om as Kleurlinge beskou te word'*]. TBK KUS B-615 File B26/2/1/3 (vol. 2): K.F.R. Budack, Etnologiese Afdeling van BAO to Hoofbantoesakekommissaris van SWA 'Nama en Namasprekende Ovaherero op Hartebeesmund 108, Warmbad' – 11 March 1971.

'unsatisfied' in the homelands, given their lifestyle by the Orange River. Willem Basson III was a boy when Coetzee and Budack came to Hartebeesmund, and he remembers their visits because it was quite rare for government officials to make the journey to see them. He recalled that his father was willing to reclassify as Coloured in order to retain access to grazing along the Orange River. At this stage, he had not been required to classify yet, and he did not see any contradiction in having 'Coloured' on his identification card and maintaining cultural membership within the Bondelswarts community; most Bondelswarts south of the river were already 'Coloureds'.[38]

The regional official for Coloured Affairs was not impressed with Budack's nuanced view on the matter, remarking: 'The ethnologist's conclusion that Nama-speakers can regard themselves as Coloureds cannot be accepted'.[39] He took steps to find a settlement in Namaland for these residents – ideally at Tses or Soromas – but both places were suffering from ongoing drought and could not accept any new grazing pressures.[40] The evictions fell by the wayside, and the Bassons and other families were – at least for a while – forgotten again. It is somewhat ironic that the heterogeneous 'Coloured' classification on the South African side of the river – made up of large numbers of Bondelswarts, Herero, Damara, etc. – went unquestioned by officials, whereas the same individuals on the Namibian side simply could not be conceived of as Coloured at all.[41] Racial categorisation was often a confusing process in Namibia and South Africa, fraught with contradictions and exceptions, though it was not entirely arbitrary either. The government used classification as both a blunt tool and as sharp language to control who could live where, and local Namibians sought to influence their own classification by drawing from their own histories in order to improve their positions and maintain hard-won gains.

•••

38 Willem Basson, interview with Bernard C. Moore and Luregn Lenggenhager (Karasburg, 11 November 2021).

39 ['*Die Etnoloog se stelling dat Nama-sprekendes hulself as Kleurlinge beskou kan nie aanvaar word nie*']. TBK KUS B-615 File B26/2/1/3 (vol. 2): Streekverteenwoordiger, KbRA 'Memorandum: Hartebeesmund: Hervestiging van Nama-Sprekende Ovaherero' – 14 June 1971.

40 TBK KUS B-615 File B26/2/1/3 (vol. 2): Streekverteenwoordiger, KbRa 'Memorandum: Verskuiwing van Namas(?) vanaf Hartebeesmund' – 19 October 1971.

41 Consult Moore, 'Swimming Upstream', and G.P. Klinghardt, 'Missions and Social Identities in the Lower Orange River Basin, 1760–1998' (PhD thesis, University of Cape Town, 2005).

The Gariep Dam was storing water by 1971, yet the administration had done nothing regarding the Orange River Coloured settlement,[42] and Coloured leaders began to petition further for 'rural areas' (*landelike gebiede*) for their people, distributed across north, central and southern Namibia.[43] These petitions – along with international pressure in the United Nations concerning Namibia – finally led the SWA executive committee to approve a full interdepartmental commission in order to investigate and plan the settlement, in late 1973.[44] Led by the SWA Department of Agricultural Credit and Land Tenure, the committee met for the first time in February 1974. No African representatives were present for any of the meetings over the following year or involved in any of the investigation work, and all documentation related to the project was stamped 'confidential' [*vertroulik*]. The chairman also requested that no publicity be made about the committee's investigations at all.

From the very first meeting, there were a few fundamental changes in the language and goals of how the Orange River Coloured settlement could go forward. First, the changed international status of Namibia meant that all discussions about a combined settlement with the South African side were abandoned. Although the Orange River was a fluid boundary, as it related to state projects, it was no longer feasible to view it as simply an internal river within South Africa. Second, the planners recognised that an irrigation settlement was not a viable option, because the irrigation potential of these particular six farms were fairly low, and the mountainous grazing would not be enough to support a larger number of Coloured livestock farmers. In 1964 when the Odendaal report was published, there were 140 Coloured family heads engaged in full-time farming, with approximately 69,000 small stock.[45] By 1974, that number was up to 257 Coloured family heads with more than 98,000 small stock.[46] One representative of the Department of Agricultural and Technical Services (LTD) urged that the commissioners completely abandon the idea that the six Odendaal farms could sustain enough stock-farming to accommodate even a small portion of this growth in livestock numbers. He

42 NAN LKG 52 File 7/3/3/30: Sekretaris van Waterwese, Pretoria: 'Derde aanvullende Verslag oor die Eerste Fase van die Oranjerivier-Ontwikkelingsprojek' (1971–1972).

43 NAN CRN 105 File 29/2/4/1: Hoofsekretaris, SWAKO to Streekverteenwoordiger, KbRa 'Beskrywingspunte van Tiende Jaarkongres van SWAKO' – 16 July 1970.

44 Department of Coloured Relations and Rehoboth Affairs, Annual Report 1973/74.

45 TBK KUS 1/3 File 1/1/6 (vol. 1): Plaaslike Bestuur en Kleurlingsake 'Memorandum Nedersetting vir Kleurlinge' – undated, likely 6 November 1964.

46 Large-stock units (GVE) were converted to small-stock units (KVE) at the official LTD rate of 6:1. NAN LKG 51 File 7/3/3/27: Streekverteenwoordiger, Dept KbRa 'Memorandum: Kleurlingveeboere in Suidwes-Afrika' – 10 June 1974.

said that the area was far too mountainous and inaccessible (*ontoeganklik*); the farms were 'on the whole physically and biologically unsuitable for farming of any sort whatsoever'.[47]

It is important to remember that, unlike the apartheid government's planning of homelands, Coloured settlements were never intended to be communal lands. The Bassons and other river-folk were able to successfully farm along the river and in the mountains because their operations were inherently transhumant: moving from river to communal areas in pursuit of rain and grass. Per the Rural Coloured Areas Act of 1963, these settlements were not to be communal or privatised but rather allocated to an individual/family on a demarcated, usufruct land-use basis.[48] They were planned to be smallholdings – economic units – that could sustain a specific annual family income. In this way, the land allocated to Coloureds depended on how this 'economic unit' – defined in quantity of small stock owned – was calculated. If the official Coloured economic unit measurement (700 small stock) were utilised, settling even one quarter of the Coloured bona-fide stock farmers would require the state to purchase nearly 490,000 hectares of grazing land in southern Namibia, much more than the Odendaal Report's recommendation of around 90,000 hectares. Even with a more modest economic unit of 300–500 small stock per family, only a little more than half of the Coloured farmers could be settled on an expanded scheme, though it would not allow for any growth in stock numbers whatsoever.[49]

After further investigations from committee members into irrigation potential along the Orange River, the outlook grew more pessimistic. By 1974, there was a total of 344 hectares of irrigated land along the six Odendaal farms, spread from Haakiesdoorn to Homsrivier.[50] The Department of Water Affairs believed that this potentially could be expanded to a maximum 1,030 hectares, but Girtis and Hartebeesmund had no irrigation possibilities at all.[51]

47 [*'in die geheel fisies en biologies nie geskik is vir boerdery van enige aard hoegenaamd nie.'*] NAN LKG 51 File 7/3/3/27: Notule: Eerste Vergadering van die Interdepartmentele Komitee belas met die Ondersoek na die Stigting van 'n Landelike Gebied vir Kleurlinge – 13 February 1974.

48 See Moore, 'Swimming Upstream', pp. 137–138.

49 NAN LKG 51 File 7/3/3/27: Notule: Eerste Vergadering van die Interdepartmentele Komitee belas met die Ondersoek na die Stigting van 'n Landelike Gebied vir Kleurlinge – 13 February 1974.

50 NAN LKG 51 File 7/3/3/27: Direkteur, LkGb SWA Streek to Hoof, LkGb Pretoria 'Besproeiingsgrond op Plase Langs die Oranjerivier' – 14 May 1974.

51 NAN LKG 51 File 7/3/3/27: Departement van Waterwese, 'Voorgestelde Besproeiingsaanleg Langs die Noordelike Oewer van die Oranjerivier op die Plase wat Strek vanaf die Stroomopgrens van Hartebeesmund tot die Stroomafgrens van Haakiesdoorn' – 2 May 1974.

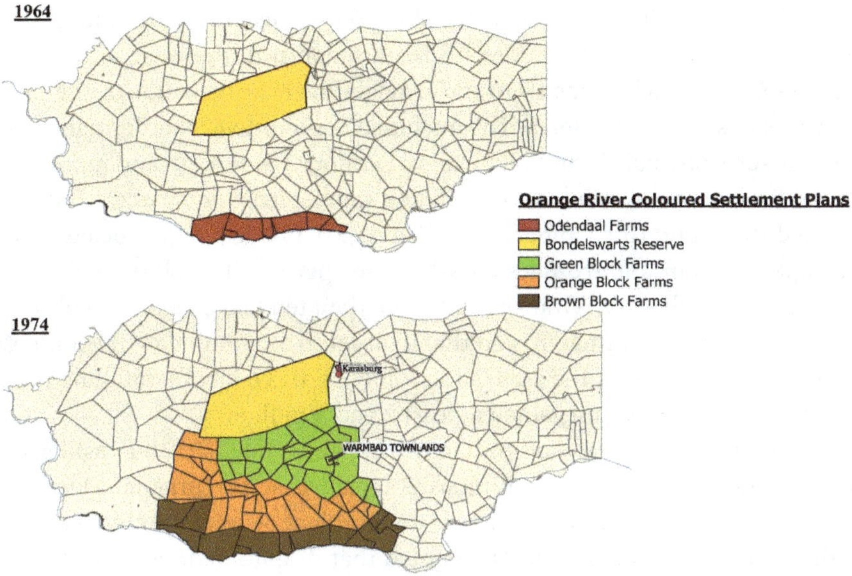

FIGURE 50 The Coloured settlement plans based on the Odendaal Report (top) compared to the Department of Agricultural and Technical Services (LTD) 1974 Ecological Groupings (bottom)
CARTOGRAPHY: B.C. MOORE

In keeping with 1970s apartheid-era considerations, economies of scale were growing more important to government planners. Although they envisioned the Coloured settlement as a smallholder scheme, they did not grant it the same leniency and subsidies that the previous Orange River 'poor-White' schemes had received. Representatives of the Department of Agricultural Economics and Marketing reasoned that even with a healthy crop mixture of sultana grapes, lucerne and alternating cotton and maize, the irrigation scheme would barely cover the costs of maintenance, operations and the diesel pump.[52] Without subsidies, they reasoned, smallholder irrigation schemes could not be profitable.

Realising that although irrigation was possible it was unlikely to flourish, the commission entrusted the Department of Agricultural and Technical Services (LTD) with the task of planning the expansion of the livestock settlement (*veenedersetting*) portion of the scheme. With the assumption that at least 50,000 small stock would have to be accommodated into the scheme, the LTD

52 Ibid.

recommended that, in addition to the six Odendaal farms, the government should buy the entire stretch of farms along the Orange River, from KumKum in the east to the state lands beyond Haakiesdoorn in the west, and all the commercial livestock farms from the Orange River in the south to the Bondelswarts Reserve in the north. This included Sandfontein, Sandfontein West, Bankwasser, Norechab, Kinderzitt, Umeis, Eendoorn, KumKum and more than fifty additional commercial stock farms and smaller irrigation parcels. The LTD grouped them into three 'blocks' (see Figure 50), divided by perceived carrying capacity, from approximately six hectares per small stock (green), eight hectares per small stock (Orange) and more than ten hectares per small stock (brown).[53] The LTD ultimately said that in order to accommodate the landless Coloured stock farmers, the state should buy all three blocks, totalling more than 580,000 hectares and valued at least ZAR 2.3 million.

The Interdepartmental Committee met again in July 1974 to consider the new proposals. The team unanimously agreed that the original Odendaal Plan was not sufficient, but there was a great deal of hesitation at the scale of the LTD's recommendation. They noted that despite their classification as 'brown block farms', Eendoorn and KumKum were actually making a fine living for some White farmers, and so, perhaps they should not buy out all the Whites. Furthermore, if the government were to assist with developing irrigation works and constructing the necessary fencing and water infrastructure to make the farms viable as Coloured smallholdings, it could cost the state up to ZAR 9 million to complete the scheme.[54] In addition, the strong rains that southern Namibia experienced in 1974 led to many White farmers departing the Livestock Reduction Scheme, revealing that the farms were perhaps not as arid as people thought in the 1960s. The committee was in a quandary. It was not eager to fund such a large-scale development scheme.

• • •

After the second meeting, the team boarded a flight from Windhoek to Karasburg. The next morning they climbed into a South African Air Force helicopter, so that the committee members may get their first aerial tour of the Orange River area

53 NAN LKG 51 File 7/3/3/27: Direkteur, LTD, SWA Streek to Sekretaris, Interdepartmentele Komitee oor Landelike Gebied vir Kleurlinge – 27 May 1974.

54 NAN LKG 51 File 7/3/3/27: Notule: Tweede Vergadering van die Interdepartmentele Komitee belas met die Ondersoek na die Stigting van 'n Landelike Gebied vir Kleurlinge – 1 July 1974.

FIGURE 51 Concrete slabs built by Chiebella in the 1970s for drying sultana grapes into raisins
PHOTO: B.C. MOORE, 2021

they had spent so many months thinking about.[55] The pilot was unable to find a single place to safely land the aircraft on Hartebeesmund, Girtis or Homsrivier, demonstrating the utter hopelessness of subdividing the Orange River farms into smallholder economic units. The team visited Haakiesdoorn, however, and met the managers of Chiebella (Edms) Bpk, who had recently acquired the farm and was in the process of expanding production of sultana grapes, lucerne and dates. The committee members understood that profitability along the Orange River necessitated either large-scale subsidies or deep pockets. As the following case of Chiebella shows, the company certainly had the latter.

The South African business tycoon, Anton Rupert, founded Rembrandt Group Ltd in 1948 with the initial goal of manufacturing cigarettes and growing tobacco, with most operations in the Stellenbosch area. Fairly soon after, however, Rembrandt branched into the wine and spirits industry, with the formation of Distillers Korporasie Bpk, which primarily produced brandy and cheaper wines, such as Tassenberg.[56] With generous loans from Sanlam and other Afrikaner-dominated financial institutions, Rembrandt and Distillers

55 NAN LKG 51 File 7/3/3/27: Addendum tot die Notule van die Interdepartmentele Komitee se Vergadering gehou op 1 July 1974 te Windhoek – Besoek aan die Oranjerivier.

56 Remgro Ltd, *2015 Integrated Annual Report*, p. 6.

thrived in South Africa through the 1950s and 1960s.[57] In 1972, much of the tobacco and other luxury goods branches of Rembrandt were spun off into the Swiss company Richemont, and by this stage Rembrandt and Distillers had become interchangeable. At that point, the company doubled down on alcohol production as well as sourcing the raw materials necessary for it.

Rembrandt sought to increase its production of grapes for base wines, both for its low-grade though high-selling Tassenberg and for its higher-grade brandies. Five litres of base wine is required for the production of one litre of brandy.[58] Rembrandt's brandy division, Oude Meester Bpk, looked to the Orange River, where land was cheap, humidity was low and irrigation waters were bountiful. Many vintners and farm managers working for Oude Meester believed that base wine grapes (often white grapes like sultanas) grew best in these arid parts of the Orange River valley.[59] Furthermore, sultanas could also be used for raisin production if weather conditions rendered the crop unsuitable for wine production.[60] The company had already purchased some properties farther upriver, in the area of Kakamas, Keimoes and Upington, and now, like most Afrikaner capital during the mid-twentieth century, it sought to enter cheaply into the newly claimed colony, Namibia.[61]

Between 1970 and 1972, SWA residents Johannes Adolf Kleynhans and his wife, Jean-Lavinia, purchased several properties along the north bank of the Orange River with the intention of entering into a sales or production partnership with Oude Meester. The properties included several unsuccessful sheep farms – like Pelladrift and Oranjefall – and some farms with outdated irrigation infrastructure – like Beenbreek and Haakiesdoorn.[62] Oude Meester wanted to purchase the properties to grow grapes and diversify into lucerne, dates and some vegetable production.[63] It formed a series of subsidiary companies

---

57 P. du Toit, *The Stellenbosch Mafia: Inside the Billionaires' Club* (Cape Town, Jonathan Ball, 2021).

58 L. Louw and M.G. Lambrechts, 'Grape based brandies: Production, sensory properties, and sensory evaluation' in J. Piggott (ed.) *Alcoholic Beverages: Sensory Evaluation and Consumer Research* (Woodhead Publishing, 2012), pp. 281–298.

59 Leon Fouche (Managing Director, Distell Namibia), interview with Bernard C. Moore (Windhoek: 7 March 2022).

60 Andries Nolte (Technical Manager, ZZ2 Group Ltd), interviews with Bernard C. Moore (Haakiesdoorn Farm: 20 September 2021 and 20 July 2022).

61 Distillers has today morphed into Distell Group Holdings, Ltd. Its Namibian subsidiary, Distell Namibia (Pty) Ltd, founded in 1968, is owned via Distell's investment vehicle South African Distilleries & Wines S.A. Ltd. See BIPA File 1968/15689: Annual Return (CM-23) – 7 November 2017.

62 See WDO: Registrasie Afdeling 'V': Pelladrift no. 102.

63 BIPA File 1973/5752: Agreement signed between Oude Meester Groep Bpk and J.A. Kleynhans, J.L. Kleynhans, P.G. Steyn – 16 April 1973.

that would operate as holding vehicles for the farm properties, the two most important being Chiebella (Edms) Bpk (which would own Haakiesdoorn) and Pelladrift (Edms) Bpk (which would own Oranjefall and Pelladrift).[64] Most of the directors were from Oude Meester, but Adolf and Jean Kleynhans would remain on the board of both operations until 1980.[65]

Haakiesdoorn was already operating as an irrigated farm when Oude Meester purchased it (Charles Weidner had also grown some citrus there in the 1920s), but the new corporate arrangement expanded production significantly. The lands directly across from Goodhouse had been already flattened for several decades, so the company hired additional workers and machinery to flatten land for several kilometres west along the Orange River. Vineyards and some date palms were planted, and a number of fifty-metre-long concrete slabs were built to facilitate drying sultana grapes. Similar land-flattening was done on Pelladrift, Oranjefall and Beenbreek, but Haakiesdoorn was the largest of Oude Meester's operations on the Namibian side. When the Interdepartmental Committee arrived at the farm in 1974, Kleynhans had more than 160 Ovambo migrant workers from northern Namibia in his employ on short-term harvesting contracts, as well as Coloureds and river-folk. He anticipated eventually increasing to 500 workers during harvests, once additional lands came under cultivation. This caused the local magistrate to raise a red flag, and he remonstrated Kleynhans on the poor-quality worker housing.[66]

The Interdepartmental Committee boarded their military helicopter for home with a deeper understanding of what was necessary to develop irrigation works along the Orange River. Perhaps their smallholder system for Coloureds was not that viable in the post-Gariep Dam era. The Orange River was now tied closer to regional and global markets, and without granting heavy subsidies to Africans (something the committee was reluctant to do), smallholdings like the old 'poor-White' schemes were increasingly an anachronism.

In May 1975, the committee completed its final report, recommending the purchase of most of the lands demarcated in the LTD's three block land arrangement.[67] After reducing the 'economic unit' measurement for Coloureds, they were able to cut off some of the proposed purchases along the eastern and western extensions, and specifically excluded buying Tantalite Valley mine and

64 BIPA File 1973/5752: Sertifikaat van Inkorporasie, Pelladrift (Edms) Bpk – 10 May 1973.

65 BIPA File 1973/5920: Inhoud van Register van Direkteure, Ouditeure en Beamptes (CM-29) – 3 December 1980.

66 NAN Archives of the Chief Bantu Affairs Commissioner, Windhoek (hereafter, BAC) 305 File (43)N.3/13/3/11: Landdros, Karasburg to Hoofbantoesakekommissaris, Windhoek 'Inspeksie: Chiebella (Edms) Bpk' – 19 December 1974.

67 NAN LKG 51 File 7/3/3/27: Verslag en Aanbevelings deur die Komitee wat Ondersoek Ingestel het na 'n Landelike Nedersetting vir die Kleurlinge in SWA (May 1975).

FIGURE 52 Remnants of Chiebella's shelters for Ovambo contract workers
PHOTO: B.C. MOORE, 2021

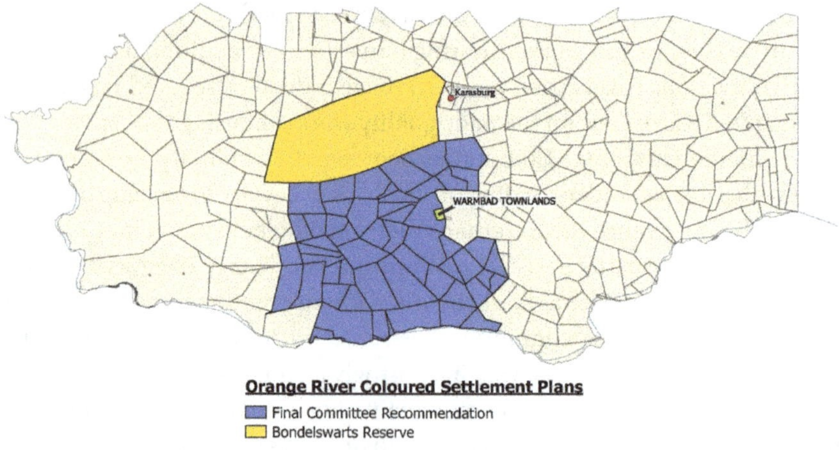

FIGURE 53 Final committee recommendations concerning the Orange River settlement, 1975
CARTOGRAPHY: B.C. MOORE

Chiebella's Haakiesdoorn, on the grounds that the capital investments on the properties would render the whole scheme unaffordable.

•••

The 1975 Coloured settlement proposal was one of the last grand schemes of apartheid-era land planning in Namibia. In a number of ways, it was fairly ambitions and held the capacity not only to improve the livelihoods of the Coloured stock farmers who sought grazing access, but it would inadvertently allow river-folk like the Bassons more secure land tenure and grazing access along the river. By this point, many Nama and river-folk had registered as 'Coloured' for apartheid-era classification purposes. Once the Bondelswarts Reserve was reproclaimed in 1978, this scheme would have finally reconnected the Bondelswarts communal areas and the Orange River, a link that had been severed by colonialism and land alienation so many decades before. Despite this promise, not a single hectare of land was ever purchased for the purpose of an Orange River settlement. Late-1970s political transformations in SWA/Namibia meant that the sixteen-month investigation into the Coloured settlement was never even laid before the government for approval.

In response to UN pressure, South African State President B.J. Vorster held a number of meetings during 1973 and 1974 with centrist and sympathetic African leaders in Namibia in an effort to build a pro-Pretoria, though multiracial, political party.[68] This eventually gave rise to the Turnhalle Constitutional Convention and the political party, the Democratic Turnhalle Alliance (DTA). Apart from gradually abolishing 'petty apartheid' laws (such as pass laws and anti-miscegenation laws), one of the critical outcomes of the 1975–1977 Turnhalle Conference was a slight transformation in the ways in which land was handled in NAMIBIA. Rather than focusing on homelands, which would achieve, per the Odendaal Plan, 'political independence alongside economic dependence' (*politieke onafhanklikheid met ekonomiese afhanklikheid*), the DTA envisioned a multinational ethnic federal system, built through institutions known as Representative Authorities (*Verteenwoordigende Owerhede*) that would see to the wellbeing of registered members of their given population group, regardless of where these members resided.[69]

Existing homelands, such as Namaland, would remain as communal lands belonging to their given Representative Authority. (New, grand apartheid planning schemes were now over; they were simply too expensive and politically risky.)[70] In many cases, this was a two edged sword. The Bondelswarts Reserve was reproclaimed in 1978, meaning no more land was to be taken

68 See TBK KUS B-667 File B39/8 (vol. 2): Notule van Adviesraadsvergadering op 26 en 27 September 1974, Kaapstad.

69 This was eventually put into law in 1980, repealing the Self-Government for Native Nations of SWA Act. See Proclamation no. A.G. 8: General Provisions Applicable in Respect of Representative Authorities Established for Population Groups (24 April 1980).

70 TBK KUS B-641 File B39/9 (vol. 2): Ekonomiese Komitee van SWA Beraad: 'Ontwikkelingsplan van SWA: Samesprekings met Mnr Mudge (L.U.K.), Dr Swarts en Mnre van Gas (BAO) en Gous en Rautenbach (SWA Admin)' – 21 July 1976.

from Africans, but the Orange River Coloured settlement plan died – no more land was to be taken from Whites either. If Representative Authorities wanted to expand grazing areas for their people, then the new dispensation meant they would have to buy the land on a willing-buyer, willing-seller basis. Under the DTA, the status quo concerning land ownership was maintained, dressing racial inequalities in the clothing of class, rather than race. It is in this context that the Interdepartmental Committee published its 1975 final report on the Orange River settlement, and for this reason, the Department of Coloured, Rehoboth and Nama Relations wrote in its 1976/77 annual report that 'in view of the constitutional development in SWA it [the 1975 Report] will not be submitted to Cabinet at this stage'.[71] The same statement was written for the next two years as well.[72] The Orange River settlement would never be mentioned again.

The Orange River of the 1970s and the 1980s was simply not the same as it was in the smallholder 'poor-White' settlement days of the 1920s and 1930s; even the crown jewel of such settlements on the Namibian side, Noordoewer, was changing rapidly. In the late 1950s, a bridge for trucks was constructed spanning the Orange River – negating the need for the Goodhouse pontoon – and a tarred trunk road was completed in 1969.[73] The opening of the government-regulated Cape Town Fresh Produce Market in the 1960s – equipped with cold rooms – allowed even Orange River farmers to gradually diversify away from monocrop lucerne production towards fruits and vegetables for regional and global markets. By 1965, more than two thirds of all irrigated lands along the Namibian side of the Orange River was planted exclusively with year-round lucerne to supply agricultural cooperatives and private sheep farmers.[74] Especially on the Namibian side, the turn of the 1970s signalled a gradual shift from irrigation as ancillary to the livestock industry to irrigation for export production. This would only accelerate in the 1980s.

Oude Meester sold off Chiebella (Edms) Bpk and their 315 hectares of irrigation in 1988 to the Thalwitzer family, owners of the Komati Fruit Group of

71 NAN Archives of the Administration for Namas (hereafter, ANA) 1 File 1/3: Department of Coloured, Rehoboth and Nama Relations: Annual Report 1976/77.

72 [*'Met die oog op ontwikkelings in SWA is hierdie aangeleentheid voorlopig agterweë gelaat'*]. NAN ANA 1 File 1/3: Department van Kleurling-, Rehoboth-, en Namabetrekkinge: Jaarverslag 1977/78 & Jaarverslag 1978/79.

73 NAN PLA 162 File P.41/48 (vol. 1): Acting Chief Roads Engineer to Director, Department of Local Government, Windhoek 'Noordoewer: Establishment of a Township' – 10 October 1975.

74 Department of Water Affairs, SWA, *Preliminary Feasibility Report on the Irrigation Possibilities along the Lower Orange River* (Planning Division: August 1987).

FIGURE 54 A pivot irrigation system on Haakiesdoorn, from the Komati days and unused since at least 2011

Tzaneen, who bought Haakiesdoorn in part to grow citrus and grapes but also to further develop existing lucerne capacities.[75] They moved away from flood-furrow irrigation and invested heavily in centre-pivot sprinkler irrigation systems, now that the farm was connected to the South African ESKOM power grid.[76] As for Oude Meester's eastern holdings, particularly Pelladrift (Edms) Bpk, the company and assets were sold in 1983 to Gerrit Luttig, head of one the old *oewerboere* families in the region, who had been farming on both sides of the Orange River since before the land-surveying boom of the 1920s.[77]

Gerrit Luttig's father had been a sheep farmer for a number of years on Kum-Kum, and Luttig himself had owned Kambreek, Pelgrimsrust and Khaais farms since 1971, as well as additional parcels on the South African side near Pella and Onseepkans.[78] Luttig ran an ESKOM powerline across the Orange River from Pella Mission Station and along several kilometres of narrow flattened Pelladrift river frontage, where pumps irrigated lucerne, tomatoes, peas and other vegetables. The powerline then went up over the Spelonkberge mountains to another portion of irrigation on the border with Oranjefall, crossing the river

75 BIPA File 1973/5920: Inhoud van Register van Direkteure, Ouditeure, en Beamptes (CM-29) – 7 December 1988.

76 BIPA File 1973/5920: F.V.U. Thalwitzer to Departement van Handel en Nywerheid, Windhoek 'Insake skrapping van Chiebella (Edms) Bpk' – 17 December 1991.

77 BIPA File 1973/5752: Contents of Register of Directors, Auditors, and Officers (CM-29) – 1 July 1983.

78 WDO: Registrasie Afdeling 'V': Pelgrimsrust no. 100 and Kambreek no. 104.

again to his estate at Luttigshoop, Onseepkans.[79] Gerrit Luttig embraced the emergence of a regional market for Orange River produce, and built a private pontoon to cross from Pelladrift to Luttigshoop, equipping it to carry a sixteen tonne truck that would proceed on to Cape Town.[80] In total, Luttig expanded Oude Meester's original production by up to 140 hectares.[81]

Irrigation farmers on the South African side of the Orange River also expanded production during the 1970s and 1980s, targeting regional and global markets. In response to declining karakul production and thus reduced demand for lucerne, Gertjie Niemöller made plans to transform some of his irrigated fields from fodder to Medjool date palms. He travelled to Arizona in 1978 and 1979, from where he imported more than 4,000 seedlings to plant on Kambreek-Zandfontein.[82] With financial and technical assistance from the South African state-owned Nywerheidsontwikkelingskorporasie, Niemöller's date operations along the Orange River and farther inland eventually became the largest date plantation in the southern hemisphere.[83]

Indeed, as the Orange River became more valuable as a place to grow export crops, a greater incentive grew to police farm boundaries and the river crossing more strictly. This made life for the Basson family and other river-folk even more difficult, since northbound transhumance had already been severed by the Odendaal Plan. Now their east – west mobility was increasingly threatened by the lure of irrigation profits on Pelladrift, Haakiesdoorn and Kambreek-Zandfontein. Willem Basson III's father eventually had a falling out with Niemöller – who regularly called the man 'Willem Oorlog' ('War Willem') – because some of his herders inadvertently allowed the livestock to stray from Hartebeesmund and KumKum into Niemöller's lucerne fields across the river.[84] Once Niemöller switched to more valuable date-farming, he formally rented the Orange River islands from the RSA state, in order to monitor the cross-border mobility of the river-folk.[85]

79 Nico Strauss, interview with Luregn Lenggenhager and Bernard C. Moore (Velloor farm, Karasburg District: 10 November 2021).

80 Pieter Luttig, interview with Bernard C. Moore and Luregn Lenggenhager (Tigerberg farm, Karasburg District: 10 November 2021).

81 Department of Water Affairs, SWA, *Preliminary Feasibility Report on the Irrigation Possibilities along the Lower Orange River* (Planning Division: August 1987).

82 GNPA: R.G. Niemöller, 'Die Dadelbome van Klein Pella' – undated.

83 See Moore, 'Swimming Upstream', pp. 136, 141.

84 GNPA: R.G. Niemöller, Pofadder to Stasiebevelvoerder, SAP Karasburg – 29 March 1972.

85 GNPA: Departement van Openbare Werke en Grondsake, Tydelike Huurkontrak ten Opsigte van Vakante Staatsgrond: R.G. Niemöller – 11 December 1984.

When South Africa took over Namibia from Germany, the Orange River ceased to be a formal international border, though it still was relevant for veterinary and administrative purposes. With the independence of Namibia in March 1990, the international border returned, and the legal definition was again the high-water mark on the north bank of the river. Niemöller's islands in the middle remained South African territory. After centuries of living in this transriverine zone, for the Bassons and other river-folk it would take a while for the border to become meaningful. This was in part because neither government strictly policed agricultural endeavours along the river. Gerrit Luttig wrote letters to Windhoek and Pretoria requesting instructions on how to proceed with his (now cross-border) business – because Pelladrift was accessible from the south bank only – and no one requested that he stop his pontoon.[86]

Niemöller actually did not care if Willem Basson II [Willem Oorlog] moved his livestock around, as long as it stayed on the Namibian side or on the islands. For Niemöller, the new regime of passports, IDs and veterinary certificates needed to be followed, but he was asking for a radical change from how things had operated for centuries: even when the Germans controlled Namibia, movement across the border in these remote corners was rarely policed. Niemöller simply could not grasp that someone whose primary residence was in Namibia on Hartebeesmund Farm could occasionally work at Aggeneys Mine and Pofadder town and be a member of the community at Pella.[87] Perhaps Niemöller took it personally that not all the livestock belonged to his enemy Willem Basson II, but also to Willem's sons, Petrus and Gertus Basson, who had been in Niemöller's employ on his farms and mines for years.

The independence of Namibia in 1990 and the return of the Orange River as an international border changed the situation drastically for the Bassons and other river-folk. Facing a rapidly expanding irrigation sector along the river, east – west transhumance was severed. The failure of the Orange River Coloured settlement meant that movement northbound to the Bondelswarts Reserve was blocked by White livestock farmers. The international border with South Africa meant that obtaining grazing on 'Coloured' communal areas like Pella and Witbank was technically illegal without livestock export permits, passport stamps and an arduous journey to the official border posts at

86 Pieter Luttig, interview with Bernard C. Moore and Luregn Lenggenhager (Tigerberg farm, Karasburg District: 10 November 2021).

87 [*'Die familie woon dus in Namibië, maar die vee loop op die eilande wat aan Suid-Afrika behoort. Hoe kan hulle as die gemeenskap van Pella beskou word?'*]. GNPA: R.G. Niemöller, Pofadder to Streekverteenwoordiger, Departement van Openbare Werke, Kimberley 'Huur van Eilande in die Oranjerivier teenoor die Plase Hoogoor en Kambreek' – 5 July 1995.

Noordoewer or Onseepkans. Their original capitalist patrons – Weidner and Niemöller – were now either gone or relations with them had soured. Finally, their longest-standing patron – the Roman Catholic Church – was forced to sell off Homsrivier Farm and mission station in 1992, to raise funds to build the offices for the new, post-independence, Diocese of Keetmanshoop.[88] While the rest of Namibia was enjoying the fruits of national independence, the Bassons and the river-folk were now out of patrons, and they were trapped.

88 WDO File T402/1992: Remainder of Farm Homsrivier no. 133, Deed of Transfer – 31 January 1992. A.P.J. Beris CMM, *Making the Desert Bloom* (*vol. 4*) (Keetmanshoop, RC Diocese, 2014), p. 38.

PART 3

Global Capitalists, Global Conservationists

CHAPTER 8

A Mountain Retreat

On 21 March 1990, after decades of political and armed struggle, the Republic of Namibia earned its political independence from South Africa. For the newly elected government in Windhoek – led by the South West Africa People's Organisation (SWAPO) – it was quite important that it could formally declare in Article 1 of the post-apartheid Constitution that there was now an international border near the Orange River. For those who lived by the river, however, this new round of the politics of passports and permits was yet another move in reducing their pastoral and transhumant mobility. At the same time, independence and the end of apartheid legislation brought fundamental changes to those who lived by the river. They were now citizens of a new and independent country, and they were supposedly able to benefit from the privileges of that status. However, political independence brought challenges: independence came with deeply neoliberal policies, and the new government did very little to address the land question, especially during the first five years.

Most of the farms that stretched from the Orange River up to Warmbad and beyond remained in White hands, though most of these Whites were now Namibians. Many Whites, however, did not wish to live in a post-apartheid Namibia, and there was a rush of farm sales in the years between 1988 and 1995. Land in Namibia was sold for pennies on the dollar, the sellers returning to South Africa where apartheid would not let go for a few more years. Until the promulgation of Namibia's Agricultural (Commercial) Land Reform Act of 1995, there were still no laws restricting foreign ownership of farmland, so many of these sales went to foreign individuals or businesses. Some of these new owners continued to run their properties as livestock or irrigation farms; others sought to diversify into nature conservation, tourism and trophy-hunting, sectors that were expected to grow dramatically in the post-apartheid years.

Driving southbound on district road 292, the road gradually rises over the Orange River mountains before descending for about 10–15 kilometres to the irrigated fields at Gaidip and Homsrivier Mission Station. At the very top of those mountains – where the road crosses from the southern boundary of Gaobis Farm to Ramansdrift Farm – there is a small steel placard on the left-hand side, pointing eastwards to a disused public road and a place called 'Great River Game Ranch'. You might miss it if you drive too fast. Owned for many years by an eccentric Danish engineer, named Steen Severin, the Great River Game Ranch was a small tourism and trophy-hunting company and, as described at

FIGURE 55 A three-metre-high game fence built along the southern border of Homsrivier Farm
PHOTO: B.C. MOORE, 2021

the beginning of this book, it was allegedly responsible for kraaling ostriches in Homsrivier Church. Nevertheless, its establishment in the 1990s set in motion a fundamental shift in southern Namibia; new owners introduced a new regime of land use and value. Steen Severin was not a farmer, and his modest operation brought a new set of challenges to river-folk along the Orange, many of which they still face today.

• • •

Steen Severin was born on 14 April 1939 in Denmark to a prominent family, who were major shareholders in Det danske Stålvalseværk A/S, which operated the largest iron and steel foundry in Copenhagen.[1] The Danish steel industry was closely tied to the boatbuilding and shipping industries in both Copenhagen and Odense. Severin and his father ran a Copenhagen-based company called Silikadan A/S, which manufactured and sold industrial products and machinery.[2] Severin trained as an engineer, and for a final project as part of his studies he designed a new kind of industrial boiler for steam generators,

1 Peter J. Cox, interview with Bernard C. Moore (telephone, 15 January 2023). Cox believed that the Severins were somehow connected to the Danish royal family, which is possible, because Prince Axel was a shareholder in Det danske Stålvalseværk.

2 Ministeriet for Handel, Industri og Søfart, *Registreringstidende for Aktieselskaber, Forsikringsselskaber og Foreninger* (no. 10 of 1955), pp. 472–3.

which made substantial profits for the company, especially in sales to Japanese factories in the post-WWII era.[3] Severin and his father travelled around Europe during the early 1960s, marketing boilers and other products manufactured by the family company. In a chance encounter in 1962, at a Zürich restaurant named l'Eiffel near the airport, Steen Severin met his future wife, Franka.

Born in 1942 in the small Italian village of Loneriacco, near the Slovenian border, Franka Cussigh lived a modest life. Her father had been a soldier in Mussolini's army, but the family moved to Paris in 1952 because many demobilised soldiers who had fought for the old fascist regime were finding it hard to get work in the new Italy.[4] Her father got a job as a stone mason, and Franka completed school, learning French along the way; she was then employed by the Printemps department store. Through her work, she became acquainted with a photographer named Raoul, who took her to the French Riviera in 1959. Recognising her beauty, he urged her to enter the Miss Côte d'Azur and Miss Film Festival modelling competitions, and to her surprise, Franka won.[5] Sporting the new name Franka Bell – which Raoul considered more cosmopolitan than her own provincial surname – Franka moved to the Riviera, where she dined with the likes of Pablo Picasso, Salvador Dali, Le Corbusier, the Onassises and the Kennedys. Via connections with producer and publicist Angelo Rizzoli, she travelled for a year throughout Europe, acting in small films. She eventually invested her savings in the restaurant l'Eiffel in Zürich.

According to Franka, 'after that meeting in Zürich' with Steen Severin she fell pregnant with his child. A few years later, in 1965, they got married, and Franka moved to Copenhagen with the Severin family.[6] Steen Severin continued with his family business while Franka raised their child, took Danish language classes and studied fine arts. Through the family's longstanding connections with the Danish Steelworks, from the late 1960s Severin's main business was shipping raw materials from South America and Africa to foundries and factories in Denmark.[7] By 1970, Severin had purchased two major shipping vessels, which would make return journeys from Rotterdam to Cape Town; he named them the *Tinou Bell* and the *Franka Bell*, after his wife. According to one of Severin's former business partners, he used them to ship coal from South Africa and engaged in 'sanctions busting' by shipping Rhodesian chrome to

3 Peter J. Cox, interview with Bernard C. Moore (telephone, 15 January 2023).

4 Hélène Fincker (ed.), *Franka: Une Femme Affranchie: Chronique d'une vie* (Paris, Éditions Balland, 2018), p. 35.

5 Ibid, pp. 57–60.

6 Ibid, pp. 99–102.

7 Severin's ships would also import animal feed to Denmark. See US Department of Agriculture, 'Exports', *Journal of Commerce* (1988), p. 12B.

FIGURE 56 Gittelil, later renamed Tinou Bell after Severin purchased it
PHOTO: H. PETERSEN, 1965. REPRODUCTION COURTESY OF M/S MARITIME MUSEUM OF DENMARK

Europe.[8] Severin clearly took risks: in late 1971 the *Franka Bell* was nearly sunk in the North Sea after colliding with an English vessel while his boat was carrying eighty tonnes more than capacity.[9] On a journey back from Cape Town in March 1973, the *Tinou Bell* suffered a hull breach off the coast of Gabon. The ship was abandoned, and the crew was thankfully saved by local coastguards.[10]

After their second child, Sabine, was born in 1972, Franka began to involve herself more in Severin's businesses. They founded Steen Severin A/S in 1975 to consolidate their shipping and trading operations.[11] During the 1970s, Western Europe was facing a shortage of metallurgical coking coal and thermal coal for energy production. In the wake of the OPEC embargo and 1973 oil crisis, the International Energy Agency (IEA) initiated policies to get Europe to transition to coal again, as well as promoting scientific research into liquification and gasification schemes.[12] Coal use expanded across the developed world throughout

8 Peter J. Cox, interview with Bernard C. Moore (telephone, 15 January 2023).
9 *Dansk Søulykke-Statistik, 1971* (København, Handelsministeriet, September 1972).
10 Museet for Søfarts Billedarkiv File 51700: M/S *Gittelil.*
11 Danske Erhvervsstyrelsen File A/S60693: Steen Severin Aktieselskab.
12 H. Türk, 'From Oil to Coal? The International Energy Agency and International Coal Policy since the End of the 1970s', in L. Bluma et al. (eds.) *Boom-Crisis-Heritage: King Coal and the Energy Revolutions after 1945* (Berlin, De Gruyter Oldenbourg, 2021), pp. 81–92.

the late 1970s and 1980s, and shortages abounded. Severin sought new sources to fuel the Danish foundries.

While Steen may have been the engineer, Franka had the personality and skills to drive business relationships, and she was often the face of their operations.[13] Franka helped negotiate a partnership with Les Houillères du bassin du Nord et du Pas-de-Calais, a division of the French national coal company, Charbonnages de France (CdF), in which Steen Severin A/S would supply CdF through its southern hemisphere suppliers, as well as buy coking coal and chemical fertiliser from CdF.[14] This arrangement was successful for a few years, though by 1977, the coal reserves in northern France were increasingly depleted and CdF was looking to increase imports from other countries. Steen Severin A/S was given a contract to investigate possible coal exports from South America and South Africa. Steen and Franka decided that, given their existing connections in southern Africa, they would forge additional relationships with the coal mines in South Africa's KwaZulu-Natal province to supply CdF.

By the end of the decade, the Severins wanted 'complete control' over the coal production and shipping arrangements – they wanted their own mines. In her biography, Franka reflects on the family's decision to move to the Republic of South Africa. It was driven by high Danish taxes and, 'for political reasons which are beyond me', the Danish government would not allow a Danish company to invest in South Africa: 'I am not hostile to a form of solidarity', Franka wrote, concerning anti-apartheid sanctions, 'but it seemed abusive to me'.[15] Just before leaving for Johannesburg, Franka bought a show apartment in Monaco (the Monte Carlo Star), presumably as a fall-back option in case things did not work out. Nevertheless, in 1979, the Severins emigrated to South Africa with their children, with the intention of buying a coal mine.

•••

Having purchased a villa in Sandton, in Johannesburg, and enrolled their children at the French school, the Severins got busy scoping out mining opportunities. They hired a geologist, Louis Carroll, to assist them in surveying and acquiring a coal mine, offering him payment in shares in their new South African company, Severin Minerals Corporation (Pty) Ltd.[16] With a ZAR

13 Peter J. Cox, interview with Bernard C. Moore (telephone, 25 March 2023).

14 Fincker, *Franka*, pp. 111–113.

15 Ibid, p. 114.

16 Ibid, p. 120.

35 million loan from Standard Bank to boost their personal payment, Severin bought St George's Colliery, in Dundee, in today's KwaZulu-Natal province. The sulphur content of the coal matched what was required by Charbonnages de France, and additional production went into briquettes, which the company supplied to the South African public utility Eskom for electricity production.[17] By the end of 1981, 7.9 per cent of French coal imports came from South Africa, and this rise was in part due to the Severins and their relationship with CdF.[18]

Steen Severin was not a geologist, but his skills in engineering and metallurgy – related to his work with Det danske Stålvalseværk – made him a fairly successful mining engineer. With steady cash flow coming in from their coal sales as well as from their operations in Denmark, he took a tour of recently decommissioned mines in South Africa to see what he could get on the cheap. Many of South Africa's older gold mines were no longer profitable using traditional extraction methods, and some of the shafts were nearly a century old by that stage. Severin visited Rand Leases, west of Johannesburg. Although the mine shafts were not likely to be profitable, Severin was impressed with the tailings and overburden on the property.[19] He was about to do something very different in the mining sector: hydrometallurgy and heap leaching.

At its most basic, hydrometallurgy is the process of using liquid acids and alkalis to extract metal from ore; in prior years this was done in the intense heat of furnaces. Cyanidation of metallic gold has been known about since at least the eighteenth century. However, the application of hydrometallurgical principles to leach ore on a larger scale is a moderately new phenomenon, developed after the Second World War as metallurgists came to understand better the properties of acids, reagents and lixiviants, and the pressurised systems for treating ore.[20]

Once finely crushed ore is removed from the shaft or open pit mine it is piled into massive tailings or 'heaps' laid on an impermeable, often plastic surface. Sprinklers and pumps spray a lixiviant solution (depending on the metal, this could be sulphuric acid, sodium cyanide, hydrochloric acid or even just water) onto the heap, which permeates the crushed ore and chemically reacts with the metals in the material, absorbing some of the gold, copper, zinc, etc. into the liquid itself. The pregnant leach solution then seeps down to vats at the bottom of the heap, where it is either repumped to the top to repeat the process

17 Ibid, p. 124.

18 A.L. Giraud, 'Energy in France', *Annual Review of Energy*, 8 (1983), pp. 165–191.

19 Peter J. Cox, interview with Bernard C. Moore (telephone, 25 March 2023).

20 F. Habashi, 'A Short History of Hydrometallurgy', *Hydrometallurgy*, 79 (2005), pp. 15–22.

and absorb more metal or is pumped to a plant to extract the metal from the solvent via mechanical or chemical processes, or else by electrowinning.[21]

Heap leaching is a far more profitable extraction of metals from low-grade ore than via mechanical and thermal means.[22] As it relates to average gold leaching operations, heap leaching can make ores of less than 0.5 grams of gold per metric tonne viable (though this does not include the economic and environmental costs of between fifteen and thirty billion litres of water per year needed for both the lixiviant and dust suppression).[23] Furthermore, it has the capacity to drastically reduce the physical labour needed to mine valuable minerals like gold, copper, lithium, uranium and others. Additional electricity costs are needed in the metallurgical plant, however. Given the fact that the Rand gold mines were quite low grade to begin with, and viable only because of the apartheid regime's cheap labour polices, heap leaching was a means for Severin to overcome some of these limitations.[24]

When Severin bought Rand Leases in 1985 – registering it under his new company, Severin Mining & Development (Pty) Ltd – it had been barely mined at all since 1971. Many observers in the industry welcomed the decision to dig open-cut mines and reopen two of the mine shafts to extract more ore,[25] though it was clear that the main purpose of Rand Leases was to focus on the heap leaching of the existing tailings, which dated back to the early 1900s. Severin spent more than ZAR 16 million on the hydrometallurgical plant on site to treat the pregnant leach solution.[26] The mine resumed operations, and between 1985 and 1986 the workers extracted approximately 35,000 tonnes of ore per month, which were then added to Severin's heap system.

Steen Severin hired Peter J. Cox as one of his geologists in 1986; Cox was not long out of university and had earlier worked for the mining company, Johannesburg Consolidated Investment Co. Ltd (JCI). 'When Steen Severin arrived in South Africa, the mining industry was incredibly centralised: JCI, Anglo American, RandGold, Goldfields, AngloVaal dominated everything', Cox said, reflecting on those years. 'Severin was basically the first "small mining company" in the region which was really going fairly successfully, and Severin's

21 M. Nicol, *Hydrometallurgy: Volume 2, Practice* (Amsterdam, Elsevier, 2022), pp. 22–26.

22 M. Nicol, *Hydrometallurgy: Volume 1, Theory* (Amsterdam, Elsevier, 2022), pp. 1–13.

23 D.I. Bleiwas, *Estimated Water Requirements Gold Heap Leach Operations* (US Geological Survey, 2012), p. 11.

24 See, J. McCulloch, *South Africa's Gold Mines and the Politics of Silicosis* (Oxford, James Currey, 2012).

25 West Wits Limited, ASX Initiation Report (June 2022), p. 7

26 Savannah Environmental (Pty) Ltd, *Final Basic Assessment Report: Rand Leases Extension 15, Industrial Township* (November 2014), p. 5.

efficient heap leaching was one of the reasons. Today, all these people know about heap leaching, but very few know the name of Steen Severin'.[27] Although his adoption of new hydrometallurgical mining techniques was a major reason for his success, it is important to remember that Severin was willing to participate in the economy of 1980s southern Africa, a region increasingly targeted by international and corporate divestment and sanctions. He bought his mines cheaply, in part because they were 'low grade' and in part because foreign companies were willing to sell them at low prices.[28]

The US company General Electric (GE) had bought Utah International in 1976 for more than USD 2.1 billion (see chapter 6). This gave General Electric overseas exploration and production assets in South America, Australia and southern Africa. The GE acquisition of Utah – which was done partly as a means to access uranium and other rare-earth minerals to supply its nuclear division and partly as a hedge against inflation by obtaining Australian coal assets and thus Australian dollars – was the largest corporate merger in American history at the time.[29] But not long afterwards, rotating chairmanships led to changing priorities, and GE sought to sell assets and cut labour. The first task was to sell the Utah division. At the same time, the Australian mining giant, Broken Hill Proprietary (BHP), was seeking to expand beyond down under, not only to increase profits and foreign reserves but also to try and buy back domestic coal assets owned by foreign mining ventures. Between August 1982 and January 1983, GE and BHP negotiated a USD 2.4 billion merger. BHP would take most of GE's mining assets in Australia, Chile and the USA, as well as its coal shipping vessels.[30] The deal would include Utah/GE's southern Africa assets, bundled under the subsidiary company Southern Sphere Mining and Development (Pty) Ltd, which owned a coal mine in KwaZulu-Natal, Eersteling Gold Mine near Polokwane and Tantalite Valley Mine in southern Namibia.[31] BHP completed the merger with GE's Utah by April 1984.[32]

27 Peter J. Cox, interview with Bernard C. Moore (telephone, 25 March 2023).

28 Sabine Anderson (née Severin), interview with Bernard C. Moore (telephone, 2 January 2023).

29 P. Thompson and R. Macklin, *The Big Fella: The Rise and Rise of BHP-Billiton* (Sydney, William Heinemann, 2010), p. 149.

30 BHP-Billiton, 'The Utah Acquisition' (28 January 1983), https://www.youtube.com/watch?v=sKzBsAE7_lw

31 Thompson and Macklin, *The Big Fella*, p. 160. See also A. Cooper and M. Tusenius, *International Investment in South Africa* (Washington DC, Investor Responsibility Research Center, 1987). R. Knight, *Unified List of United States Companies Doing Business in South Africa* (New York, The Africa Fund, 1990).

32 'BHP now a "Moderate Sized International Company": Takeover Deal for Utah Completed', *The Canberra Times* (4 April 1984).

Southern Sphere Mining and Development (Pty) Ltd proved to be a liability for BHP. Within two years of concluding the GE/Utah merger, BHP finally got on board with anti-apartheid divestment schemes, such that by 1986 they were looking to sell their South African assets very quickly.[33] Arthur Anderson, one of the financial managers at Rand Leases, informed the Severins of the forthcoming sale of Southern Sphere from BHP and urged them to buy, especially given Eersteling Gold Mine's inclusion in the package.[34] At this recommendation, the Severins bought Southern Sphere on 27 February 1987, and they took steps to spin off some of the assets into separate companies. Eersteling Gold Mining Corporation Ltd would be eventually listed on the Johannesburg Stock Exchange in 1987, and the official reopening of the mine in January 1988 featured Foreign Minister Pik Botha as keynote speaker. Like Rand Leases, Eersteling would combine a modest ore extraction programme with a vigorous heap-leaching operation.

Severin's heavy investment in Rand Leases, Eersteling and heap-leaching technology meant that it took him quite a while to even visit the smaller assets held by Southern Sphere, which included some coastal diamond prospecting licences in Namaqualand and Tantalite Valley Mine in southern Namibia.[35] Peter J. Cox was seconded from JCI to join Southern Sphere, and in November 1987 he and Severin took their first visit to Tantalite Valley. 'There was really very little going on at the time', Cox reflected. 'The mine had been closed for a while, and a guy named Boet Adriaanse was the caretaker there on Umeis and Kinderzitt. There were some folks renting grazing for sheep, but that was it'.[36] Severin was not that impressed with the mine, and he did not feel that it had much potential except during times when tantalite prices were quite high.

According to Cox, Steen Severin was far more impressed, even *enamoured*, with the landscape of Tantalite Valley, and for that reason he looked to southern Namibia as his personal retreat. Steen and Franka Severin, Rand Leases geologist Louis Carroll and their top mining engineer, Ian Preston, were all appointed directors of Peter Weidner's old company, Tantalite Valley Minerals (Pty) Ltd, in place of the BHP directors.[37] At the time, the assets of Tantalite Valley Minerals included the mining licence, the equipment and the actual land: Kinderzitt and Umeis farms. By 1989, Severin changed this relationship

33 Peter J. Cox, interview with Bernard C. Moore (telephone, 15 January 2023).

34 Fincker, *Franka*, pp. 147, 153.

35 Ibid, pp. 147–148.

36 Peter J. Cox, interview with Bernard C. Moore (telephone, 15 January 2023).

37 BIPA File 1952/0590: Contents of Directors, Auditors, and Officers (CM–29) – 20 January 1988.

by founding Namibian company Kindermeis Holdings (Pty) Ltd, appointing his Geneva-based accountant as the sole director of the company, which would represent Severin's interest in the land.[38] Although Severin was not certain that he would ever mine the pegmatites at Tantalite Valley, he knew from his very first visit that he wanted to live there.

•••

At the turn of the 1990s, when the gold price was volatile, life in South Africa had changed for the Severins. Their daughter, Sabine, had finished school and gone to the University of the Witwatersrand to study mining engineering – one of the first women to be able to enrol in that degree. Rand Leases was far less profitable than it used to be, and Franka was feeling increasingly unsafe in crime-ridden Johannesburg.[39] She wanted to focus more on her art and painting; the bohemian Franka wanted a somewhat simpler day-to-day life. Severin would sell Rand Leases to Brett Kebble of JCI for just ZAR 0.01 per share, bringing in just ZAR 40,000 in revenue. In reality, the ZAR 4 million valuation of the mine was a gross overestimation, because the company was riddled with debt and owned old, unproductive shafts, tailings and heaps that had been treated with as much cyanide as possible, rendering them an ecological ticking timebomb.[40] Rand Leases was a giant liability, and Steen was happy to scrap it.

For the Severins, Namibia appeared an acceptable place for Franka to practise her art and for Steen to retreat into the stunning landscape. Sabine Severin remembered that, even from the late 1980s, the family would take holidays in Namibia, where they would hike in the mountains, ride Yamaha motorbikes and escape from the hectic life of Johannesburg.[41] These visits would increase during the early 1990s, when Steen and Franka made the decision to establish a permanent residence at Tantalite Valley. In May 1991, together with geologist Louis Carroll and the controversial Swiss finance manager Pierre Hafner, the Severins founded a Namibian company Tantalite Mining & Exploration (Pty)

---

38 BIPA File 1989/088: Inhoud van Register van Direkteure, Ouditeure en Beamptes (CM–29) – 8 March 1989. His two Geneva-based associates were Raymond Felix and Pierre Hafner. Severin would technically become a director of Kindermeis only in 1999. See BIPA File 1989/088: Contents of Register of Directors, Auditors and Officers (CM-29) – 8 April 1999.

39 J-P. Ceppi, 'Portait: Ils ont largué les amarres Franka Severin: Croqueuse de pépites du Tyrol à l'Afrique du Sud, elle a été comptable, ou starlette, avant de chercher de l'or au pays de l'apartheid', *Libération* (17 August 1996).

40 B. Sergeant, *The Kebble Collusion* (Auckland Park, Jacana Media, 2012), p. 23.

41 Sabine Anderson (née Severin), interview with Bernard C. Moore (telephone, 2 January 2023).

Ltd, which per the articles of association would 'carry on the business of exploration and mining of tantalite and other minerals'.[42]

According to Severin's geologist and long-time business partner, Peter Cox, although Severin intended to prospect for tantalite, the price was very low at the time, and outside of Tantalite Valley itself, the other mineral reserves in the area simply were not economically viable; the company's purpose shifted simply to land speculation. Cox reflected, 'the Severins would regularly visit Europe, and both moved in elite circles. Steen apparently got into a competition with Prince Rainier of Monaco to determine who owned the largest farm'.[43] The Monégasque royal had apparently been accumulating land in Tunisia, and Severin tried to counter with land deals in Namibia. Umeis and Kinderzitt – owned by Tantalite Valley Minerals (Pty) Ltd – together covered more than 16,000 hectares, but that would not be enough to surpass the assets of European royalty.

In the immediate aftermath of independence in Namibia, many White farmers were selling land cheaply in order to move some of their assets to South Africa, and this was certainly the case in southern Karasburg District. Severin went on a shopping spree. He bought several farms from Pauline L. du Toit (née Baard), whose father was one of the largest karakul farmers in the area. In April 1991, she sold Severin both Sandfontein and Sandfontein West for ZAR 288,515, and they were officially transferred to his new company in September of that year.[44] Du Toit's other parcel at the Orange River, Girtis, was sold for only ZAR 5,000; the property was registered in Franka's personal name.[45] With the independence of Namibia, the Catholic church had to establish a new diocese in Keetmanshoop – it was no longer appropriate to have the southern Namibian church still reporting to Keimoes in South Africa. This required cash to build the Pastoral Centre and hire staff, and Severin therefore offered to buy Homsrivier in October 1991 for ZAR 610,000, transferring it to his company early the following year.[46] In less than a single year, Severin had increased his Namibian landholdings to more than 73,000 hectares, sufficient for his wife to allege that they had defeated Prince Rainier, and she was told that she had acquired the

42 BIPA File 1991/0173: Memorandum of Association (CM–2) – 13 May 1991.

43 Peter J. Cox, interview with Bernard C. Moore (telephone, 15 January 2023).

44 WDO File T3413/1991: Sandfontein no. 131 and Sandfontein West no. 148, Deed of Transfer – 16 September 1991.

45 WDO File T284/1992: Girtis no. 109, Deed of Transfer – 27 January 1992.

46 WDO File T402/1992: Remainder of Farm Homsrivier no. 133, Deed of Transfer – 31 January 1992.

world's largest private property in the hands of one individual.[47] This was also the largest consolidation of southern Namibian farmlands since the properties were originally surveyed and demarcated in the German and early South African era. This was something completely new.

Although the size of the consolidation of land ownership was indeed new, what was most significant to most of the residents of southern Namibia – especially the Bondelswarts and the river-folk – was that these new owners were not interested in farming, or indeed in using the land at all. The Severins were gradually establishing a residence at Tantalite Valley, likely staying in Weidner's old compound, while still travelling between Europe and Johannesburg. The bulk of their 73,000 hectares simply existed on paper; Severin rarely ever saw them. While the caretaker Boet Adriaanse was supposed to be maintaining the fences and patrolling the boundaries, this was a farm larger than even Prince Rainier of Monaco could manage, so people began to cross Severin's properties with their livestock.

During the mid-1990s, Severin was touring some of his western properties, possibly Homsrivier or Sandfontein West, and he discovered that local Black pastoralists – probably river-folk – were grazing herds of goats on his properties and had taken up temporary residence there. He warned these 'squatters' to leave the farms immediately, otherwise he would kill the goats. He returned a few days later to find that the herders had not trekked the animals away, and, as the story goes, Steen Severin shot dead an entire herd of goats.[48] This quickly sent the message that absentee (or at least semi-absentee) landlordism did not mean that these private properties would become usufruct Bondelswarts communal lands. Severin had other plans.

One of Severin's mining engineers working in South Africa with the Southern Sphere assets, a Frenchman named Michel J. DuRocher, would regularly come to Tantalite Valley and the other properties during the 1990s on holiday. According to Steen's daughter, Sabine, 'Michel fell in love with the place', and although DuRocher was technically a miner, he urged Steen and Franka to build up some sort of tourism operation on the properties.[49] DuRocher began surveying the grounds for the kinds of game that were available to hunt, and allegedly he was the one who kraaled ostriches in the former Catholic church at Homsrivier. Signage described Severin's properties as Great River Game

47 ['*On m'a dit que j'avais acquis la plus grande propriété privée au monde aux mains d'un particulier*']. Fincker, *Franka*, p. 147.

48 Peter J. Cox, interview with Bernard C. Moore (telephone, 15 January 2023).

49 Sabine Anderson (née Severin), interview with Bernard C. Moore (telephone, 2 January 2023).

Ranch, and a few roads were cut connecting a small newly constructed hunting lodge on Sandfontein with Severin's other farms.

A year after DuRocher's tragic death in a car accident on Sandfontein Farm in May 1999,[50] Severin founded a new company, Great River Game Ranch (Pty) Ltd. It was to carry out Du Rocher's vision, being officially a tourism and hospitality company that offered traditional trophy-hunting trips and photographic safaris, targeting primarily the Scandinavian market. Severin partnered with a fellow Dane and hunting enthusiast, Günther Hadsbjerg, who was for many years the Johannesburg representative for the Swedish electrical company Allmänna Svenska Elektriska Aktiebolaget. He had presumably come into contact with the Severins as chairperson of the Danish Society of Johannesburg during the 1980s.[51] Hadsbjerg would also join the board of Tantalite Mining & Exploration (Pty) Ltd.[52]

By the late 1990s, Steen Severin wanted to make Tantalite Valley his home for good, and therefore needed to build a house. It would, in the words of Peter Cox, have to 'make a statement'. Franka lived a very social life, and even though she was moving away from the business and mining life of Johannesburg, she did not intend to give up her lifestyle, and neither did Steen. They saw Tantalite Valley becoming a cosmopolitan place where people could visit them, socialise, hunt, anything. In 1999, they hired the South African architect, Conrad Margoles, to build Casa Severin, a villa nestled in the mountains on Umeis Farm about five kilometres south-east of the Homestead pegmatite outcroppings at Tantalite Valley. The architects and builders were tasked to complete construction before the end of the year. Timber and joinery was brought in to Karasburg by train, then transported to Umeis by truck. Severin had his private quarters and office on the south side of the building separated from many guest rooms by a vast foyer and lounge. A twenty-metre lap pool runs along the north side of the house, and date palms grow within the compound.[53] This was

50 Republic of Namibia, *Government Gazette* (no. 2416 of 29 September 2000) and (no. 2424 of 13 October 2000). Also see *Acte de décès en Namibie, 1999*. It appears that donations were made to the village school in Warmbad in his name, hence its name, Michel DuRocher Primary School.

51 'Günther Prien Hadsbjerg: Utilities Executive', *Marquis: Who's Who Database* (Accessed 13 April 2023).

52 BIPA File 1991/0173: Contents of Register of Directors, Auditors and Officers (CM-29) – 27 August 2004.

53 Conrad Margoles, 'Casa Severin: Tantalite Valley, Namibia'. https://conradmargoles.com/portfolio/casa-severin/ (Accessed 1 May 2024).

FIGURE 57 Casa Severin on Umeis Farm
PHOTO COURTESY OF CONRAD MARGOLES ARCHITECTURAL STUDIO

FIGURE 58 Steen Severin sips coffee in the sitting room of his new villa
PHOTO COURTESY OF CONRAD MARGOLES ARCHITECTURAL STUDIO

Severin's mountain retreat. According to Peter Cox, 'one could sit with Steen on the veranda for long hours and talk about life, God, anything'.[54]

'Franka was a very social person, and she always sought to build relationships which would add value', said Peter Cox. 'That house was intended as a

54 Peter J. Cox, interview with Bernard C. Moore (telephone, 25 March 2023).

place to bring guests, which were often prominent people: European ballerinas, the editor of *Le Monde*, all sorts of people'.[55] Franka even invited the late Kapteine of the Bondelswarts, Anna Katrina Christian, to come visit Tantalite Valley; for Franka, all sorts of relationships could 'add value'. Indeed, bits of Tantalite Valley even made their way to Europe: Franka once took with her a *Pachypodium* succulent on a flight from Namibia to Monte Carlo, unaware that this particular variety was an endangered species, until customs officials seized the plant.[56] Franka had intended it as a gift to Prince Rainier of Monaco; eventually she was able to negotiate for the plant to be kept in the Exotic Garden of Monte Carlo.[57]

From 1992, Sabine Severin brought her Wits University friends to Tantalite Valley for regular holidays – including Finn Behnken, Heye Daun and her then-boyfriend Sean Gilbertson. They would stay at Tantalite Valley and explore the farms that Severin had recently bought. In Gilbertson's words: 'I was blown away – I loved it'.[58] Though he and Sabine stopped dating in 1994, Gilbertson kept in touch. He got along well with Steen Severin, and he would still occasionally visit. In Peter Cox's assessment, the relationship between Gilbertson and Severin gives credence to the lasting relationships that Severin formed, for whom 'a handshake was as good as a contract'.[59] Casa Severin was completed in time to ring in the year 2000, an event that had close to eighty people staying at the villa and on the property. Sabine, Steen, Sean, the university mates – everyone was there. For them, the twenty-first century started in southern Namibia.

All was not necessarily well at Tantalite Valley, however. In Gilbertson's words, 'Severin's health had been deteriorating, and he was increasingly estranged from his wife'.[60] Franka describes the situation more bluntly in her autobiography. Steen Severin had apparently been drinking heavily for quite a long time, and Franka had warned him in 1998 that he had to make a choice – drink or family. She gave him two years to get his act together. He was unable to break his habits, and in 2000 Franka divorced him and returned to Monte Carlo.[61] In her words, 'our masquerade of married life was no longer bearable to me'.[62] These problems clearly had been going on for some time, since Franka had resigned her directorship with Tantalite Valley Minerals (Pty)

---

55 Peter J. Cox, interview with Bernard C. Moore (telephone, 25 March 2023).

56 Sabine Anderson (née Severin), interview with Bernard C. Moore (telephone, 2 January 2023).

57 Fincker, *Franka*, p. 170.

58 Sean T. Gilbertson to Bernard C. Moore (7 January 2023), letter in the authors' possession.

59 Peter J. Cox, interview with Bernard C. Moore (telephone, 25 March 2023).

60 Sean T. Gilbertson to Bernard C. Moore (7 January 2023), letter in the authors' possession.

61 Fincker, *Franka* p. 110.

62 ['*notre mascarade de vie de couple ne m'est plus supportable*']. Ibid, p. 169.

Ltd in October 1994.[63] She resigned from Tantalite Mining & Exploration (Pty) Ltd in November 2000.[64]

•••

The year 2000 was clearly a difficult year for the Severins. At age 27, Sabine moved to Tantalite Valley to join her father, presumably to look after him in his ill-health, but also because she had been assigned by the family to 'dispose of the assets' in Namibia: namely the Tantalite Valley mining licence.[65] In January 2000, the price of tantalite oxide was about USD 20.50 per kilogram: a very low rate. However, the price rose throughout the year, eventually hitting USD 136 per kilogram by the end of 2000. In this context, Sabine thought it was the time to get a buyer for the mine, and she was in discussions with Marubeni Corporation of Japan, which was apparently sourcing raw materials for Japanese capacitor manufacturers, such as Hitachi and Samsung.[66] Marubeni wanted an offtake agreement, though they were willing to front some of the costs to refurbish the mining equipment at Tantalite Valley. After all, the mine had not been fully operational in almost two decades.

Sabine and Steen partnered with Peter Cox, Severin's long-time friend and former geologist at Southern Sphere, to get the machinery up and running and train workers. Cox remembered the situation somewhat differently – that the mine was reopened only after threats by the Namibian state to withdraw the mining licence after such a long time of inactivity.[67] Nevertheless, Sabine was appointed managing director of Tantalite Valley Minerals,[68] and they formed a new subsidiary, Tantalite Valley Mining (Pty) Ltd, to handle the expected 1.5 tonnes of Ta_2O_5 (tantalite oxide) concentrates per month over eight years of mine life. Tantalite Valley mine was officially recommissioned by then Namibian President Sam Nujoma in July 2001, after nearly two decades of closure. After reopening, Tantalite Valley mine employed about 120 people, of whom

63 BIPA File 1952/0590: Contents of Directors, Auditors and Officers (CM-29) – 22 November 1995.

64 BIPA File 1991/0173: Contents of Register of Directors, Auditors and Officers (CM-29) – 17 April 2001.

65 Sabine Anderson (née Severin), interview with Bernard C. Moore (telephone, 2 January 2023).

66 E. Brandt, 'Tantalite Mine Opens on Back of Favourable Market', *The Namibian* (1 August 2001).

67 Peter J. Cox, interview with Bernard C. Moore (telephone, 15 January 2023).

68 BIPA File 1952/0590: Contents of Directors, Auditors and Officers (CM29) – 28 August 2000.

eighty per cent came from impoverished Warmbad. Within just a few weeks, however, the mining operations were stopped, and more than a hundred employees were fired by the end of August 2001. The mine was officially closed again by December, citing the reason as the varying price of tantalite oxide, which fell to USD 22–32 per kilogram.[69]

Sabine explained that the price fluctuations in Ta_2O_5 were scaring away the Japanese buyers, who were no longer speaking to the Severins. Also, when the Severins had declared that they would reopen the mine, she started getting letters and phone calls from other mines elsewhere in Africa, asking if Tantalite Valley would market their production as Namibian tantalite. 'There wasn't really a global oversight process for tantalite at the time, leading to a lot of "conflict tantalite". A lot of these mines were in Mozambique, and I suspected that this could be the case'.[70] Reopening the mine, while helping the company keep its mining licence, was clearly not successful or well-orchestrated to any reasonable degree. Sabine remarked that 'it was a very sad sort of period' when the decision was made to close the mine. The workers were given very little notice, only one month, and she did not return to Tantalite Valley for a long time after that because she felt unsafe, given the workers' grievances.

Sean Gilbertson did not really believe in Tantalite Valley mine, but he did believe in Steen Severin and his daughter Sabine. He and Heye Daun learned in early 2000 of the offer for sale of Norechab Farm, located just north of Tantalite Valley, and their love of the southern Namibian landscape was reason enough for them to buy it in December 2000.[71] According to Peter Cox, it was also conditioned by the fact that Sabine had joined her father at Tantalite Valley, and Sean wanted to be near them.[72] Each of the excursions of his Norechab Sand & Tent Club would start and end at Tantalite Valley airstrip and Casa Severin. Steen's luxurious villa and swimming pool was a welcome relief from days in the bush, and additional time spent with Steen and Sabine was always enjoyable.

But there would not be any more mining in Tantalite Valley for the Severins. In August 2001, just a few days after declaring that the mine would be closed, Tantalite Mining & Exploration (Pty) Ltd officially changed its company name to Sandfontein and Houmsrivier Properties (Pty) Ltd, dropping all mention of mining from its articles of association and declaring the main purpose of

69 E. Brandt, 'Tantalite Mine in South Closes "Temporarily",' *The Namibian* (9 January 2002).

70 Sabine Anderson (née Severin), interview with Bernard C. Moore (telephone, 2 January 2023).

71 BIPA File CC/1996/0628: Amended Founding Statement (CC-2) – 1 December 2000.

72 Peter J. Cox, interview with Bernard C. Moore (telephone, 25 March 2023).

FIGURE 59 The Norechab Sand & Tent Club would routinely dine on the terrace at Casa Severin, taking a refreshing dip in Severin's lap pool
PHOTO COURTESY OF CONRAD MARGOLES ARCHITECTURAL STUDIO

the company to be 'property holding and any related business'.[73] Tantalite Valley Minerals (Pty) Ltd did the same in November, becoming Tantalite Valley Estates (Pty) Ltd, also a property-holding company.[74] In July 2002, Tantalite Valley Mining (Pty) Ltd was placed into provisional liquidation; Tantalite Valley Estates purchased the lands and the mining licences over the next few years.[75]

Around 2003, Franka Severin thought that it was time to sell off the properties, given Steen's declining health, and Sean Gilbertson agreed to help facilitate the sale of Great River Game Ranch.[76] Together with Norechab Club member Murli Bhamidipati, Gilbertson took out an advert in *The Economist*, with the heading 'Exceptional Game Ranch, Southern Namibia'. It described the property as '57,700 hectares, licenced and fenced with 18

73 BIPA File 1991/0173: Certificate of Change of Name of Company (CM-9) – 11 September 2001.

74 BIPA File 1952/0590: Application for Change of Name of Company (CM-9) – 14 November 2001.

75 Government Gazette of the Republic of Namibia, no. 2780 (2 August 2002). Government Gazette of the Republic of Namibia, no. 3820 (5 April 2007).

76 Sean T. Gilbertson to Bernard C. Moore (7 January 2023), letter in the authors' possession.

kilometres of river frontage. New and operational lodge, over 2,100 animals. Unrivalled potential'.[77] The minimum asking price was USD 2.5 million. Sean and Murli made a promotional website for the sale, describing the lodge as perfect for 'hunting, 4 × 4, photographers, etc', declaring vaguely that the 'interests of Namibian nationals pertaining to employment and maintenance of indigenous sanctity have been fully complied with'.[78] They mentioned the quantity of game, as well as the guest lodge and chalets built by DuRocher in 1999 on Sandfontein, and the website declared that the owners would be interested in discussing including neighbouring properties in the sale, to increase the size to nearly 100,000 hectares.[79] This presumably meant including Umeis, Kinderzitt and Norechab. 'Sadly', reflected Sean, 'despite quite an effort in 2003 and 2004, we could not attract any serious interest'.[80]

Great River Game Ranch would continue to operate for a few years as a hunting lodge, gradually marketing itself as 'Sandfontein Game Ranch'. Günther Hadsbjerg remained closely involved with Steen Severin, and it is clear from the web advertisements that the ranch continued to target the Scandinavian trophy-hunting market.[81] They described the land as follows:

> This vast area is unsullied by fences or any sign of human habitation, except for some lonely Bushman grave cairns, and 'The Ghost Town' on the banks of the Orange River – once a thriving Catholic Mission Station, which was abandoned when the Orange River became the border between South Africa and Namibia, and the congregation was no longer free to cross the river to attend services … The photographer or hunter must use all his wiles to get close enough for a good shot. Cover is sparse and the game is wary.[82]

77 'Businesses for Sale', *The Economist* (17 January 2004).

78 IAWB: 'Details about Great River Game Ranch in Namibia' (21 January 2004). https://web.archive.org/web/20040121033318/http:/www.greatrivergameranch.com:80/prop_details.htm

79 IAWB: 'Location of Great River Game Ranch in Namibia' (21 January 2004). https://web.archive.org/web/20040121033336/http:/www.greatrivergameranch.com:80/location.htm

80 Sean T. Gilbertson to Bernard C. Moore (7 January 2023), letter in the authors' possession.

81 IAWB: 'Sandfontein Game Ranch: Contact Us' (7 December 2004). https://web.archive.org/web/20041207064111/http://www.sandfontein.co.za:80/contact_us.htm

82 IAWB: 'Sandfontein Game Ranch: About Us' (7 December 2004). https://web.archive.org/web/20041207063631/http://www.sandfontein.co.za:80/about_us.htm

The lodge would continue to operate as a small-scale hunting property through 2005/2006, occasionally catering to battlefield tourists from South Africa looking to explore First World War sites.[83]

Although the church in the 'Ghost Town' described on the Sandfontein website in 2004 was not holding Mass anymore – perhaps it was still holding ostriches – the Orange River area was never a land of ghosts, even after Severin consolidated all this farmland. Perhaps frightened off by Severin's massacre of a herd of goats in the mid-1990s, or perhaps because grazing was better elsewhere, the Bassons and other river-folk were moving a little farther upriver, near KumKum, Kambreek, Pelladrift and Pelgrimsrust.[84] They still recognised Severin's properties as their ancestral lands, but they chose the path of least resistance and moved a bit upriver for the time being.[85] However, Severin's purchase of Homsrivier was not the last time that the river-folk would pass by their old church building or seek grazing north of the river.

83 IAWB: 'Camelthorn Adventures: Sandfontein Lodge' (15 May 2006). https://web.archive.org/web/20060515233746/http://www.camelthorn.net/Sandfontein.htm

84 LAC BCF: Bondelswarts Traditional Authority Councillor to Director for Land Reform, Ministry of Lands and Resettlement – 31 July 2006.

85 LAC BCF: B.E. Simboya, Pella 'Verblyf teen die Oranjerivier' – 16 July 2004.

CHAPTER 9

From Project Egg to Sandfontein

By the turn of the twenty-first century, the Bassons and other families along the Orange River were increasingly in conflict, not with the original local White farmers descended from colonial settlers but rather an internationally connected and hyper-wealthy business elite, many of whom spent little time in Namibia at all. These new landowners were also seeking to use the land in very different ways from those of the karakul sheep farmers before them. Steen Severin and his family were the first of this new wave of incoming wealthy residents seeking rural leisure, but we have already seen that Sean Gilbertson and his Wits University mining finance friends were not close behind. After several chapters examining the lives and worldviews of the Namibian river-folk, this chapter returns to Sean Gilbertson and his father, Brian, to situate his purchase of Namibian farmland within the broader trajectory of his family businesses outside Namibia and South Africa. Doing so enables us to better understand the purposes and reasons for him seeking the luxury of Namibia's rural spaces.

In March 2003, Sean Gilbertson stepped down as CEO of globalCOAL.[1] Through his longstanding familial connections with globalCOAL's major shareholders E.On, Rio Tinto, Glencore and others, under Gilbertson, the company expanded its share of coal brokering, reaching 420,000 tonnes within just its first month of transactions. 'It's all driven by the deregulation of the utility industry', said Sean at the time. 'These utilities want the ability to match their fuel mix to whatever the current electricity price is doing.'[2] Sean kept his shareholding in the company, and he spent a bit of time planning the largest trip of the Norechab Sand & Tent Club to date, involving eleven people flown in three Cessna aeroplanes.[3] At the end of his holiday, though, it was time for Sean to join his father in developing the family business.

Within just a few years of leading the merger of the Dutch mining conglomerate, Billiton, and its South African counterpart, Gencor, Brian Gilbertson entered into discussions with the Australian company, Broken Hill Proprietary (BHP), about a possible merger – one that would be the largest in Australian

1 UKCHA #04007764: Terminating Appointment as Director or Secretary (288b) – 10 March 2003.

2 'Global Coal's Gilbertson on New Online Coal Market', *Bloomberg News* (5 June 2001).

3 'The Twin Rivers Trip [October 2003]', *Norechab* http://www.norechab.com/oct2003.htm (Accessed 1 May 2024).

FIGURE 60 Brian Gilbertson on holiday in Tasmania, August 2003, a few days after leaving BHP
PHOTO: GETTY

history. In March 2001, when the merger plans were announced by the BHP board in Melbourne, it had been only about fifteen years since BHP had divested itself of its South African assets, yet this was now the new millennium and expansion into newly-democratic South Africa was in vogue again. Brian Gilbertson took over as CEO of the newly merged conglomerate, though his 'autocratic management style' did not mesh well with the majority-Australian board, and Brian's plans to forge another merger with Rio Tinto were not accepted.[4] After six months with the company, he was instructed to resign, and cited 'irreconcilable differences' with the company.[5] He was able reap a GBP 4-million payout from BHP along with a GBP 600,000 pension, though.

Brian Gilbertson was a dealmaker, and he lived the high-stress life of corporate politics, mergers and acquisitions. He also apparently sought to maintain his independent leadership and use the perks of the executive position. While with Billiton, he would travel by helicopter to the Johannesburg offices, and when he arrived, staff would joke that 'the ego has landed'.[6] Although his name

4 P. Thompson and R. Macklin, *The Big Fella: The Rise and Rise of BHP-Billiton* (Sydney, William Heinemann, 2009), pp. 358–376.

5 Ibid.

6 H. Connon, 'Mining a Very Rich Seam', *The Guardian* (10 July 2004).

may have been taboo in Australia, he quickly landed a position as chairman of the board of Vedanta Resources Ltd, the London-registered Indian base-metals mining company. However, after just seven months, he was fired by the CEO for refusing to terminate his advisory position with Sibirsko-Uralskaya Aluminium Company (SUAL), one of the larger aluminium companies in Russia and a direct rival of Vedanta.[7] Rather than leave SUAL, Brian took his GBP 7 million in Vedanta shares with him as he negotiated a rumoured USD 50 million employment package with the Russians.[8] In August 2004, he was appointed Executive President of SUAL.

SUAL was a product of the Yeltsin-era privatisation schemes, enabling Viktor Vekselberg – the Ukrainian-born Russian oligarch with close ties to the Kremlin – to obtain large aluminium concessions in Siberia.[9] In a lot of ways, Brian Gilbertson and Viktor Vekselberg saw the 1990s and early 2000s in a similar light: each was trying to get their large, yet politically isolated mining empires internationalised and interconnected with the rest of global mining finance. Just as the end of apartheid allowed Gilbertson to bring South African mining capital and expertise to the world, so did the fall of state communism in the Soviet Union facilitate Vekselberg's plans. Between 2004 and 2007, Gilbertson and Vekselberg would merge SUAL's assets with RUSAL [Russkiy alyuminiy] and the aluminium assets of the Anglo/Swiss giant Glencore to form United Company-RUSAL, IPJSC – at the time, the largest aluminium producer in the world.

•••

When Brian Gilbertson arrived at SUAL, he quickly realised that Viktor Vekselberg had an expensive, yet patriotic habit: Vekselberg liked Russian art. In February 2004, two months before one of the largest Sotheby's auctions of antique Fabergé eggs – belonging to the Forbes family – Vekselberg showed up at Sotheby's New York offices and bought the entire collection of nine eggs on the spot for more than USD 100 million, well above what it was anticipated to bring at auction. 'He likes to keep a low profile', said Sotheby's chief executive Bill Ruprecht, 'but he had an intensity and focus and commitment to repatriating these things that made this a great transaction for everyone involved'.[10]

---

7 T. MacAlister, 'Gilbertson Carpeted at Vedanta', *The Guardian* (5 July 2004).

8 S. Shah, 'Gilbertson Quits Vedanta for $50 Million Russian Job', *The Independent* (7 July 2004).

9 C. Walsh, 'A Big Moment in the Life of Brian', *The Guardian* (2 September 2006).

10 C. Vogel, 'Russian Industrialist Buys Fabled Forbes Fabergé Eggs', *New York Times* (4 February 2004).

Vekselberg, and other Russian oligarchs like Alexander Ivanov, developed a hobby of diversifying their assets by acquiring Russian art overseas, registering it in the names of offshore companies in Panama, Cyprus and other tax havens, and 'repatriating' it to private museums in Moscow or St Petersburg.[11]

When Brian Gilbertson began his tenure with SUAL in 2004, the Renova Group of companies – controlled by Viktor Vekselberg – was one of the largest shareholders in the giant aluminium company and Russia's main oil producer. Renova was also rapidly expanding into other emerging markets, such as South African manganese mining.[12] Although Brian had been hired to grow SUAL, he clearly spent just as much time with his boss discussing the latter's art purchases and how to grow this hobby into a larger corporate pursuit. In 2005, Vekselberg held a private meeting with Brian and Sean Gilbertson to discuss acquiring the rights to the Fabergé luxury goods brand and gemstone egg designers, which Unilever had owned since 1989. This was conceived of as 'a potential investment project'. Vekselberg was enthusiastic about the imminent partnership with the Gilbertsons, and he termed the operation Project Egg.[13] Before discussing the Gilbertsons, Fabergé and how this connects to Namibia, it is important that we talk about companies broadly or else the complexity of these dealings will be unrooted.

Limited liability companies and global variants are nothing new on the international stage, nor necessarily is complexity within corporate structures. Proprietary companies (Pty Ltd) have existed in South Africa and Namibia for more than a century, with German-style GmbHs preceding them. However, since the collapse of the Bretton Woods system and the birth of the era of floating exchange rates in the early 1970s, there have been some significant shifts in how global corporations have structured themselves as legal entities.

---

11 S. Walker, 'Russian Billionaire Feathers his Fabergé Nest Egg', *The Independent* (10 December 2010).

12 M. Creamer, 'Russia's Renova Mulling Uplift of South African Manganese Business', *Mining Weekly* (2 December 2011). Renova holds 49 per cent of United Manganese of Kalahari (Pty) Ltd in the Northern Cape.

13 Unless mentioned specifically, information relating to 'Project Egg' comes from the following Cayman Islands court documents and legal advisory opinions: 'Renova Resources Private Equity Ltd v. Gilbertson and Four Others', *Cayman Islands Law Reports* (judgement date: 14 April 2009), pp. 268–298. 'Renova Resources Private Equity Ltd v. Gilbertson and Four Others; Gilbertson and Autumn Holdings Asset Inc v. Vekselberg, Kuznetsov, Renova Holding Ltd, and Renova Private Equity Ltd', *Cayman Islands Law Reports*, no. 1 of 2010 (judgement date: 28 April 2010). 'Renova Resources Private Equity Ltd v. Gilbertson and Four Others', *Cayman Islands Law Reports* no. 2 of 2012 (judgement date: 5 November 2012), pp. 416–486. 'Gilbertson and Autumn Holdings Asset Inc. v. Renova Resources Private Equity Ltd', *Cayman Islands Law Reports* no. 2 of 2017 (judgement date: 12 September 2017), pp. 136–333.

Large corporations increasingly took advantage of new computer technologies to form and operate shell companies in tax havens and accommodating jurisdictions across the globe – something that was once reserved for Switzerland alone.[14] More and more, corporate acquisitions and operations used limited liability entities for privacy and to cap investors' and directors' losses, but not their gains.[15] In addition, the use of transfer pricing – particularly for 'intangible' assets like brand names and intellectual property – by 'tax mitigation' specialists within the corporations allowed their grounded operations outside of accommodating jurisdictions to appear less profitable than they truly were.[16]

There is nothing inherently illegal about owning and operating shell companies in tax havens, nor is structuring corporate acquisition and operations in such a complicated manner that attributing beneficial ownership becomes obscure or impossible. That is entirely the point. Tax havens are more than just about allowing corporations and wealthy individuals to avoid tax burdens, which is why researcher Alain Deneault refers to them as 'accommodating jurisdictions': each territory operates as an inverse glove of accepted legal behaviour concerning a given action. Some, like Liberia or the Marshall Islands, specialise in foreign ship registry; Ireland specialises in intellectual property; the Cayman Islands is known as a regulatory haven for high-risk finance.[17]

They are not interchangeable. Each territory finds its own ways to legalise what would be illegal or more highly regulated where sales and profits are obtained, such as in Paris, Frankfurt or Washington. The use of alternative jurisdictions involves (1) avoiding tax burdens, (2) avoiding legal ramifications or regulation, or (3) ensuring privacy and restricting information. These three aims ride alongside one another, and the first two are ultimately dependent on the third. Fragmented transnational corporate structures can be designed to conceal beneficial ownership of a particular company; obscurity is obtained through complexity.[18] There need not be any illegal operations or malicious intent when these companies and these structures are built – in many or

14 See G. Zucman, *The Hidden Wealth of Nations: The Scourge of Tax Havens* (Chicago, University of Chicago Press, 2015), pp. 8–34. V. Ogle, 'Archipelago Capitalism: Tax Havens, Offshore Money, and the State, 1950s– 1970s', *American Historical Review*, 122, no. 5 (2017), pp. 1431–1458.

15 A. Knobel, 'Rethinking Limited Liability: Beneficial Ownership Transparency to Reform the Liability System', Tax Justice Network Working Paper (November 2021).

16 A. Deneault, *Legalizing Theft: A Short Guide to Tax Havens* (Halifax, Fernwood Publishing, 2018).

17 Ibid, pp. 20–23.

18 A. Knobel, 'Complex Ownership Structures: Addressing the Risks for Beneficial Ownership and Transparency', Tax Justice Network Working Paper (February 2022), pp. 6–9.

most cases, there are not – but maintaining strict privacy renders it difficult or impossible for authorities or citizens to find out the movement of funds and who ultimately owns the assets. Brian Gilbertson and Viktor Vekselberg watched this world emerge as their careers rose, and Sean Gilbertson was born into it.

After Sean left globalCOAL, he worked for his father managing the family offshore investments and trusts, as well as establishing a new mining private equity fund. Between 2005 and 2007 he founded the UK-based Pallinghurst Resources LLP and at least nine other holding companies and 'investment vehicles' in the Cayman Islands, known as the Pallinghurst Structure.[19] These were founded alongside the Cayman Islands company Project Egg Ltd, which would be a wholly owned subsidiary of Pallinghurst, and which would be the shelf company for the future Fabergé arrangements. The Gilbertsons envisioned Project Egg Ltd (as part of the Pallinghurst Structure) acquiring the Fabergé brand using investments into the structure from, inter alia, Vekselberg's Renova Holdings Ltd and the Gilbertsons' Jersey-based family trust.[20] Pallinghurst would negotiate with Unilever to buy the brand, beginning in May 2006 with a USD 20-million offer, though Unilever was unwilling to budge.

From the beginning of 2006, Brian Gilbertson sensed potential profits on the horizon. During the 1960s, De Beers signed agreements with the government of the Soviet Union to sell certain quantities of Russian diamonds within its international sale networks. The ALROSA Group of Companies – the Yeltsin-era inheritor of the privatisation of Russian diamond mining – maintained this sales agreement with De Beers until February 2006, when the European Commission's Competition Board (an anti-trust entity) made legally binding a three year phaseout of De Beers' purchases of Russian diamonds.[21] By 2009, ALROSA would have to have a separate buyer facilitating sales in the European Union, and if De Beers were to violate the decision, the European Commission

19 UKCHA #OC315855: Application for Incorporation of Limited Liability Partnership (LLP2) – 20 October 2005. See also BIPA File 2016/0546: Annual Return (CM-23) – 31 July 2017.

20 For the Gilbertsons, this was Autumn Holdings Asset Inc, a British Virgin Islands company which was a wholly owned subsidiary of Fairbairn Trust Ltd (the Jersey-based branch of Nedbank Private Wealth), acting as trustees of the Gilbertson family trust. At this time, it appears that Autumn/the Gilbertson trust was the controlling party in the Pallinghurst Structure. See, 'Autumn Holdings Asset Inc' (British Virgin Islands company), *Panama Papers* (https://offshoreleaks.icij.org/nodes/10116588) and 'Autumn Holdings Asset Inc.' *The Paradise Papers* (https://offshoreleaks.icij.org/nodes/80031926 [Accessed 1 May 2024]). See, concerning Fairbairn, 'Offshore with Sharks', *Noseweek* (June 2010), pp. 28–30.

21 European Commission, Press Release IP/06/204: 'De Beers' commitment to phase out rough diamond purchases from ALROSA made legally binding by Commission decision' (22 February 2006).

FIGURE 61 Renaissance, a Fabergé egg made for Alexander III of Russia in 1894. It is owned by Viktor Vekselberg
WIKIMEDIA COMMONS: FABERGÉ MUSEUM, SAINT PETERSBURG

would impose a ten per cent fine on its worldwide turnover, effectively removing De Beers from the European market. Brian Gilbertson retained an interest in expanding his mining investments in base metals (after all, he was Executive President of SUAL at the time) but moving into gemstone marketing and production was a quick way to tap into these new opportunities. Vekselberg, the Gilbertsons likely reasoned, had the connections in Russia; Fabergé had the prestige and the name recognition to bring ALROSA's diamonds to luxury boutiques abroad.[22]

The Gilbertsons knew that Fabergé held a special place in Vekselberg's identity as a Russian – he allegedly dubbed himself 'Mr Fabergé' – and they knew that he had already invested more than USD 100 million in gemstone eggs.[23] Sean Gilbertson, however, was becoming increasingly frustrated with Vekselberg's negotiating strategy as well as Vekselberg's constant micromanagement of

22 See, regarding intentions, D. McKay, 'Gilbertson Eyes Russian Diamonds', *News24* (19 August 2008).

23 C. Vogel, 'Russian Industrialist Buys Fabled Forbes Fabergé Eggs', *New York Times* (4 February 2004).

the Pallinghurst corporate structure. Brian wrote to his son on 8 August 2006 echoing these sentiments, saying 'Buy the Egg, and I'll pull the plug on 'em [i.e., Vekselberg's team]'.[24] It appeared that Unilever would eventually be willing to agree to about USD 40 million for the rights, and a provisional verbal agreement was made on 15 December 2006. Because Unilever demanded a higher price than initially anticipated – which caused several other Pallinghurst founding investors to pull out – there was now a problem of where the funds would come from. Vekselberg proposed making up the difference using some of his own personal funds, rather than relying just on his and the Gilbertsons' initial investments into the Pallinghurst structure. This meant that he was going to want the rights to be held by one of his own companies, which would simply lease them to Pallinghurst. Vekselberg envisioned that 75 per cent of the shares in Project Egg would be held by his 'Lamesa Group of Companies', domiciled primarily in Cyprus and Panama.[25]

This was unacceptable to Brian Gilbertson, who apparently viewed the Fabergé project as his personal endeavour. From the beginning of December 2006, according to the Cayman Islands court records, Brian and Sean Gilbertson looked to old friends from the coal industry – Hans Mende and Fritz Kundrun, who controlled the American coal giant AMCI Group and Milan Jelinek of the Czech coal sector. Together, they intended to purchase Fabergé without Vekselberg and his Renova and Lamesa companies, and on New Year's Day 2007, they negotiated a USD 38-million cash deal with Unilever. Brian Gilbertson was clear with his coal mining partners that the goal was to use the Fabergé brand to negotiate for Russian diamonds, and perhaps coloured gemstones. He wrote an email to his team, reminding them what was at stake:

> This opportunity could be worth serious money for us after only a few months: We need those few months—and the brand name—to negotiate

24 *Cayman Islands Law Reports* no. 2 of 2017.

25 See, 'Lamesa Group Inc' (Panama company no. 155672687), *OpenCorporates*. 'Lamesa Group Holding S.A.' (Panama) *D&B Business Directory*. See also, 'Russian Billionaires Dispose of Assets Ahead of US Sanctions', *UAWire* (2 January 2018). The particular company that would hold the assets was 'Lamesa Arts Ltd'. Sean Gilbertson wrote to us in 2024, describing the situation as follows: 'When the other founding investors in the Pallinghurst Structure withdrew their support in December 2006, Mr Vekselberg, as the sole remaining funder, sought to exploit that position by insisting that his funds would only be forthcoming if the Fabergé brand was owned by one of his personal vehicles (Lamesa Arts) rather than being owned by the Pallinghurst Structure, and with the Pallinghurst Structure having only a 10-year management contract instead of outright ownership, as had been the original deal'. Sean T. Gilbertson, personal communication with Bernard C. Moore & Luregn Lenggenhager (email, 28 October 2024).

> with Alrosa, and/or to develop the non-diamond-angle, and we would add substantially to the current brand valuation which Unilever has BADLY mis-managed for decades. We cannot lose. Viktor will be willing to buy us out at the $38m+ at any time. (I told him some months ago that he/we would have to pay $100M for the name: he winced but said he could live with it. Remember that he paid $120M for eggs. The brand gives him serious cred in Russia/the Kremlin).[26]

The Gilbertsons understood that the name Fabergé carried a significant amount of weight, such that there were rumours that ALROSA itself had considered buying the brand from Unilever a few years earlier. Gilbertson and Vekselberg had joked that they would market Russian diamonds using the James Bond tagline, 'From Russia with Love', a clear nod to the De Beers 'Diamonds are Forever' marketing campaign.[27] However, it appeared that Brian Gilbertson wanted to be in control of the situation. Registering the Fabergé brand with Lamesa and then licensing it to Pallinghurst was not sufficient. If he were able to secure ownership of the Fabergé brand outside the auspices of Vekselberg's companies, then Brian could simultaneously negotiate a diamond marketing deal with ALROSA on his own *and* negotiate better terms with Vekselberg as a means to hedge his bets.[28] Purchasing the rights from Unilever without Vekselberg's knowledge would enable him to negotiate 'from a position of strength' – it would also ensure that Vekselberg's team would not separately negotiate with either Unilever or ALROSA, as all parties were waiting on Brian.[29] Gilbertson moved USD 38 million from the Gilbertson trust in Jersey to pay back his coal partners, and Fabergé was sold to the Cayman Islands company Project Egg Ltd.

While Brian Gilbertson was running these negotiations with his business partners and with Unilever, his son Sean was keeping Vekselberg's team busy, speaking on the telephone with them on 2 January 2007 to discuss technical aspects of the Lamesa deal, never mentioning that his father was arranging a completely separate purchase. By the end of the day on 3 January 2007, the Russian brand was owned by Project Egg, which itself was 99.11 per cent owned by the Gilbertson trust. As a result of the dilution of shares after the loan from

26 *Cayman Islands Law Reports* no. 2 of 2017, p. 187.

27 J.L. Thompson and J. Day, 'Towards a Model of Entrepreneurial Behaviour: An Evaluation of the History of Fabergé from 1842 to 2017', *Journal of Management History*, 29, no. 1 (2023), pp. 134–155. Ironically, neither James Bond film featured Fabergé eggs; this would be reserved for *Octopussy* in 1983, starring Roger Moore.

28 *Cayman Islands Law Reports*, no. 2 of 2017, pp. 187–188.

29 *Cayman Islands Law Reports*, no. 2 of 2012, p. 447.

Mende, Kundrun and Jelinek, Vekselberg was left with only a single share in the company. In March 2007, Project Egg Ltd officially changed its name to Fabergé Ltd, and the Gilbertsons and their colleagues had nearly full control of the brand.

Unsurprisingly, Vekselberg was furious and took it as a personal offence. He sacked Brian from his SUAL position and immediately pressed charges in the Cayman Islands court. Over the course of ten years, lawsuits and counter-lawsuits were flung between Vekselberg and Brian Gilbertson. The court ruled that Brian Gilbertson had breached his fiduciary duties as director of Project Egg Ltd, initially ordering him to pay Vekselberg USD 2.3 million plus interest, but the judges in time dismissed the fine and all further lawsuits between the two parties, identifying the cases as arising primarily from personal vendettas.[30] Viktor Vekselberg would never negotiate with the Gilbertsons again.[31]

In the months immediately following their purchase of Fabergé, Sean Gilbertson reported that Pallinghurst was intending to raise between USD 1 billion and 3 billion to buy diamond and gem producers, gradually integrating them with jewellers and the Fabergé brand.[32] This would create a vertically integrated system, from mine to polisher to boutique. From the beginning it was clear that the Gilbertsons had drawn lessons from the way De Beers had handled the diamond market in previous years.[33] This went beyond just gemstones and diamonds, however. The Gilbertsons negotiated a USD 1.5-billion investment from South Africa's Investec Group, US-based AMCI Group and 'an unnamed Asian steel company' to target base metals, such as manganese in Australia and Ghana.[34] Pallinghursts base metals operation would go on to

30 B. Butler, 'Caymans Court Lambastes former BHP Chief Brian Gilbertson', *Sydney Morning Herald* (28 July 2014).

31 Sean Gilbertson wrote to us in 2024, declaring that the Project Egg Ltd team subsequently offered Vekselberg the opportunity to join the business arrangement at the original terms 'i.e. that Project Egg Ltd as owner of the Fabergé brand would be wholly owned by the Pallinghurst Structure with Mr Vekselberg as one of the investors in the Pallinghurst Structure', but Vekselberg declined. Sean T. Gilbertson, personal communication with Bernard C. Moore & Luregn Lenggenhager (email, 28 October 2024). Since the Project Egg events, Vekselberg and the many Renova Group companies are currently under US, UK and Australian sanctions for their connections with the Russian invasion of Crimea. These sanctions were strengthened in March 2022, resulting in travel bans for Vekselberg and many of his colleagues, as well as the seizure of some assets held in foreign accounts.

32 S. Bailey, 'Gilbertson's Pallinghurst Raising $1 Billion for Resource Fund', *Bloomberg News* (28 May 2007).

33 Thompson and Day, 'Toward a Model of Entrepreneurial Behaviour', p. 146.

34 V. Wessels and S. Bailey, 'Gilbertson, Investec Start $1.5 Billion Resource Fund', *Bloomberg News* (28 May 2007).

include platinum in South Africa, graphite and lithium in Canada and nickel in the USA. This was Brian's arena, however; Sean remained focused on gemstones, and with the gradual retirement of his father, Sean would become the new face of the Gilbertson mining empire, albeit with a more subdued personality.[35]

The Gilbertsons set their eyes on a company called Gemfields. In 2000, businessman Rajiv Gupta – who operated emerald-polishing facilities in India – founded Gemhouse Inc. of Canada. He and his business partners acquired a number of gemstone mines and exploration licences in east and southern Africa, particularly Zambia. In June 2004, the company was moved to the United Kingdom and reformulated as Gemfields Resources plc, being listed on the London AIM stock exchange. Meanwhile, in 2007 Pallinghurst acquired a seventy five per cent ownership in Kagem Mining Ltd, which held the largest emerald mine in Zambia, and which would become the largest coloured-gemstone mine in the world.[36] As these licences were ultimately controlled by Pallinghurst and/or the Gilbertson family, the ownership structure was complex and global.[37]

In December 2007, Sean Gilbertson negotiated a sale of the Kagem corporate structure to Gemfields, and Pallinghurst granted Gemfields a fifteen-year exclusive marketing licence with the Fabergé name for all non-diamond endeavours (Pallinghurst received royalties of 2–6 per cent on sales). In exchange, Pallinghurst (via its subsidiaries) would be granted 59 per cent of the share capital of Gemfields, and both Sean Gilbertson and his Wits University mate Finn Behnken would be placed on the board of directors.[38] The deal

35 'Pallinghurst Group Announces Retirement of Chairman Brian Gilbertson', *The Pallinghurst Group* (Press Release, 28 April 2021).

36 See S. Gilbertson, Keynote Speech at Zambia Mining Investment Conference, London (29 June 2015) (https://youtu.be/zRJkgMb5lqs).

37 The Kagem licence was obtained in 2001 as part of the Zambian government's scheme to privatise mining assets; the government held on to 25 per cent of Kagem Mining Ltd. The remaining 75 per cent was owned jointly by Hagura Mining Ltd (England) and Hagura Mining Ltd (BVI), which were both wholly subsidiary on a 50/50 basis to Greentop International Inc (BVI) and Krinera Group S.A. (Panama); the latter featured Sean Gilbertson and Finn Behnken as directors. As part of the purchase Greentop and Krinera became, in turn, subsidiaries of the Cayman Islands registered 'holding vehicles' Rox Ltd and Rox Conduit Ltd, which were ultimately controlled by Pallinghurst and the Gilbertsons. See 'Krinera Group, S.A.' (Panama company no. 46750) *OpenCorporates*. On Rox Ltd, Rox Conduit Ltd and Sean Gilbertson's relations to them, see BIPA File 2016/0546: Annual Return (CM-23) – 31 July 2017.

38 Gemfields Resources plc, 'Letter to Shareholders from the Chairman of Gemfields' in *AIM Readmission Document* (13 May 2008), pp. 16, 24.

FIGURE 62 Sean Gilbertson at the Gemfields Ruby book launch event (28 November 2017) in London
PHOTO: GETTY

was finalised in July 2008, and it was tantamount to a Pallinghurst takeover of Gemfields with the acquiescence of the latter's directors.[39]

For some time, the Gilbertsons believed that there was still the possibility that Pallinghurst itself could win contracts to sell ALROSA diamonds on the world market, which is why it granted Gemfields only a fifteen-year licence to the Fabergé name and restricted it to non-diamond designs.[40] This would never come to fruition, however, and plans were initiated during 2012–2013 for Gemfields to officially take over the Fabergé rights permanently, making Pallinghurst the largest shareholder in Gemfields by far.[41] Via their direct and indirect interests, Sean Gilbertson and his father were significant shareholders in Gemfields as well, though some of this shareholding was through their offshore family trusts. Sean was eventually appointed CEO of Gemfields in March 2018; his friend Finn Behnken was on the board throughout 2017.[42]

Gemfields had become exactly what the Gilbertsons had probably wanted from the start: a vertically integrated coloured-gemstone company, specialising in emeralds, rubies and sapphires. Gemfields had investments from prospecting to mining licences to mining itself, to polishing, luxury retail and offshore investment vehicles. By early 2020, Gemfields Group had at least 52 subsidiaries across mining, polishing, sales and finance, with domicile in mining countries such as Zambia, Mozambique, Madagascar, Ethiopia, South Africa, Brazil, Canada and Sri Lanka, or in tax havens and alternative jurisdictions like the Cayman Islands, British Virgin Islands, Mauritius, Delaware, Switzerland and Guernsey.[43] Sean Gilbertson himself was a director of more than half of them.[44]

39 Ibid, p. 11.

40 D. McKay, 'Gilbertson Eyes Russian Diamonds', *News24* (19 August 2008). See also, 'Fabergé Seeks Partnership with Alrosa', *Rough Polished* (28 May 2008).

41 Gemfields plc, 'Proposed Acquisition of Fabergé Ltd' (6 December 2012).

42 Gilbertson was still CEO of Gemfields in May 2024, though he stepped down as CEO of Fabergé in April 2023, remaining as of October 2024 the chairman of Fabergé, however.

43 List derived from numerous Gemfields reports, particularly: Gemfields Group, Ltd. 'Appendix to the Pre-Admission Announcement: Further Information on Gemfields Group Limited in Connections with its Proposed Admission to Trading on AIM' (17 January 2020).

44 See Ibid. Data on Gilbertson's directorships are also found in BIPA File 2016/0546: Annual Return (CM-23) – 31 July 2017. Gilbertson wanted to remind us that under his leadership, Gemfields sought to initiate accountability policies to 'assist the citizens of mineral-resource-hosting nations to better understanding the share of natural resource wealth being paid to the host nation's government'. Sean T. Gilbertson, personal communication with Bernard C. Moore & Luregn Lenggenhager (email, 28 October 2024).

By 2023, Sean personally owned approximately 1.5 per cent of Gemfields as a whole, and through his trust the family owned more.[45] As CEO, Sean received a 'fee' of USD 548,000 per annum, as well as a bonus of the same amount, depending on company performance.[46] Although Brian Gilbertson had clearly played a role in the rise of Pallinghurst, the acquisition of Fabergé and the formation of the new Gemfields, the growth of the latter was ultimately Sean's project. Alongside his rise in gemstone mining and finance, Sean nevertheless sought relief and relaxation from corporate life – a place where he could, in his words, 'listen to the silence of the wind'. For Sean, this place was southern Namibia.

•••

FIGURE 63 Sean Gilbertson (left) and friends receive 'bush pilot' training on Sandfontein in 2006. The training involved landing aircraft in the dry bed of the Hom River
PHOTO: MARKUS MÖLLMANN, USED WITH PERMISSION

45 Gemfields Group Ltd, *Annual Report, 2023*, p. 120. Gilbertson personally owned 17,548,327 shares at the end of 2023.

46 Gemfields Group, Ltd. 'Appendix to the Pre-Admission Announcement: Further Information on Gemfields Group Limited in Connections with its Proposed Admission to Trading on AIM' (17 January 2020), pp. 31, 38. He also earned dividends through his interest in globalCOAL and in Pallinghurst's non-Gemfields investments.

Sean Gilbertson's efforts to sell Steen Severin's properties in 2003 and 2004 had failed, and Severin and his Scandinavian partners continued to run the properties as a modest hunting lodge. While Steen Severin was in poor health and getting ready to move back to Denmark, and Sean's ex-girlfriend Sabine was no longer in Namibia, he still had his eye on Sandfontein. 'Franka [Severin] kept sharpening the price until, in 2005, I thought the price looked pretty attractive', Sean reflected. 'And I tried to put together a consortium of mostly foreign (individual) investors to buy it'.[47] Over the course of 2005–2007, Sean's 'consortium' (including his 'Namibian business partner', Willem G. Agenbach) purchased Severin's company, Sandfontein and Houmsrivier Properties (Pty) Ltd, which owned the farms Sandfontein, Sandfontein West and Homsrivier.[48] Sean would also purchase Girtis Farm directly from Franka Severin in 2009, paying her ZAR 750,000. He registered the farm under the name of his other company, Marwilben Farming cc, which already owned Norechab.[49] Finally, he would also purchase in 2013 a small portion of Bankwasser Farm, to the west of Sandfontein West, giving his properties access to the government-serviced D208 road to Karasburg.[50]

With the promulgation of the Agricultural (Commercial) Land Reform Act of 1995, there was a drastic increase in the transfer of Namibian farmland to limited liability companies rather than individuals. There are a number of reasons for this, from the most benign to nefarious. First, as subdivision of farmland was no longer possible, it was one of the few options for farm partnerships (even among Namibians) to be put into action. Second, as with other uses of limited liability, it reduced the possibility that land assets could be taken through divorce, lawsuit, etc., because the company would technically own the farmland. Third, it reduced government property transfer duties from future land sales, because shares in the company could be sold rather than the property itself. Finally, as with limited liability companies more broadly, it could be a means to build complex corporate structures for land ownership that could hide the beneficial owner. Between 1995 and 2000, farm sales to corporations increased by 400 per cent per annum on average.[51] In 1995 alone, more than 2.2 million hectares of Namibian farmland were simply *donated* from private

47 Sean T. Gilbertson to Bernard C. Moore (7 January 2023), letter in the authors' possession.

48 BIPA File 1991/173: Contents of Register of Directors, Auditors and Officers (CM-29) – 24 April 2007.

49 WDO File T3293/2009: Girtis no. 109, Deed of Transfer – 13 July 2009.

50 WDO: Registrasie Afdeling 'V': Bankwasser no. 139 gedeelte A.

51 B. Fuller and G. Eiseb, 'The Commercial Farm Market in Namibia: Evidence from the First Eleven Years', Institute for Public Policy Research, Working Paper no. 15 (November 2002), pp. 12–14.

owners to limited liability companies, quite possibly with the original owner as a director.[52]

Under section 61 of the 1995 legislation, a Namibian must sign an affidavit stating that he/she is the owner of 50.01 per cent or more of the property, and that he/she is not a 'nominee owner' standing in for a foreigner. The signatory can sign as a representative of another 50.01 per cent Namibian company, however, which makes fragmented ownership of Namibian farmland both possible and comparatively easy. It is difficult for the Ministry of Agriculture, Water and Land Reform to examine and investigate every affidavit, which probably means that many 'nominee owners' exist, facilitating foreign ownership of farmland in Namibia. The opaqueness of limited liability companies makes this situation possible. Also, public land registries become less effective unless combined with other data from elsewhere, and interministerial cooperation is difficult.

After the merger of Gencor and Billiton, Sean's father made use of the company's private Dulini Game Lodge within the Sabi Sands Private Game Reserve, which borders the Kruger National Park, to host informal business meetings, impress clients with African 'nature' and eventually win potential business.[53]

FIGURE 64 Aircraft lands on the runway which Sean Gilbertson constructed on Sandfontein, 2023
PHOTO COURTESY OF ROSITA SMEENK AND MAURITS MATERS, USED WITH PERMISSION

52 Ibid, p. 12.
53 Thompson and Macklin, *The Big Fella*, p. 304.

Although Sean framed his appreciation of Sabi Sands as a model of economic development and employment possibilities, something he thought Sandfontein could achieve to only a limited extent, it is hard to imagine that he was not impressed with his father's use of Dulini Game Lodge and what it could mean for mining business.[54] Sean had already toured Severin's properties during his visits with Sabine and more extensively with the Norechab Sand & Tent Club, going as far as the Homsrivier Catholic Mission in November 2005. He clearly had some ideas for how, in Sabine Severin's words, he could 'embellish' what her father had started.[55]

The Norechab Club was no longer meeting as regularly for their air-and-bush vacations, but Sean remained a keen amateur pilot of Cessna aircraft. In October 2006, while he was in the process of purchasing Severin's properties – and while he was presumably in the process of cutting Vekselberg out of the Fabergé deal – Sean hosted a 'bush pilot' training course on Sandfontein. This involved bringing in expert flight instructors from Johannesburg to train Sean and some friends on irregular take-offs and landings, such as in the dry bed of the Hom River that passed through his new properties.[56] Over the first few years of operation, Sean upgraded the main lodge building, built a massive swimming pool, constructed additional manager's and staff housing, and built a graded gravel airstrip to the southwest of the lodge itself.

Although Sean Gilbertson had purchased Severin's western properties, he was not at that point an owner of Umeis, Kinderzitt or Tantalite Valley mine itself. In 2006, the Australian company Magnum Mining and Exploration Ltd entered into discussions with Steen Severin to take over his companies, his mining licences and his remaining lands. This was motivated by, inter alia, a gradual rise in the price of tantalum pentoxide and an announced gradual phaseout of tantalum production on several of Australia's largest pegmatite belts.[57] However, since the time that Severin had purchased the mine, there were now new land reform laws, forbidding foreign ownership of farmland, so new arrangements had to be made. Mining Licence no. 77 – covering Tantalite Valley mine – was sold by Severin to Tameka Shelf Company Four (Pty) Ltd,

54 Sean T. Gilbertson to Bernard C. Moore (7 January 2023), letter in the authors' possession. Gilbertson would later write to us in 2024 the following: 'I have never hosted potential business partners or clients at Sandfontein (other than those directly associated with Sandfontein and its purposes)'.

55 Sabine Anderson (née Severin), interview with Bernard C. Moore (telephone, 2 January 2023).

56 'The Bushpilot Training Course – An African Adventure', *African Pilot* (January 2007).

57 N.A. Mancheri et al., 'Resilience in the Tantalum Supply Chain', *Resources, Conservation, and Recycling* 129 (2018), pp. 56–69.

FIGURE 65 Sandfontein Lodge in 2011
PHOTO: KAREN ASH, WIKIMEDIA COMMONS

a 100 per cent subsidiary of Magnum, such that the Australians now held the mining licence.[58] As for the land, this was more complicated. The Australians needed Namibian business partners, and for this they turned to Severin's former and business partner, Peter J. Cox.

Like Gilbertson, Peter Cox had a love for the southern Namibian landscape, and although he lived in Johannesburg, he wanted to continue to travel to Tantalite Valley for pleasure (staying the night in Casa Severin) or business (mining the pegmatites). When the first push to sell Severin's properties occurred in 2003, Cox registered a Namibian company, Warmbad Investment Holdings (Pty) Ltd, presumably to play a role in acquiring some of the property and assets held by Severin's company Tantalite Valley Estates (Pty) Ltd.[59] After Magnum expressed interest in buying the mining licence, two Namibians were appointed alongside Cox to Warmbad Investment Holdings – attorney Caroline S. McLeod and secretary Heidie Veldskoen – which must have been part of

58 Magnum Mining and Exploration Limited, Annual Financial Report for the Year Ended 31 December 2006. All the machinery and equipment related to the mine was registered under a different subsidiary of Magnum, Namibia Tantalite Investments (Pty) Ltd.

59 BIPA File 2003/0136: Memorandum of Association (CM-2) – 4 February 2003.

a process to make the company 51 per cent Namibian.[60] The plan would involve Warmbad Investment Holdings purchasing 51 per cent of the share capital in Tantalite Valley Estates from Severin's holding company, Kindermeis Holdings (Pty) Ltd, with the remaining 49 per cent being acquired by Magnum.[61] The deal was finalised in July 2007, transferring ownership of the land to Peter Cox's companies and Magnum Mining and Exploration Ltd.[62] However, the economic downturn of 2008 would quash all plans to reopen Tantalite Valley mine, and the project would remain mothballed for a number of years.[63]

While the price of tantalum pentoxide was rising during the mid-2000s, the price of uranium oxide was increasing even faster. In January 2001, uranium was selling around USD 2.95 per kilogram; by January 2005, its price had risen to USD 9.57 per kilogram.[64] This was during the run-up to the great uranium bubble, caused broadly by the rise of the Chinese nuclear power sector and technical and environmental problems at some of the largest Canadian and Australian mines.[65] The veteran Canadian geologist, Gennen McDowall, who had worked during the 1970s as a prospecting geologist for Langer Heinrich Uranium near Swakopmund, returned to Namibia at this time to investigate prospecting opportunities. 'I had known Sean Gilbertson for a long time', McDowall said. 'My wife used to babysit for Brian when I was working in Johannesburg in the 1970s'.[66] Possibly at the suggestion of Sean, who was in the process of purchasing Severin's properties, McDowall visited the geological survey offices to see what uranium deposits existed within the pegmatites of southern Namibia. He recollected that he had found a report by Southern Sphere, which was at the time considering purchasing Tantalite Valley from Weidner. Through a partnership with a Springbok-based American prospector,

60 BIPA File 2003/0136: Contents of Register of Directors, Auditors and Officers (CM-29) – 21 February 2007.

61 Magnum Gold NL, Notice of General Meeting and Explanatory Memorandum – 21 June 2006. Note: Magnum Gold was the former name of Magnum Mining and Exploration Ltd before it went into tantalum production. See also BIPA File 1989/088: Contents of Directors, Auditors and Officers (CM-29) – 6 October 2008.

62 BIPA File 1952/0590: Contents of Directors, Auditors and Officers (CM-29) – 2 September 2011.

63 Magnum Mining and Exploration Limited, Annual Financial Report for the Year Ended 31 December 2008.

64 Uranium prices based on *Trading Economics* databases (https://tradingeconomics.com/commodity/uranium).

65 See T. Johnson, 'Global Uranium Supply and Demand', *Council on Foreign Relations* (14 January 2010).

66 William Gennen McDowall, interview with Bernard C. Moore (telephone, 1 February 2023).

George E. Swanson, it had discovered leucogranite 'pegmatic alaskites' on Homsrivier Farm, which likely contained uranium.[67] Although the uranium mineralisation of the Pofadder Shear Zone might have been insufficient to warrant exploration and exploitation during the twentieth century, the situation was different during the uranium bubble.

With financial backing from Xemplar Energy Corporation of Vancouver and a Namibian partnership with the veteran attorney Peter F. Koep – who had long represented mining companies, including Steen Severin's – McDowall founded Namura Mineral Resources (Pty) Ltd, with the goal of applying for reconnaissance and exclusive prospecting licences (EPLs) on Homsrivier, Sandfontein and Sandfontein West.[68] Namura was a wholly owned subsidiary of Xemplar, of which McDowall was CEO, and Koep was appointed to the board of directors.[69] Namura was granted a reconnaissance licence for the farms, and during 2006 did some radiometric surveying, boldly describing the Homsrivier pegmatites as similar 'to the uraniferous alaskitic pegmatites found at Rössing'.[70] By the end of 2006, Namura was granted EPLs for the properties, and they planned drilling for the following year.[71]

As EPL-holders, Gennen McDowall and Namura had to negotiate a land-use agreement with Gilbertson and his company. This generally involves rent payments to access portions of the property to drill, costs related to water and electricity use, as well as fees paid for accommodation or the rights to build a mining camp. Namura rented from Gilbertson the former Homsrivier Catholic mission station as their base of operations.[72] Staff were housed in the accommodation there, and the core samples and uranium dust were stored in the church building itself.[73] It is quite possible that when McDowall, Koep or the other Canadian directors visited, they were housed in Sandfontein Lodge itself.

Sean Gilbertson admits that this land-use agreement brought in significant income during the first few years of his ownership of the properties. Even for

67 MME File 1234/EPL919: A Review of Southern Sphere Mining and Development Company (Pty) Ltd and the Company's Policy Towards Exploration in South West Africa, 1978.

68 BIPA File 2005/032: Memorandum of Association of a Company Having Share Capital (CM-2) – 12 January 2005.

69 Xemplar Energy Corp. 'Press Release: Peter Koep Joins Board of Directors' – 18 December 2006.

70 MME File 4876/ERL65: Namura Mineral Resources (Pty) Ltd. 'Final Assessment Report for Exclusive Reconnaissance Licence 65' – 16 May 2006.

71 Xemplar Energy Corp. 'Business Acquisition Report' – 10 October 2006.

72 MME File 4090/EPL3568: Namura Mineral Resources (Pty) Ltd. 'Biannual Environmental Report on EPLs 3567 and 3568, Warmbad' – January 2010.

73 William Gennen McDowall, interview with Bernard C. Moore (telephone, 1 February 2023).

FIGURE 66 Slinging mining equipment from Xemplar's helicopter on Homsrivier Farm, 2009
PHOTO COURTESY OF THE MINISTRY OF MINES AND ENERGY, WINDHOEK [FILE 4090/EPL3568]

Gilbertson, though, he could see that Namura and Xemplar were speculating in the bubble and that it was going to burst. 'Mining companies can (1) mine the ground or (2) mine their investors and/or run their businesses as "lifestyle businesses". In my career, I have seen both types', reflected Sean. 'Xemplar brought much income to Sandfontein before 2010 when they claimed to have found uranium on our properties, but it turned out to be a "pump and dump" operation'.[74] Indeed, the uranium bubble peaked in December 2006, when the price reached USD 63.50 per kilogram, and by December 2007, the price had fallen to USD 40.05 per kilogram. Although the price might not fall to the lows of five years earlier, it was going to fall further.

Over the first few days of January 2008, a rumour was leaked to the British tabloid *Sunday Express* that Rio Tinto was considering buying Xemplar in order to obtain its Namura subsidiary and its uranium EPLs on Gilbertson's properties.[75] This caused the Xemplar stock (listed on the Namibian and

74 Sean T. Gilbertson to Bernard C. Moore (7 January 2023), letter in the authors' possession.

75 C. Matthews, 'Rio Tinto Linked to Offer for Xemplar', *The Namibian* – 9 January 2008.

Canadian exchanges) to rise by twenty two per cent[76] and Rio Tinto's stock fell by about four per cent.[77] When asked about this, McDowall reasoned that it was likely 'a promoter looking to short a stock sale after he learned about our meeting with Rio Tinto'.[78] McDowall maintained that this was about seeking a kind of technical partnership rather than a buyout, but (he holds) regardless of the stock price increase, the drama soured whatever relationship he could have built with Rio Tinto. Nevertheless, citing medical reasons, McDowall quickly left the CEO position of both companies by the end of the year.[79]

In March 2010, after drilling intermittently for two years, Xemplar scaled down its activities on Sandfontein and Homsrivier, closing down all operations by June 2010.[80] Both Gennen McDowall and Peter Koep maintain that the fall in the price of uranium oxide on the world market was the reason for the closure of operations.[81] There is some validity to this – the price fell in February 2010 to only USD 18.75 per kilogram. Peter Cox, who was watching the situation closely from Tantalite Valley, views it rather differently. While not accusing the Xemplar team of fraudulent pump-and-dump schemes like Gilbertson, Cox read their drill reports, noticing that the uranium – although present – was excessively deep, mostly between 300–500 metres below the surface-level outcroppings, with some of the more mineralised portion lying more than 700 metres deep.[82] This makes small-scale mining of uranium very difficult and costly even in boom times.

Despite their failure to actually develop uranium mining, McDowall and Koep were able to maintain a strong relationship with Sean Gilbertson and Willem Agenbach – probably because of the stable income they were providing. As per Namibian mining law regarding EPL land-use agreements, the duty and cost of rehabilitating the mining and prospecting area would fall on Xemplar. Interestingly, the company did not mandate that Xemplar rehabilitate the vehicle tracks or most of the drill sites. In the report to the Ministry of Mines and Energy, the company declared:

---

76 L. Hill, 'Uranium Explorer's Shares Jump on Rio Bid Speculation', *Engineering News* (RSA) – 7 January 2008.

77 N. Fletcher, 'British Energy is Hot Nuclear Favourite', *The Guardian* – 8 January 2008.

78 William Gennen McDowall, interview with Bernard C. Moore (telephone, 1 February 2023).

79 L. Hill, 'Uranium Explorer Xemplar Appoints President, *Mining Weekly* (4 December 2008).

80 MME File 4090/EPL3568: Namura Mineral Resources (Pty) Ltd. 'Biannual Environmental Report on Cape Cross, Aus/Garub, and Warmbad Exploration Projects' – July 2010.

81 Peter F. Koep, interview with Bernard C. Moore (telephone, 11 January 2023).

82 Peter J. Cox, interview with Bernard C. Moore (telephone, 15 January 2023).

> due to the busy, mountainous landscape, the landowners feel that the remaining protruding casing does not pose a visual impact since it is hardly noticeable [and] due to the undulating topography and limited access on the farm, he did not feel the need to rehabilitate any of the newly created Namura tracks, since it allowed him better access to remote lookout places with his guests.[83]

Photos from the environmental report show that the Xemplar team was escorted throughout the premises in official Sandfontein Toyota Land Cruisers. Despite obtaining the property from Severin with the intention to run a luxury nature reserve in the wilderness, Sean Gilbertson was cultivating a decidedly human landscape on Sandfontein.

•••

'I used to work on Sandfontein, back when Agenbach was there', said Johannes Gertze as he rolled a cigarette. 'It was a hunting farm [*jagplaas*], and there were people coming from South Africa, Europe, many places. I'm not sure if it still is. The pay wasn't much better than I get here, but we could sometimes get tips, which was nice'.[84] Gertze was in his late fifties, and he worked as a shepherd

FIGURE 67 Imported giraffes on Sandfontein, April 2016, with high-voltage power lines in the background
PHOTO: KAREN ASH, WIKIMEDIA COMMONS

83 MME File 4090/EPL3568: Namura Mineral Resources (Pty) Ltd. 'Biannual Environmental Report on Cape Cross, Aus/Garub and Warmbad Exploration Projects' – January 2011.

84 Johannes Gertze, interview with Bernard C. Moore (Arus Farm, 16 October 2022).

for Henry Pretorius, particularly looking after the newborn lambs that Pretorius kept on his rented grazing on Arus Farm, just to the east of Sandfontein and Umeis. Gertze remembers helping guests to track game for the hunt and to prepare the carcasses after an animal was bagged. He implied that about ten or fifteen years earlier, there had been a lot of hunting in the area, much more than today.

After the takeover of Fabergé and Gemfields, Sean Gilbertson did not have quite as much time to travel to Namibia as in earlier years with the Norechab Sand & Tent Club. His income had likely grown significantly, but so had his responsibilities. It was useful, therefore, that he had a business partner in Willem Agenbach to look after the place while he was in London. However, Sean would learn that partnering with local individuals is far from a simple negotiated business arrangement. As mentioned, foreign ownership of farmland was banned in Namibia, and it was necessary for people like Sean to ensure that they had a local partner. He was used to this and had bypassed the regulation with his mate Heye Daun's 51 per cent shareholding in Marwilben Farming cc, which owned Norechab and Girtis. When Sean bought Severin's properties, his British Virgin Islands property-holding company Arianna Investments Ltd bought 49.98 per cent of the shares in Sandfontein and Houmsrivier Properties (Pty) Ltd, while 50.02 per cent of the shares were allocated to Namibian citizen, Willem Agenbach.[85]

In some ways, Willem Agenbach and his wife Rodica were successfully integrating Sandfontein into the high-class conservationist circles of Namibia. In June 2011, for example, Sandfontein partnered with N/a'ankusê – a large Namibian ecotourism company with a parallel wildlife foundation – to resettle a family of cheetahs onto Sandfontein's properties. A film crew from Animal Planet accompanied the convoy of specialised trucks, which drove the cats more than 700 kilometres to the fenced camp where Agenbach thought best to acclimatise them. N/a'ankusê described it as the 'Release of a Lifetime'.[86] The cheetahs were collared with GPS tracking devices, which unfortunately reported their demise a year later. The cats had strayed across the property lines to one of the irrigated portions of Ramansdrift next door, where they were shot by the farm manager and were allegedly 'fed to the farm workers'.[87]

85 RoHC File HC-MD-CIV-MOT-GEN-2016/00262: Founding Affidavit: Sean Thomas Gilbertson and Arianna Investments, Ltd – 17 August 2016. According to Gilbertson, Agenbach purchased the shares outright. Sean T. Gilbertson, personal communication with Bernard C. Moore & Luregn Lenggenhager (email, 28 October 2024).

86 'The Release of a Lifetime', N/a'ankusê Carnivore Conservation Research Update (11 July 2011).

87 'Bedreigde jagluiperds geskiet, vleis vir werkers gegee', *News24* (3 July 2012).

FIGURE 68 Pedro Ampuero with one of his bowhunting kills in 2014
PHOTO COURTESY OF P. AMPUERO, USED WITH PERMISSION

Sean Gilbertson identifies himself as a 'non-hunter', and while he remained friendly with Severin during the years of Great River Game Ranch's hunting operations of the early 2000s, it is unclear precisely if he condoned hunting at the time. His website in August 2015 described Sandfontein as a 'Luxury and Hunting Lodge' – perhaps indicating the presence of trophy-hunting on the property – though perhaps this was a holdover from the Severin days.[88] Regardless, Sean described his fallout with Agenbach as arising from when Willem 'went bananas, keeping his own livestock in the nature reserve (illegal) and started shooting all manner of animals, including leopard (illegal)'.[89] Agenbach was apparently partnering with a neighbour, Willem F. Mans, who was running a hunting business on neighbouring KumKum, allowing his hunters to come onto Sandfontein (see chapter 10).

It appears that Agenbach was also beginning to formally advertise Sandfontein as a hunting lodge, hiring a Berlin-based film and design agency to produce films, brochures and business cards describing himself as the Managing

88 IAWB: 'Luxury and Hunting Lodge, Nature Reserve in Namibia, Africa' (11 August 2015). https://web.archive.org/web/20150811161956/http://www.sandfontein.com

89 Sean T. Gilbertson to Bernard C. Moore (7 January 2023), letter in the authors' possession.

Director of Sandfontein's 'hunting and fishing safaris'.[90] Agenbach brought in the Spanish bowhunter, writer and internet influencer, Pedro de Ampuero Castellanos, for a week of hunting during June 2014. Castellanos's wife took film and photographs, making quite a professional promotional film representing Sandfontein as the best of Namibian hunting.[91] With the assistance of a truck-full of workers, like Johannes Gertze, Castellanos was able to bag at least one gemsbok female and one kudu bull. They filmed his trophy kills alongside images of luxury food in the lodge, refreshing swims in Gilbertson's massive pool and fishing along the Orange River. While Castellanos' trip was perhaps the grandest, it was also clear that Gilbertson's apparent goal of a high-end lodge for wealthy tourists and businessmen – as his father used Sabi Sands – clashed with Agenbach's vision of Sandfontein as a hunters' paradise, using the naturally large numbers of game to bring in trophy revenue alongside photographic tourism for the non-hunter.

Their difference probably lay in where the money would come from. Agenbach wanted to profit from Sandfontein itself, whereas Gilbertson sought something else, and possibly he did not care if the Sandfontein company made money at all. The sociologist Justin Farrell explored this phenomenon in his study of Teton Country, Wyoming, noting how the flood of multimillionaires and billionaires into an isolated corner of the United States was accompanied by a general interest in 'conservation', broadly defined. Farrell found that many of these corporate executives who set up second residences in Wyoming often used conservation initiatives – normally defined by simply buying up land 'for conservation' – as a 'special dispensation for purchases and practices that may otherwise be viewed as morally suspect, opulent or greedy'.[92] Especially for those who made their fortunes in polluting industries like mining or oil, Farrell found that many of his respondents understood their later 'conservation' work as wiping the coal soot from their environmental record, as well as offering a 'du jour leisure activity providing social integration into elite circles, with the implicit assumption that such conservation activities are also wholly altruistic, publicly spirited and ecologically progressive'.[93]

Sean Gilbertson likely would have fit in nicely in Teton County. In fact, on 17 August 2016 he was precisely in Jackson, Wyoming – perhaps on holiday,

90 Anne Prinz and Mickael Brock's *Astronaut Studio*, http://www.anneprinz.de/astronautstudio/work/sandfontein/ (Accessed 1 May 2024).

91 P.A. de Castellanos, 'Bowhunting in Namibia' (20 June 2014). https://pedroampuero.com/2014/06/20/bowhunting_namibia/ (Accessed 1 May 2024).

92 Farrell, *Billionaire Wilderness*, pp. 100–101.

93 Ibid, p. 79.

perhaps for a business meeting – when he visited the local commissioner of oaths to sign the founding affidavit of his legal proceedings to liquidate the company he owned with Willem Agenbach (against the latter's will). According to Sean's affidavit, Agenbach apparently informed Gilbertson in September 2015 that he was no longer interested in working as the general manager for Sandfontein, and that he was 'disinterested in the running of the business and noted repeatedly that he did not want to interact with visiting clients or guests'.[94] Sean negotiated that Agenbach would remain manager until the end of April 2016, at which point he would vacate the manager's house and make way for his replacement.

Based on the available documents from the Registrar of the High Court, it appears that Sean was in a difficult situation, because at the end of the day, Willem Agenbach was the 50.02 per cent owner of Sandfontein, not Gilbertson. As one of only two directors of the company, Agenbach's voice on how Sandfontein could be used – whether for hunting, livestock, tourism, whatever – weighed more than Sean's 49.98 per cent. It appears that the most effective strategy to get Agenbach out was to hold him to his fiduciary duties. After all, although Agenbach was not a poor man, he could not front all the expenses to run a lodge that, Gilbertson admits, was bleeding cash: its monthly expenses were twice that of its reported revenue generated from all activities.[95] As an isolated lodge which sleeps only ten people, and which, following Gilbertson's wishes, would not allow hunting, revenue generation would be quite difficult, even when room rates at that time approached USD 1,500 per night.[96] It is unclear what the financial situation of Sandfontein was in 2007, but by 2016 Sean Gilbertson was providing most, if not all, of the cash needed for renovations, upkeep and salaries (which Agenbach and his parents received monthly). In the period after June 2009 – well after all the farms were purchased – Sean's company Arianna Investments Ltd fronted at least USD 2.94 million to cover Sandfontein expenses and Sean himself paid out another R788,000.[97] Likely around 2014, Sean stopped making the cash available, and he began to formally request pro-rata contributions from both directors, which included Agenbach.

The increased costs of running a lodge and nature reserve with imported charismatic fauna like giraffes and black rhinoceros – without hunting,

94 RoHC File HC-MD-CIV-MOT-GEN-2016/00262: Founding Affidavit: Sean Thomas Gilbertson and Arianna Investments, Ltd – 17 August 2016.

95 Ibid, para 45.

96 See RoHC File HC-MD-CIV-MOT-GEN-2016/00262: S.T. Gilbertson to W.G. Agenbach 'Ongoing Use of General Manager's House' – 31 May 2016.

97 RoHC File HC-MD-CIV-MOT-GEN-2016/00262: Annexure STG10: 'Arianna Loan Funding to Sandfontein'.

farming or a large hotel – made sufficient revenue generation nearly impossible, especially given the strained relations between Agenbach and Gilbertson. Sean stopped making payments from Arianna Investments and allowed the company's bank accounts to decrease dramatically in June 2016. He made a formal request to his fellow director that urgent funding of NAD 500,000 be provided to Sandfontein, and Agenbach would have to put forth his 50.02 per cent share. Sean wrote to Agenbach:

> Please confirm urgently whether or not you are prepared to provide your share of the funding. There is of course no obligation to provide the funding, but failure to do so will – as your legal advisor will confirm – trigger certain legal obligations on the part of the company's directors given that the company will have insufficient funds to continue its business and pay its liabilities ... You are reminded that you remain a director of the company and that you accordingly owe a fiduciary duty to the company and to its shareholders (including Arianna Investments Ltd) and must accordingly act in their best interests. As your legal advisor will confirm, it is an offence for a director to allow a company to continue doing business if it is reasonably suspected that the company will not be able to pay its debts as they fall due.[98]

Agenbach finally wrote to Sean after nearly three weeks, ignoring the issues of fiduciary duties, stating 'I would still like it very much if you and me can sort this mess out between us regarding Sandfontein. I do not want to sell my shares or give them to anybody. I would like to make a deal and contract with you to manage and run it without me as if you were 100 per cent owner'.[99] Agenbach must have realised that Sean had perhaps revealed the tensions of post-apartheid Land Reform legislation, and he simply could not supply his share of the funds to run what had become an astronomically expensive operation. His proposition to sign a nominee owner contract with Gilbertson, if followed through, would also have been illegal by law. Sean had to send a cash influx from Arianna to cover general expenses like diesel, worker salaries, etc, which required that Agenbach as the local director sign papers at the bank. Agenbach wrote to Sean:

98 RoHC File HC-MD-CIV-MOT-GEN-2016/00262: S.T. Gilbertson to W.G. Agenbach 'Urgent: Funding the Company' – 10 June 2016.

99 RoHC File HC-MD-CIV-MOT-GEN-2016/00262: W.G. Agenbach to S.T. Gilbertson 'Sandfontein Salaries' – 30 June 2016.

> I do not oppose Arianna transferring money into the Sandfontein account ... I do not want anything to do with the Sandfontein account. I will ASAP sign the papers to give you the signing power over the account alone. The question that you tend to avoid all the time is what do you suggest we do now to let Sandfontein continue to be run as a Game and Nature Reserve.[100]

Agenbach likely understood that his and Gilbertson's definitions of a game and nature reserve were very different. They involved a different use of the land, a different source of profits and a fundamentally different clientele.

But Agenbach would have no place in this future Sandfontein. In his affidavit to the High Court, Sean requested that since neither shareholder was willing to provide funds to the company, the company should be closed and a liquidator be appointed to sell and distribute the assets among the creditors (being Sean Gilbertson and Arianna Investments).[101] The court naturally sided with Gilbertson, and Sandfontein and Houmsrivier Properties (Pty) Ltd was provisionally wound up in September 2016,[102] completing the process the following month.[103]

In order to hold onto the properties, Sean had to repurchase all the farms at a public auction – after they were waived for sale by the Ministry of Land Reform.[104] Sean quickly founded a new company the following month, called Sandfontein Lodge and Nature Reserve (Pty) Ltd.[105] Perhaps in order for it to meet land reform laws, he turned to his old Wits University mate and Namibian citizen Heye Daun, who was already the 51 per cent owner of Marwilben Farming cc's farmland. The new Sandfontein company became a 100 per cent subsidiary of Sean and Heye's Lukin-Heydebreck Holdings (Pty) Ltd, named after the two generals in the Battle of Sandfontein during the First World War.[106] Heye Daun was appointed director of both companies, and presumably,

100 RoHC File HC-MD-CIV-MOT-GEN-2016/00262: W.G. Agenbach to S.T. Gilbertson 'Excess on Account' – 8 July 2016.

101 RoHC File HC-MD-CIV-MOT-GEN-2016/00262: Founding Affidavit: Sean Thomas Gilbertson and Arianna Investments, Ltd – 17 August 2016.

102 RoHC file HC-MD-CIV-MOT-GEN-2016/00262, Government Gazette of the Republic of Namibia no. 6128 (16 September 2016).

103 RoHC File HC-MD-CIV-MOT-GEN-2016/00262, Government Gazette of the Republic of Namibia no. 6153 (21 October 2016).

104 Sean T. Gilbertson to Bernard C. Moore (7 January 2023), letter in the authors' possession.

105 BIPA File 2016/0546: Contents of Register of Directors, Auditors and Officers (CM-29) – 3 October 2016.

106 BIPA File 2016/0546: Annual Return (CM-23) – 31 July 2017.

FIGURE 69 Swimming pool at Sandfontein Lodge, 2023
PHOTO COURTESY OF ROSITA SMEENK AND MAURITS MATERS, USED WITH PERMISSION

Arianna Investments held 49 per cent of the share capital again.[107] This time, however, Heye Daun was a mining executive – CEO of Osino Gold – and he could meet his 'fiduciary duties'.

In September 2018, Sandfontein Lodge and Nature Reserve (Pty) Ltd successfully repurchased the farms at public auction for NAD 29 million plus fees to the state amounting to about NAD 3.5 million.[108] While useful in building his property holdings, partnering with Agenbach was an expensive mistake – even for Sean Gilbertson – who ultimately lost at least NAD 7.1 million in the liquidation process.[109]

For the Bassons and other river-folk along the Orange, there was not a big difference between Gilbertson's and Agenbach's strategies, because both still required policed, exclusive use of the farms. Around 2009, Agenbach and the Sandfontein company built their first game fence along the southern portion of the farms, running along Homsrivier and Girtis. According to testimonies at

107 BIPA File 2017/0099: Annual Return (CM-23) – 7 September 2019. Gilbertson later wrote to us that Sandfontein Lodge & Nature Reserve (Pty) Ltd is now 49% owned by his BVI company Arianna Investments Ltd.

108 WDO File T7793/2018: Deed of Transfer – 6 September 2018.

109 WDO File T7793/2018: S.T. Gilbertson, 'SaHRP Liquidation Payout Waterfall'.

the time, Agenbach also formally claimed that the company owned the state land Hartebeesmund, east of Girtis. This restricted the pastoralists' abilities to move with their livestock along ancestral trek-paths, like the Hom River.[110] Agenbach also built and locked gates along farm roads 217, 284 and 311, all of which pass through Homsrivier, Sandfontein, Sandfontein West, Girtis or Norechab (see chapter 14).

For those living by the river, the conflicts between Agenbach and Gilbertson about what a 'game reserve' looked like were problems for those with much deeper pockets; they were problems for those whose livelihoods did not come from subsistence farming. And for the Bassons and other river-folk, their influence again had been dramatically reduced in the balance of power in the region. The number of commercial farmers had diminished, and there was little need for the casual tenant labour that the Bassons could provide. Mining had nearly collapsed, and Tantalite Valley was sitting dormant. In fact, the owners of the Sandfontein company and its assets no longer resided on the properties at all; they were international mining executives jet-setting to conferences and board meetings across the globe.

---

110 LAC BCF: Lesle Jansen, notes from consultation with Willem Basson – 20 July 2010.

CHAPTER 10

Rewilding the Rhinoceros

On 29 April 2013, a sixteen-kilogram parcel was delivered to Michael E. Nacol at his home in Georgetown, Texas. The cargo manifest described the contents as 'hard dried skins and bony souvenirs of a hunt'.[1] It was shipped by Willem F. Mans of KumKum Game Ranch in southern Namibia. Nacol had travelled to KumKum in 2012 for his second big hunting trip in Namibia, having hunted farther north in 2004. Reflecting on his first impression, he said, 'I thought it was very sharp and rocky as hell'.[2] Nacol has been a long-time active member of the Dallas Safari Club and Safari Club International, and he would learn about hunting opportunities at their conventions and charity auctions. He and a friend purchased a one week hunt on KumKum from a Dallas Safari Club auction, and they went particularly in hopes of shooting leopard.

Originally from South Africa, Willem Mans arrived in the region at the same time as Steen Severin, and he purchased the 15,656-hectare KumKum Farm in December 1987 for the bargain price of ZAR 290,000.[3] Mans increased his land assets over the following years, purchasing Norechab for only ZAR 115,000 in 1996 and a 3,000-hectare portion of Arus Farm in 2004 for NAD 100,000.[4] Mans was not very interested in farming but rather in hunting, and on a large scale. Building on the abundance of natural wildlife in the area, he built game fencing and imported additional game from central and northern Namibia, stocking the farm heavily to ensure that his foreign hunters left with their trophies and he with their US dollars. In addition to the daily hunters' rate of USD 300 per day, Mans charged for each animal bagged: from USD 500 for a caracal or ostrich to USD 900 for a kudu, USD 950 for a Hartmann's zebra, USD 1,800 for an eland and USD 8,000 for a twelve-day leopard hunt.[5] These prices were tempting to nearby landowners, who at times permitted Willem Mans to escort hunters across their farm borders for a cut, making it possible for Mans to advertise hunting grounds of 100,000 hectares, which was much larger than

1 Bill of Lading no. MAEU559718413; Container no. MSKU5054010.

2 Michael E. Nacol, interview with Bernard C. Moore (telephone, 5 January 2023).

3 WDO File T4309/1987: Deed of Transfer: KumKum no. 105, consolidated no. 413.

4 WDO File T4842/1996: Deed of Transfer: Norechab no. 129. WDO File T5116/2004: Deed of Transfer: Remainder of Farm Arus no. 111 – 25 August 2004. KumKum, Eselruh and the Arus portion were combined in 2004, making the total area of KumKum 18,597 hectares.

5 IAWB: KumKum Game Ranch, 'Pricelist 2009' (11 November 2009). https://web.archive.org/web/20091111012449/http://www.kumkum.com.na/pricelist_pg.htm

 | DOI:10.1163/9789004712454_011

FIGURE 70 Black rhino on Sandfontein, after feeding on lucerne
PHOTO COURTESY OF BIRGIT KÖTTING, USED WITH PERMISSION

his actual property holdings.[6] Together with his American bluetick coonhound and a team of Nama workers, Mans targeted the American hunting market, bringing in high-value clients from Idaho, Montana, California, North Carolina and Texas.[7]

Michael Nacol and his hunting partner stayed in the small chalets beside the farmhouse, where they would dine on their hunting while Mans 'chain-smoked more cigarettes than I've ever seen in my life', Nacol reflected.[8] On KumKum, he shot a gemsbok and a few springbok. Some of the meat was eaten each night, though most of it was sold by Mans to nearby farmers for workers' rations and to shops in Karasburg. Nacol recalled that Mans 'dried some of it into jerky [biltong], using all these Biblical spices, like coriander. It was nice'. Most of Nacol's time was spent with Mans and his workers tracking a leopard, which he successfully shot, and for which Mans received USD 8,000. On Nacol's last day of hunting, Mans took them to 'a nearby conservation area', where they

6 IAWB: KumKum Game Ranch, 'Hunting Information' (11 November 2009). https://web.archive.org/web/20091111012420/http://www.kumkum.com.na/huntinginfo_pg.htm

7 IAWB: KumKum Game Ranch, 'Photo Gallery 2008' (5 October 2008). https://web.archive.org/web/20081005113249/http://www.kumkum.com.na/photo_gallerypg.htm

8 Michael E. Nacol, interview with Bernard C. Moore (telephone, 5 January 2023).

had 'an arrangement' with the owner: Willem Agenbach of Sandfontein. Nacol bagged an ostrich from their herd, which presumably was included in his parcel of skins shipped a few months later. He recalled that 'they had this herd of rhinoceros on the property. You couldn't hunt them, though. That place was really beautiful, for sure'.

Up to 2012 – and even beyond – trophy-hunting was a prominent feature of the farms to the east of Sandfontein (and, as chapter 9 showed, on Sandfontein itself). KumKum's hunting operations allowed Willem Mans to retire comfortably in South Africa. Throughout these years, the biggest difference between Willem Mans's KumKum operations and Gilbertson/Agenbach's Sandfontein operations – apart from window-dressing – was that the latter had a charismatic herd of black rhinoceros. This chapter considers the history, political economy and ecology of black rhino on Sandfontein and how the allure of rhinos shaped 'conservation' initiatives near the Orange River.

•••

In November 2021, we left Henry Pretorius's farm, Eendoorn, with spare keys to the gate to farm road #227, which runs south to Pelgrimsrust Farm and the Orange River farms of Kambreek and Pelladrift. We were surprised that we needed keys in the first place, because locking gates on public roads is illegal by Namibian law. However, Pretorius noted, as long as all the owners beyond the gate had keys, it was permitted; concerning those farms in question, Pete Morkel owned them all. We did not really know anything about Morkel at the time and our goal was not to speak with him but rather to find our way along the jeep tracks to Oude Meester and Gerrit Luttig's old irrigation fields on Pelladrift. Henry Pretorius looked at our 4 × 4 vehicle and shook his head. 'That plastic thing won't be able to handle those mountain roads,' he said. 'Maybe Pete will let you use his stronger bakkie'.

After unlocking the gate near the abandoned farmhouse on Keimas, we drove for a few more kilometres of 4 × 4 track until we reached the large, modernist, two storey house on Pelgrimsrust that belonged to Pete Morkel. Built of concrete and glass, it was more brutalist than boer. We parked the car and a sunburnt man greeted us in Afrikaans and invited us inside for some coffee and a chat. Morkel was friendly and talkative, and he clearly appreciated having an audience after living for quite a while out of regular reach of a cell phone signal. Although we were familiar with the situation on Sandfontein, it quickly became clear to us that Pete Morkel was a nodal point around which the entire conservation plans for the Orange River area revolved – plans much bigger than Sandfontein.

FIGURE 71 Pete Morkel, photographed after successfully translocating three black rhino from Port Lympne Zoo in Kent, England, to Mkomazi National Park in Tanzania, 2012
PHOTO: GETTY

Peter van der Byl Morkel was born in 1960, in Umtali (today Mutare), Zimbabwe.[9] The Morkel family goes back to the early 1700s, when Philip Morkel – an artillerist on a Dutch East India Company warship en route to Holland – decided to stay in Cape Town.[10] The Morkels became a large slave-owning family on their farms Onverwacht and Voorburg, near Cape Town.[11] Part of the Morkel family emigrated to Zimbabwe during the early 1900s. Pete Morkel was sent by his father to complete his secondary schooling in South Africa, and then managed to graduate in veterinary studies from the University of Pretoria, utilising his 'virtually non-existent' Afrikaans. In 1984, Morkel was called up for compulsory military service, and volunteered to go to Rundu, in Kavango. He described life in the guerrilla war zone as 'two wonderful years for us there in northern Namibia; they were probably the best two years of my life'. He met his wife, Estelle, who was working as a radiologist at Rundu Hospital.

In June 1986, Morkel was hired as a veterinarian for the SWA Division of Nature Conservation as part of their game capture team. At the time, they were relocating lechwe from Caprivi to other parks and private farmlands in Central Namibia, and from 1989 to 1991 he worked on relocating black rhino between Damaraland, Etosha National Park and White-owned farms. In 1991, Morkel 'went private', leaving his post in the Namibian government (he was by then a naturalised Namibian citizen) to work on a contract basis for farms, lodges and private and public game reserves in the Northern Cape, and eventually in East Africa.

In 2002, Morkel took a job with the Frankfurt Zoological Society as director of their 'rhino project' in Ngorongoro Crater, in Tanzania. This work involved relocating rhino throughout this section of Tanzania, but also into Kenya. Morkel was already connected to the East African conservation networks through his years as a private wildlife veterinarian. During the 1990s, he worked with a number of British and British/Kenyan businessmen and directors of non-profit organisations to relocate eastern black rhino from Addo National Park in South Africa to Mkomazi Game Reserve in Tanzania.[12] Morkel was also long involved with NGOs – like the South African organisation, Back To Africa – which relocated rhinoceros from zoos in Europe to reintroduce them to African national parks and to 'rewild' the species into landscapes that had purportedly

9 Unless otherwise noted, biographical material for Pete Morkel comes from two biographies, written by his brother. See M. Morkel, *Wildlife Conservation and Pete Morkel* (self published, December 2016); M. Morkel, *Diamonds on the Soles of His Feet* (self published, 2019).

10 P.W. Morkel, *The Morkels: Family History and Family Tree* (self published, 1961), ch. 1.

11 E. Rhoda, 'The Origin of the Rhoda Family of the Strand' (March 2001).

12 See T. FitzJohn,'25 Years of Mkomazi', *The Horn* (2014), pp. 22–23. Mkomazi has been described by Brockington as the epitome of 'fortress conservation'.

FIGURE 72 Prince William, Duke of Cambridge, presents Pete Morkel with the Prince William Award at the Tusk Conservation award ceremony, November 2018
PHOTO: GETTY

supported rhinoceros in the past. It is fair to say that Pete Morkel is one of the best-known veterinarians of black rhinoceros and specialists for their translocation living today. In 2018 he was given the Prince William Award for African Conservation by the English royal himself.[13]

Given all this, what was Pete Morkel doing on this isolated, mountainous farm in southern Namibia? After Morkel left government service in 1991, he remained a prominent face in nature conservation and rhinoceros affairs in Namibia, and in 1993 helped to design Namibia's Black Rhino Custodianship Programme (BRCP), which has been led since 2006 by control warden Birgit Kötting. During the Namibian liberation war, there was a widespread recognition that charismatic fauna – especially in northern Namibia – were under threat from poaching. There is evidence that elephant and rhinoceros numbers declined during the late 1970s and early 1980s, often at the hands of White South African Defence Forces soldiers and officials, though specific figures are hard to come by.[14] The real perceptions of poaching were further amplified by the large-scale operations of the SWA Administration's game-capture team, which Morkel would eventually join.

To understand the purpose of the game-capture programme, we need to look back at nature conservation in earlier years. When the 1964 Odendaal Commission recommended reducing the size of Etosha National Park in order to add necessary grazing to the Damaraland homeland,[15] conservationist organisations protested, declaring that 'non-Europeans are not preservers of wild animals; thousands of animals could perish'.[16] In the months before the implementation of the plan, the SWA Administration approved the formation of a game-capture team, purchasing sophisticated equipment and hiring additional staff. The team went into the homelands, particularly Damaraland and Kaokoland, emptying the areas of eland, tsessebe, roan antelope and, crucially, black rhinoceros. The Administration purchased additional farmland in Naukluft and Waterberg areas to hold the captured animals before they could be sold to private farms and national parks in South

13 See 'Winner of the Prince William Award For Conservation in Africa 2018', *Tusk Conservation Awards*. https://www.tuskawards.com/pete-morkel-2018/

14 M. Bollig, *Shaping the African Savannah: From Capitalist Frontier to Arid Eden in Namibia* (New York, Cambridge University Press, 2020), pp. 196–233.

15 NAN AP 4/1/13: Report of the Commission of Enquiry into South West African Affairs, 1962–1963 (1964), pp. 89–93.

16 BAB PA.24 VI.B.5: H.G. zu Castell-Rüdenhausen, 'How the Recommendations of the Commission of Enquiry into SWA Affairs would Affect Game and Game Reserves' – n.d., most likely mid-1964.

Africa and abroad, as the captured species were too many to be supported by Etosha National Park alone.[17] Throughout the 1980s, the SWA Division of Nature Conservation continued to capture and sell black rhino to private farms and South African national parks, particularly Augrabies.[18] The reality of poaching in north-western Namibia was complemented by the reality of large-scale emptying of the homelands' game by the government, spreading the black rhino population away from communal areas into national parks and private lands.

The inauguration of the Black Rhino Custodianship Programme continued this trend in the post-independence years, albeit in a more democratic way. There was still an interest in decentralising the critically endangered black rhino population by involving freehold landowners in the process. This was viewed as a way to protect against poaching and to spread potential economic costs and benefits. As per nature conservation legislation, all black rhinoceros were officially the property of the Namibian government, so each farmer-applicant needed to apply to the Ministry of Environment and Tourism and sign a Memorandum of Understanding.[19] In the early days, only private freehold landowners were involved, and in 1993, the first two custodians received the first BRCP herds, totalling eleven animals.[20]

After thirty years, there are almost thirty freehold landowners in the BRCP, looking after about one third of the state's black rhino population across more than one million hectares of farmland.[21] It is the responsibility of the landowner to build a steel-reinforced boma (kraal) for the herd of rhinos; veld-to-veld transfer has rarely been practised, and it is often necessary to gradually introduce the rhino to the local vegetation, while supplementing their diet with lucerne. The Namibian government pays for most veterinary expenses and transport costs. In general, those costs that the government cannot afford are supplemented with funds donated by the World Wildlife Fund, the Save

---

17 NAN ADG 2/4: Direkteur, Natuurbewaring en Toerisme 'Memorandum: Toedeling van die Regeringsfunksies ten opsigte van Natuurbewaring en Toerisme aan die Gesagowerhede op die Eerste en Tweede Regeringsvlakke' – 15 February 1980.

18 NLN TXX 0742: Department of Agriculture and Nature Conservation: Annual Report, 1985.

19 See S. Sullivan et al. 'Historicising Black Rhino in Namibia: Colonial-Era Hunting, Conservation Custodianship and Plural Values', Future Pasts Working Papers no. 13 (December 2021), p. 16. Because white rhinoceros can be privately owned – and are less critically endangered – they were not part of the custodianship programme.

20 B. Kötting, 'Namibia's Black Rhino Custodianship Programme', *Conservation Frontlines*, 2, 2 (2020).

21 Birgit Kötting, interview with Bernard C. Moore (telephone, 20 January 2023).

FIGURE 73 Black rhino on Sandfontein, suffering from drought and lack of vegetation. They would recover after additional lucerne was provided. Note the overturned melkbos to the right of the animals.
PHOTO COURTESY OF BIRGIT KÖTTING, USED WITH PERMISSION

the Rhino Trust and the United States Fish and Wildlife Service, which heavily subsidise the BRCP.[22]

Pete Morkel was hired in 2009 by Birgit Kötting and the BRCP to facilitate the transfer of a small herd of black rhino from Etosha National Park to Sandfontein.[23] Kötting reflected that this was the first such scheme that far south: 'the landscape at Sandfontein is a bit volatile, and it's not really an ideal locale for black rhino'.[24] The far south of Namibia is on the fringes of what the black rhino's natural habitat would have been, even in the most generous assessments of prior rainfall and vegetation growth. The ideal range is really from Windhoek north-west to Omaruru, and then farther into Kunene region. 'In the early 1990s, the government had transferred a breeding pair of black rhino to Naute Dam Park [near Keetmanshoop], but the browse was insufficient, leading to the rhino eating too much unusual plants in the riverbeds, and the animals died', said Kötting. 'So, for a number of years, we had a policy that no

22 Ibid.

23 Pete Morkel, interview with Bernard C. Moore and Luregn Lenggenhager (Pelgrimsrust farm, Karasburg District, 10 November 2021).

24 Birgit Kötting, interview with Bernard C. Moore (telephone, 20 January 2023).

black rhino would be transferred south of Windhoek, and Sandfontein got in right after we changed that'.

Kötting remarked that the rhino intended for Sandfontein were first kept in a kraal and fed on lucerne for a while before being transported to the south. They then were kraaled for three additional months on the property, gradually being introduced to local vegetation along with lucerne fodder.[25] Furthermore, black rhino prefer to bush browse rather than graze on grassy plains – which predominate on Sandfontein – so only about one fifth of Gilbertson's property holdings can actually be used for the BRCP. It is either too grassy or too bare and mountainous. Much of the bush is melkbos (*Euphorbia gregaria*) – poisonous for most mammals but edible for rhino – which is acceptable as a supplement to lucerne or other browse, but it is insufficient for the entirety of the black rhino's diet. The rhino can eat the latex-filled stalks, but they prefer eating the root system, which means they dig up the plants. In this way the melkbos cover on Sandfontein was reduced dramatically.[26] The south of Namibia suffered a long drought between 2013 and 2020, which resulted in the health of the black rhino on Sandfontein deteriorating significantly (see Figure 73). However, because the herd was accustomed to lucerne, their diet could be supplemented for some time, keeping them from starving until vegetation began to regrow in 2020 (see Figure 70).[27]

So, given that Sandfontein and southern Namibia are not ideal black rhino habitat, and given the expenses and effort involved in maintaining the growing number of black rhino on Sandfontein (which Kötting said had risen to about fifteen or sixteen), why would Sean Gilbertson and his business partners go to the trouble of partnering with the BRCP to have the animals translocated to their properties? Birgit Kötting divides rhino custodians into two types. There are the conservationists, plain and simple; they value the species and want to spend money to protect them. Then, you have businessmen, who run tourism or hunting operations and find that having black rhino draws guests to their lodges. Although the rhino itself cannot be hunted, there is an allure to hunting other species with black rhino nearby.[28] Michael Nacol certainly thought this when he hunted ostrich on Sandfontein in 2012. When the rhinos were first brought to Sandfontein in 2009, it is likely that Sean Gilbertson was the first type of custodian: he was a wealthy person with an interest in conservation who wanted to play a part in protecting a species. Given Agenbach's interest in

25 Ibid.

26 Ibid.

27 Kötting, 'Namibia's Black Rhino Custodianship Programme'.

28 Birgit Kötting, interview with Bernard C. Moore (telephone, 20 January 2023).

hunting, it is likely that he was the second type, looking to use rhino as a pull factor to bring in additional safari hunting tourists to shoot other animals on Sandfontein.

There is, however, a third type, whereby custodianship of the government's black rhino are perhaps a smokescreen to secure land tenure; after all, farmers do not own these species. The black rhino is really the last vestige of the old 'royal game' or 'state custodianship' model of conservation, of the MacKenzian Orthodoxy (see chapter 1). Virtually all other game in Namibia can be owned by the landowners themselves, even white rhinoceros. When Pete Morkel met Gilbertson in 2009, he was impressed with the lodge and what the Sandfontein company had built there. 'You know, Sean Gilbertson is a miner, not a conservationist', Morkel said. 'He appreciates conservation, but that's not why he bought all that land. He originally bought the farms because he wanted a big property for himself.'[29] Morkel emphasised that black rhino are not actually very useful for tourism: they are reclusive, rarely move in groups and can become aggressive easily. He implied that while the existence of black rhino on a property might be a draw for tourists or hunters, their behaviour made it difficult to easily integrate them into activities. 'I don't have much evidence for this, but I always suspected that a lot of black rhino custodians went into the programme as a way to protect against land expropriation', Morkel said. 'If they're guarding the government's rhinos from poachers, then they're probably less likely to have their lands taken by the new government.'[30] Indeed, although custodians are required to hire anti-poaching security – broadly defined – most of the benefits of the BRCP accrue to them. However, if Morkel's suspicion is correct – that the BRCP is a hedge against possible land expropriation – then that may be the largest benefit of all.[31]

• • •

Morkel stayed at Sandfontein for a little while after he delivered the black rhino, in part to monitor their behaviour and health and likely also to have a well-deserved holiday. While travelling through the area, he found that there were a few properties for sale to the east of Sean's holdings. By 2009, the

29 Pete Morkel, interview with Bernard C. Moore (telephone, 19 October 2022).

30 Ibid.

31 This echoes Yuka Suzuki's account of White property owners in post-independence Zimbabwe shoring up their land tenure by converting livestock farms into game or hunting reserves. Y. Suzuki, *The Nature of Whiteness: Race, Animals, and Nation in Zimbabwe* (Seattle, University of Washington Press, 2017). See also M. Spierenburg & S. Brooks, 'Private Game Farming and its Social Consequences in Post-Apartheid South Africa: Contestations over Wildlife, Property and Agrarian Futures', *Journal of Contemporary African Studies*, 32, no. 2 (2014), pp. 151-172.

FIGURE 74 Willem F. Mans's old farm sign still stands on the road to KumKum, long after its proprietor departed
PHOTO: B.C. MOORE, OCTOBER 2022

irrigation farmer Gerrit Luttig was getting quite old – he was well into his seventies – and was spending more time at his son Pieter's farm near Ariamsvlei. He had owned Kambreek and Pelgrimsrust farms since 1971, but the family was now looking to sell. Pete Morkel had a vision to expand his rhino conservation operations into Namibia, perhaps mimicking some of the private conservation operations in East Africa. He met Gerrit Luttig and negotiated the sale of the two farms in April 2010 for a price of just over NAD 1 million.[32]

'Gerrit Luttig was as crooked as a dog's leg', said Morkel, 'but by the time he died, he had accumulated quite a lot of property and retired a very wealthy man'. Morkel intended to transform Luttig's properties into a conservation area:

> I would have liked to have done something like this in my native Zimbabwe, but southern Namibia is as far from communities as possible, so we could attempt something at a large scale. Plus, without karakul these lands aren't so profitable. I would have felt horribly guilty attempting something like this with prime farmland like Erindi [central Namibia, near Okahandja].[33]

Gilbertson and Agenbach were trying to run Sandfontein as a conservation operation – albeit each having a very different definition of this – and their operations were marketed simply as contemporary choices of how to use the land. Morkel's beliefs and strategies, however, were not the same. He believed that he had *historical* justification for his choices.

When we met Morkel in November 2021, on learning that we were historians he asked if we had read H.J. Wikar's travelogues. We were initially surprised that Morkel had an old copy of this obscure eighteenth-century primary source. 'Wikar described this land as it should be', Morkel said, 'filled with game species in the mountains, hippos at the Orange River'.[34] Livestock farming and settlement, reasoned Morkel, was the cause of the decline in wildlife, especially black rhinoceros. 'I've got these ancient rhino rubbing rocks on Pelgrimsrust, so I know that there must have been a lot of rhinoceros here in the past. I want to rewild these lands to the way Wikar saw them'.[35]

Since we too had read Wikar, we asked Morkel what he thought of the fact that most of Wikar's travelogue is not really about the wildlife along the Orange River at all, but rather the river-folk communities and people he met along the way, with whom he traded, hunted, ate and drank. Morkel dismissed those facts as not applicable today at all. For Morkel, 'rewilding' appears to entail using the

32 WDO File T4912/2010: Kambreek and Pelgrimsrust, Deed of Transfer – 5 October 2010.

33 Pete Morkel, interview with Bernard C. Moore (telephone, 19 October 2022).

34 Pete Morkel, interview with Bernard C. Moore and Luregn Lenggenhager (Pelgrimsrust farm, Karasburg District, 10 November 2021).

35 Pete Morkel, interview with Bernard C. Moore (telephone, 19 October 2022).

lands only for establishing herds of specific species (like black rhino), housing those veterinarians and staff who see to those species, and allowing paying guests to experience the wilderness as Wikar allegedly once experienced it.[36] Based on Morkel's reading of Wikar, it would appear that those who use the land in other ways – hunting, commercial farming, subsistence herding – have no place in his rewilded lands.

Once Morkel purchased Kambreek and Pelgrimsrust – both of which border KumKum – he immediately had problems with Willem Mans and his hunters. 'Willem Mans was a complete miser', said Morkel, 'he barely ever paid his staff, and he instructed them to beg for tips from the trophy hunters. We've got one guy working for us now, Gabriël, who used to work for Mans, and he said that was the tip of the iceberg of Mans' strategies to make profits'. Morkel caught Mans and his workers routinely crossing over the farm boundaries illegally to hunt on Pelgrimsrust, Kambreek and even Pelladrift. For a price, Mans might have received permission. 'Mans could have bought these farms for cheap, but he was a miser and chose to do things illegally. But he made a tremendous amount of money, and he succeeded in southern Namibia where few had been able to before.'[37]

In 2011, not long after Morkel purchased his properties, he brought an internationally renowned expert in conservation to visit his farms and nearby Sandfontein, the Kenyan Ian Craig.[38] As the following chapter explains in more detail, Ian Craig was one of the pioneers of private wildlife conservation in East Africa, founding the Lewa Wildlife Conservancy out of his family's old cattle ranch in the Laikipia Highlands near Mount Kenya. He is also the founder of the Northern Rangelands Trust (NRT), a private non-governmental entity in Kenya that links private conservation businesses with community conservancies to expand nature conservation tourism and wildlife utilisation schemes. He would, in 2016, be awarded the Order of the British Empire by Queen Elizabeth II for his 'services to conservation and security to communities in Kenya'.[39] Pete Morkel got to know Ian Craig through their membership of the IUCN's African Rhino Specialist Group, and while Morkel was based at Ngorongoro in Tanzania, the two worked together to relocate rhinoceros throughout East Africa.[40] 'We worked in parallel careers', reflected Ian Craig. 'He invited me

36 In chapter 12, we cover in detail the political ecology of rewilding.

37 Ibid.

38 Sean T. Gilbertson to Bernard C. Moore (7 January 2023), letter in the authors' possession.

39 Lewa Wildlife Conservancy, 'Ian Craig Awarded Order of the British Empire' (10 June 2016).

40 Pete Morkel, interview with Bernard C. Moore (telephone, 19 October 2022).

and my wife to Namibia for a visit, as Pete had recently sold his land in South Africa, and had set himself up near the Orange River.'[41] Craig understood that the visit was not just a holiday with old friends, but rather the start of a business relationship.

Morkel toured Kambreek and Pelgrimsrust with Craig, introducing him to the mountainous landscape of southern Namibia and the Orange River cutting through it. He brought Craig to Sean Gilbertson on Sandfontein and Homsrivier to see what an ongoing 'conservation' operation in Namibia looked like. Then, he introduced Ian Craig to Willem Mans, in the hope that they could negotiate a purchase of KumKum, which would bring Morkel's lands closer Sandfontein.[42] Willem Mans wanted to sell KumKum at the maximum price possible. 'It's true that Mans was selling KumKum as a going concern, which needed to be taken into account, but he calculated the price of every piece of game in the sale price, down to the dassie!'[43] Mans had already sold Norechab a few years earlier, to Gilbertson, and he wanted to sell KumKum for USD 1 million. He got very near that price from Ian Craig: the property was sold for NAD 7.8 million.[44]

Morkel and Craig also met Pieter Luttig, son of the elderly Gerrit, to try and purchase Pelladrift Farm, right along the Orange River, which was owned by the company Pelladrift (Edms) Bpk. Luttig arranged to have the company sold to Craig and Morkel once his father had passed away, and the two obtained their directorships on 23 January 2012, the day Gerrit Luttig died.[45] When we asked Pieter Luttig what Morkel's interest in the mountainous and inaccessible farm was, he laughed and told us that they 'wanted to put rhinos on it'.[46]

Morkel had paid for his two properties from his own account. However, he and Ian Craig were clear that the latter – a Kenyan citizen – had paid for KumKum and Pelladrift. Nevertheless, it was the Namibian citizen Pete Morkel who signed the Land Reform affidavit stating that Pelladrift's shareholding was at least 51 per cent Namibian. By the end of 2012, the Pelladrift company owned

41 Ian Craig, interview with Bernard C. Moore (telephone, 18 October 2022).

42 Ibid.

43 Pete Morkel, interview with Bernard C. Moore (telephone, 19 October 2022).

44 WDO File T6534/2012: Kumkum no. 472, Deed of Transfer – 29 November 2012.

45 BIPA File 1973/5752: Contents of Register of Directors, Auditors and Officers (CM-29) – 18 May 2016. This record erroneously declares that Ian Craig was a Namibian citizen.

46 Pieter Luttig, interview with Bernard C. Moore and Luregn Lenggenhager (Tigerberg farm, Karasburg District, 10 November 2021). Because it was the sale of a company and the company's assets rather than the sale of property directly, the specific sale price for Pelladrift was not made public. We can reason that it was significantly less than the price for KumKum.

31,000 hectares; Morkel himself had 15,000 hectares of his own lands. Together with Gilbertson's properties, there was now the possibility that a 'conservation' project could be attempted at scale. The area under their combined ownership now reached nearly 150,000 hectares in total, and with land-use agreements with other farmers or additional sales this could be increased dramatically. The question arose as to what this large conservation project would look like. In 2013, Ian Craig brought in two young British disciples to survey KumKum and the other properties to conceptualise 'rewilding' in what they believed – erroneously – to be ideal rhinoceros territory.

•••

In the twenty-first century, nature conservation is increasingly about narratives. Conservationists spend significant time crafting stories – true or not – about processes of land degradation or regeneration. These narratives are normally interchangeable and could be used, with minimal modification, to describe conservation schemes in the American West, Brazil, Kenya or even southern Namibia. They go like this: irresponsible land use, combined with global climate change, has 'degraded' a landscape and it is the responsibility of globetrotting 'conservationists' to regenerate or 'rewild' the landscape to the way it should be. There is rarely a deep political or economic critique, and

FIGURE 75 Northern gate to KumKum Farm, branded with the Oana Adventure tourism logo
PHOTO: B.C. MOORE, 2021

these plans and these strategies are carried out in the language of morality and market forces. In this way, conservation narratives resemble Evangelical Christianity more than rangeland ecology; these actions are taken on faith, and this faith is universal.

When one reads the biographies of conservationists – established or emerging – they are eerily similar in structure to those of personalities within mining or finance, not just because they are narrating a CV of sorts, but in the way that the narrative is constructed. Conservationists 'cut their teeth' or 'find their feet' in a certain locale, often quite removed geographically from the regions and the projects where they are actually working today. In the same way that a young mining executive would name-drop working for Brian Gilbertson at Gencor or BHP, conservationists would be sure to mention having worked with the George Adamson Trust or The Nature Conservancy. Just as for the godfathers of the mining industry, there is a degree of 'hero worship' within nature conservation as well:

> Hero worship in conservation is as old as wildlife conservation itself. The subjects of this worship are invariably white men and women who are lionised for taking to a life of selfless service of the wilderness and its residents. Though they serve in environments where native Africans have lived for centuries and have done little more than simply live among animals, they have continued to receive near universal adulation. To ensure that the world does not take a second look at what it deems to be their heroism and altruism, their misdeeds, failures and true personalities are conveniently ignored so to craft as attractive a narrative as possible.[47]

Kenyan authors Mbaria and Ogada apply the 'hero worship' phenomenon to well-known figures in East African conservation, like George Adamson, Richard Leakey and, to a lesser extent, Ian Craig himself.[48]

When Ian Craig bought KumKum, he realised that Willem Mans had created a 'very carefully curated sort of hunting operation, with lots of fences and camps'. But Craig wanted an 'open landscape'.[49] He also believed that Mans's success in running trophy-hunting operations depended on overstocking the landscape with game. So, Ian Craig hired the twenty five year old Englishman, Edward 'Red' Barthorp, to travel to KumKum, survey the land and plan for

47 J. Mbaria and M. Ogada, *The Big Conservation Lie: The Untold Story of Wildlife Conservation in Kenya* (Auburn, Lens&Pens Publishing, 2017), p. 10.

48 Ibid, p. 71.

49 Ian Craig, interview with Bernard C. Moore (telephone, 18 October 2022).

future conservation projects.[50] Barthorp grew up in the idyllic Cotswolds in west England, attending the prestigious Cheltenham College and eventually completing a BA degree in Geography from the University of Newcastle. He had travelled to Kenya in 2011 and 2012, volunteering at Ol Pejeta Conservancy and the Northern Rangelands Trust (NRT), where he got to know Ian Craig. Barthorp followed Craig's and Morkel's vision that KumKum, Pelladrift and other farms needed to be 'rewilded', which involved removing human infrastructure, especially fences. After his short stint at KumKum, Barthorp enrolled in a MSc programme at the Durrell Institute for Conservation and Ecology (DICE) at the University of Kent. There, he met the future face of KumKum, Andreia Pawel, who was completing a MSc in conservation biology (on a chimpanzee sanctuary in Sierra Leone).[51] Pawel made educational wildlife films for Tanzania People & Wildlife while Barthorp conducted his MSc research with Craig's NRT.

In late 2014 or early 2015, Pawel and Barthorp visited Ian Craig and his wife, Jane, at their villa on Lewa Downs in Kenya. Craig wanted to 'brainstorm an initiative' with his two young protégés, on how to transform the land that he had bought in Namibia with Pete Morkel.[52] According to Pawel, Craig (and Morkel) bought the land specifically because the properties made a 'prime habitat' for black rhino reintroduction, a view they based on the fact that the mountainous areas on the farm was 'covered in milkbush (*Euphorbia gregaria*), which is rhino cereal'.[53] As mentioned above, although black rhino do indeed consume melkbos, they eat it in limited quantities, and it is far from the mainstay of their diet.[54] Pete Morkel's former boss, Eugène Joubert, identified melkbos as only one of the possible bushes on which black rhino browse; they generally prefer *Acacia reficiens* or *Acacia mellifera*.[55] Pawel and Barthorp saw melkbos in the hills, and Morkel saw 'ancient rhino rubbing rocks', and all assumed therefore that large herds of black rhino had once thrived in this area. Although it is likely that rhino historically visited these lands, perhaps

---

50 See Ed Barthorp, *LinkedIn* page: www.linkedin.com/in/ed-barthorp-0190965a (Accessed 30 June 2023).

51 See Andreia Pawel, *LinkedIn* page: www.linkedin.com/in/andrea-pawel-43927a2a (Accessed 30 June 2023).

52 'Andreia Pawel: Oana Namibia', *Conservation Careers* (9 December 2019). https://www.conservation-careers.com/conservation-jobs-careers-advice/podcast/oana-namibia-podcast/ (Accessed 1 May 2024).

53 Ibid.

54 Birgit Kötting, interview with Bernard C. Moore (telephone, 20 January 2023).

55 E. Joubert, *On the Clover Trail: The Plight of the World's Rhinos* (Windhoek, Gamsberg Macmillan, 1996), p. 115.

during exceptionally wet seasons, the lack of much other native bush and the omnipresence of extremely steep mountainous inclines means that the land probably never supported a large quantity of rhinoceros.

Whereas Craig and Morkel initially purchased the properties for black rhino reintroductions, this was in many ways the moral justification for a broader 'rewilding' of the region at large. According to Andreia Pawel:

> that [black rhinoceros] was the primary aim ... [but] it's actually grown far beyond a black rhino reintroduction project. There's so much scope here for other wildlife reintroductions. We want to totally re-wild the area. We're surrounded by commercial farmers, and at the moment there's the worst drought in 150 years in Namibia ... There's no choice. The government doesn't have the funds to buy up the land, and the livestock farmers are going to have to sell their land eventually ... So, what we're doing here is we're just creating a little model for conservation, and then the world will start knowing about it, the NGOs will come in, they'll support and buy land – or lease land – off the commercial farmers, and then we'll just create an amazing mosaic for wildlife.[56]

Pawel and Barthorp became the advance guard for Ian Craig's larger operations, which are explored in the next chapter. According to their understanding of rewilding, before one can 're-wild' KumKum and Pelladrift – and likely Morkel's properties – one must first 'de-human' them. This means removing human infrastructure like fences, hunting blinds, unnatural vegetation, etc.

The cheapest and most glamorous way to do this – perhaps Pawel and Barthorp reasoned – was to structure a 'gap-year' destination for volunteers from Europe and North America, to give back to nature in the same way that both of them had done in East Africa previously. In 2016, Barthorp and Pawel became minority foreign shareholders in Pelladrift (Edms) Bpk, and founded their own UK-based company, Namaqua Ltd, as a 50/50 joint operation. Presumably it was to be a means to solicit donations for conservation operations and take bookings for European clients to come to southern Namibia.[57] Ian Craig and his lawyer Peter Koep (who had formerly represented Xemplar mining, and Steen Severin in earlier years) eventually founded KumKum Adventure Tourism (Pty) Ltd, which conducted business as Oana Namibia. Its goal was to

56 'Andreia Pawel: Oana Namibia', *Conservation Careers* (9 December 2019).

57 They each became 4.5 per cent shareholders in Pelladrift. See BIPA File 1973/5752: Annual Return (CM-23) – 12 March 2018. UKCHA #09985394: 'Certificate of Incorporation of a Private Limited Company: Namaqua Limited' (2 February 2016).

'engage and promote adventure tourism, conservation and all trading activities related thereto'.[58] The Pelladrift company held the lands, while Oana represented the business of KumKum.

From its earliest days, Pawel and Barthorp advertised Oana as a way for youths from the Global North to have fun while contributing to the heroic conservationist visions of its founders:

> The renowned conservationist Ian Craig and the leading mega-fauna veterinary Pete Morkel recently secured 45,000 hectares of land for the sole purpose of conserving Namibia's wildlife. Their vision is to re-wild the land, link up with neighbouring conservancies and one day turn the greater protected area into a National Park and a home for Namibia's endangered wildlife. … [We] lead 2–5 week adventure conservation expeditions in Southern Namibia for naturalists worldwide. We offer a range of expeditions tailored at gap-year students, university students, mature nature enthusiasts, and school field trips. Our school field trips are bespoke and focus on biology and conservation or building leadership skills. This is the perfect opportunity to get off the grid and explore true wilderness with a great bunch of people. We merge cutting edge conservation techniques with local knowledge to carry out effective research and solutions to re-wild some of Namibia's most magnificent habitat and protect its endangered species. Go on … Join us![59]

Pawel and Barthorp claimed that Oana (KumKum) had been 'taken over from farmers and biltong hunters', and that it was necessary to rip up all the fencing to allow wildlife to migrate naturally. 'It's manual labour', they wrote. 'But it's a fun project where you see the start to finish and come home feeling accomplished. You will truly be restoring habitat and ultimately feeding wildlife.'[60] In 2018, they successfully brought in school groups from England and Scotland, and Oana eventually developed the motto: 'No More Livestock, No More Hunting. Purely Wild'.[61] Individuals would pay between USD 5,000 and USD 6,000

58 BIPA File 2018/2819: Annual Return (CM-23) – 29 July 2019.

59 IAWB: 'Oana Flora and Fauna' (23 June 2017). https://web.archive.org/web/20170623041156/http://oana-ff.org/

60 IAWB: 'Oana Flora and Fauna: Management' (11 February 2018). https://web.archive.org/web/20180211220335/http://oana-ff.org/management. Andreia Pawel estimates that they removed 52 kilometres of fencing from the properties. A. Pawel, Interview with Ben Goldsmith, Rewilding the World Podcast (23 January 2024).

61 IAWB: 'Oana Flora and Fauna' (24 November 2018). https://web.archive.org/web/20181124154446/http://oana-ff.org/

(plus airfare) to visit Oana for six weeks, to volunteer for a for-profit company, manually ripping up expensive farm infrastructure to 'rewild' the landscape.[62] If one did not wish to do a long expedition, the team set up the possibility for two week high school programmes 'suitable to IB/A-Level biology or geography coursework'. Or else one could drop in for a five day 'Wellness Retreat', described by Pawel and Barthorp as 'the ultimate detox',[63] hosted by Nicci Cloete, a White South African life coach and corporate yoga retreat specialist.[64]

Oana also solicited scholars and scientists to do research at KumKum, requesting that they pick from among topics prepared by the Oana staff. This involved MSc projects on horned adders or hyenas on KumKum, *Prosopis* management in the Orange River, or a PhD project on the 'biophysical baseline survey of the Oana Nature Reserve'.[65] Craig and his team provided no scholarships or research funding to the applicants, and required that accommodation, food, fuel and transport fees be paid to the company by the researcher. In addition, the data from the project was to be given to them. It appears that only one such project was completed – concerning leopards on KumKum – by a researcher from the Czech University of Life Sciences.[66]

Barthorp and Pawel were trying to straddle a blurry line between a research institute, a wildlife preservation zone and a tourism company. To an extent, they did all three – and to an extent, they did none of the above. When asked how the Oana team determined the baseline for 'rewilding', Pawel could only answer the question in the most general way: 'It's not to "re-wild" to a certain point, but rather to create resilient ecosystems'.[67] Her boss, Ian Craig, answered similarly: 'In the context of an overgrazed, abused and fenced Namibia, more space and more rain will bring it closer to what Namibia should be'.[68] Likewise, Pete Morkel could see only the 'non-human' elements of H.J. Wikar's 1779 account of his travels along the Orange River. The impression garnered from Oana's use of the term 'rewilding' is that it does not necessarily refer to precise ecological or historical baselines for how the land was used in the past. In this sense, rewilding is as much an ideological project as it is a scientific

62 'How We're Rewilding Oana', *Oana Namibia* (13 September 2019). See also, 'Wildlife Conservation Internship: Namibia', *Conservation Careers* https://www.conservation-careers.com/job/6-8-week-internship-2/ (Accessed 1 May 2024).

63 Oana Namibia, Facebook post (26 November 2019).

64 Nicci Cloete, https://metta365.com/ (Accessed 30 June 2023).

65 Oana Namibia, 'Research Projects, 2020: Call for Applications'.

66 K. Mikslová, 'Among the Leopards of Southern Namibia' (12 February 2019). https://www.ftz.czu.cz/en/r-10623-news-home/mezi-levharty-v-jizni-namibii-1.html

67 Andreia Pawel, interview with Bernard C. Moore (telephone, 10 October 2022).

68 Ian Craig, interview with Bernard C. Moore (telephone, 18 October 2022).

one. 'Rewilding' appears rather as a moral judgement rooted in the present day about how land *should* be used. It can be an emotional, even a faith-based, statement. Pawel contended that the farmers *will* sell their farms; agriculture *will* fail; tourism and conservation are the *only* economically and ecologically viable option.[69]

One wonders, however, whether – in the case of Oana – taking 45,000 hectares of agricultural land out of production in order to jet British, European and North American 'volunteers' to the far south of Namibia to rip up fencing and do yoga on the sand is a sufficient replacement. One does not need to advocate for Willem Mans's level of industrial hunting or for Henry Pretorius's large-scale karakul pelt production to view Oana's option as a far from ecologically or economically sound alternative. Willem Basson III's small-scale agricultural pursuits along the Orange River have pitted him against those who seek to 'rewild' these lands on the basis that for the Oana and Sandfontein people there should be no livestock. Ironically, perhaps, Basson would also like to return to the times of Wikar, not for the wildlife but because it was before his peoples' ancestral lands were seized by colonial settlers and postcolonial companies. 'Rewilding' for Basson would be returning to when his people could greet and negotiate with Wikar on their own terms. These tensions in the meanings of 'rewilding' are explored further in chapter 12.

For those who control Oana and Sandfontein today, 'rewilding' is perfectly compatible with the legacies, or even maintenance, of colonial capitalist structures. This has been the case in East Africa as in Namibia. Andreia Pawel and Edward Barthorp were ultimately shaped by the heroic narratives of their conservation mentor Ian Craig, with whom they first began their careers and to whose vision they were allied. To understand the broader plans for KumKum, Pelladrift, Sandfontein and the other nearby farms owned by Sean Gilbertson and Ian Craig, we must turn to the latter's operations in Kenya, particularly his property Lewa Wildlife Conservancy, and his organisation, the NRT. This was where private nature conservation in Africa started in earnest, and our story must move there in order to conceptualise what is planned for this mountainous area along the Orange River.

---

69 Andreia Pawel, interview with Bernard C. Moore (telephone, 10 October 2022). See also Pawel's interview with *Conservation Careers*.

CHAPTER 11

East African Strategies, Namibian Realities

In October 2010, Prince William proposed to Kate Middleton while on holiday in Kenya, at Ian Craig's Lewa Wilderness Lodge.[1] The English royal had long close connections to the Craig family. His friendship with Craig's daughter, Jecca – who had studied in England – was so close that it necessitated an official bulletin from the royal family denying that she and the prince were in a romantic relationship.[2] Nevertheless, she was still a guest of honour at William's twenty-first birthday party at Windsor Castle. Just a few years later, in 2016, Jecca's father himself would be honoured. Ian Craig was awarded the Order of the British Empire by Queen Elizabeth II for his 'services to conservation and security to communities in Kenya'.[3] The close connections with English royalty have brought Lewa – and the Kenyan Highlands more broadly – a boom in tourism and conservation investment, and these lodges have been able to charge immensely high prices for safari visitors, many of whom are celebrities.

In contrast, Namibia's isolation under apartheid meant that, until independence in 1990, its wildlife and tourism industry remained rather provincial. The state-owned Namibia Wildlife Resorts coordinates tourist bookings at lodges in national parks and the Ministry of Environment and Tourism regulates hunting concessions. Non-state endeavours were generally run by locally owned companies or individual White landowners, and the vast majority of the clientele leading up to independence were South Africans or West German trophy hunters.[4] This has changed somewhat in the past two decades, with the Namibian conglomerate Ohlthaver and List Group entering the hospitality sector and the meteoric rise of Gondwana Collection Namibia (Pty) Ltd, becoming one of the largest landowners in the country and dominating the fast-growing tourism industry.

Kenyan tourism has long been outward-looking, targeting the elite of the former British empire and beyond. Trophy hunting has been banned in Kenya

1 'His Royal Highness Prince William of Wales and Miss Catherine Middleton are engaged to be married', *The Royal Household* (16 November 2010) https://www.royal.uk/his-royal-highness-prince-william-wales-and-miss-catherine-middleton-are-engaged-be-married

2 R. Eden, 'Prince William's former "Fiancée" Jecca Craig calls off wedding', *The Telegraph* (11 October 2009).

3 Lewa Wildlife Conservancy, 'Ian Craig Awarded Order of the British Empire' (10 June 2016).

4 NLN TXX 0742: Departement van Landbou en Natuurbewaring: Jaarverslag, 1981.

FIGURE 76 Rhinoceros mingle amongst cattle on Ol Pejeta conservancy in Kenya
PHOTO: RIO THE PHOTOGRAPHER, © 2023. REPRODUCED WITH PERMISSION

since 1977, which structured the industry towards photographic and safari tourism. The monetary value of game lay in it being alive and clearly visible for wealthy tourists who were on tight schedules, rather than hunters slowly stalking with a rifle or compound bow. The Kenyan wildlife and tourism sector was able to fit easily within the *preservationist* side of conservation, facilitating the involvement of wealthy benefactors and international NGOs.

As the founder of Lewa Wildlife Conservancy, in Laikipia region near Mount Kenya, Ian Craig has been at the centre of private nature conservation in Kenya over the past thirty years, and he and his familial network have played a leading role in shaping Kenya as one of the premier safari destinations for the wealthy and the well connected. His relationship with external donors, benefactors and NGOs has transformed how conservation in Kenya is financed and operated, and his organisations have sought – with mixed success – to bridge the divide between White commercial landowners and African pastoralists in communal areas, framing conservation tourism as the rising tide that lifts all boats.

Ian Craig's purchase of 30,000 hectares of farmland in Namibia represented his first major venture outside of East Africa. Given the vastly different historical trajectories of the Namibian and Kenyan conservation sectors, it is worth examining Ian Craig's relationship with the latter to understand better how he and his staff are looking aim to apply East African strategies in the context of Namibia. This chapter explores the history and political economy of private nature conservation in Kenya, looking at the politics of land tenure, funding and revenue, and the relationship between White-owned private conservancies and communal lands. It helps to frame Craig's Oana/ORKCA project within a new perspective.

•••

The Central Highlands of Kenya are intricately tied to the British colonial project. Laikipia County, just northwest of Mount Kenya, is today primarily inhabited by Maasai, Turkana and other pastoralist groups who live on 'group ranches' established in the wake of Kenyan independence, and White descendants of the colonial planter and rancher class, now diversifying their business into other fields.[5] After the great rinderpest epizootic of the late nineteenth century, pastoralists like the Maasai in the Central Rift Valley, southwest of Laikipia, were suffering amidst declining cattle numbers. In addition, they increasingly were being pressured by British officials to make way for White settlement and commerce – which was expected to follow the expansion of the Kenyan Railway from Nairobi. African residents, mostly Maasai, were moved from many Rift Valley areas between 1904 and 1905, with the pastoralists being split between a reserve in the far south of Kenya, near Tanzania, and a reserve in the Laikipia highlands.[6]

However, within just a few years, growing pressures from White settlers for additional land, and a recognition that pastures in Laikipia were free from East Coast Fever, led the British to seek the removal of the Maasai again. According to a 1910 colonial report, 'as soon as Laikipia is free from the Maasai, the district will be open to White settlement'.[7] British officials abrogated existing treaties and forced the Laikipia Maasai at gunpoint to join their brethren in the south. The Laikipia removals continued from 1911 to 1913, with scores of Maasai killed in the move.

---

5 See G.R.L. Fox, 'Images, Uncertainties, and Success Stories: The Politics of Conservation in Laikipia North, Kenya' (PhD thesis, McGill University, 2018), p. 33. Long histories of in- and out-migration have rendered ethnicity in this part of Kenya very complex, such that historical ties to Maasai heritage need not correspond with one's ethnic identity today. Furthermore, while significant amounts of land are indeed owned by the White ex-colonial class, a good deal of commercial land is also owned by a rising Black elite and wealthy non-Kenyans, which complicates the land tenure situation significantly. See G.R. Fox, 'Maasai Group Ranches, Minority Land Owners and the Political Landscape of Laikipia County, Kenya', *Journal of Eastern African Studies*, 12, 3 (2018), pp. 473–493. See also, M.L. Gravesen, *The Contested Lands of Laikipia: Histories of Claims and Conflict in a Kenyan Landscape* (Leiden, Brill, 2020).

6 L. Hughes, *Moving the Maasai: A Colonial Misadventure* (New York, Palgrave, 2006), pp. 5–6.

7 Ibid, p. 38.

There were already some prominent British settlers in Laikipia, such as Hugh Cholmondeley, 3rd Baron Delamere – who occupied what would eventually be Ol Pejeta Conservancy. The number of White settlers would only grow after the First World War. The British put in place an 'Ex Soldier Settlement Scheme' – similar to Namibian laws of the same time – and the White Highlands (as they would eventually be known) became one of the key livestock ranching areas of colonial Kenya.[8] Some Maasai were able to negotiate tenancy contracts on these properties – in exchange for labour for the White settlers, but increasingly the tenants were Gikuyu cultivators.[9] In 1922, Alexander Douglas, born in South Africa to Scottish parents, secured the 16,000-hectare Lewa Downs ranch, a property bordering Laikipia County, directly north of Mount Kenya. In 1952, the property was given to his daughter, Delia, and her husband David Craig, who continued the cattle-ranching operations, eventually passing the farm on in 1977 to their son, Ian.[10]

Ian Hamish Craig was born in September 1952 in Laikipia County, just as the Mau Mau Rebellion was beginning in central Kenya. His family sent him away to Ireland to complete his studies, and he returned in the late 1970s to take over farming operations on Lewa Downs.[11] In the early 1980s, Lewa Downs was in financial trouble because of the sharp decline in the Kenyan beef industry.[12] The Craig family was fortunate to be contacted by a wealthy benefactor named Anna Merz, who was interested in partnering with them to build a black rhino sanctuary. Born in England, Merz had spent much of the past twenty years in Ghana, where her then-husband, the Swiss industrialist Ernst Kuhn, ran operations. Partly because of the pressures on cattle-farming and partly out of genuine interest in conservation, in 1983 Ian Craig sectioned off a 2,000-hectare portion of Lewa Downs, converting it into Ngare Sergoi Rhino Sanctuary. Craig supplied the land and Merz supplied the cash; she contributed USD 750,000 of her 'family inheritance' to construct 2.5-metre electric fencing around the piece of land and to fund the capture of rhinoceros from other farm parcels.[13]

8 B. Berman and J. Lonsdale, *Unhappy Valley: Conflict in Kenya and Africa* (Vol. 1) (Oxford, James Currey, 1992), pp. 88–90.

9 See T. Kanogo, *Squatters and the Roots of Mau Mau* (London, James Currey, 1997). See also, D. Anderson, *Histories Of The Hanged: The Dirty War In Kenya And The End Of Empire* (New York, W.W. Norton, 2005).

10 M. van den Akker, *Monument of Nature? An Ethnography of the World Heritage of Mt. Kenya* (Leiden, African Studies Centre, 2016), p. 143.

11 See N. Breed, *From Ox Cart to Email: The Kenya Story of Delia Craig* (Sussex, Book Guild, 2011).

12 Fox, 'Images, Uncertainties and Success Stories', p. 61.

13 See B.A. Schreiber, 'Anna Merz', *Encyclopaedia Britannica* (2015). 'Obituary: Anna Merz', *The Telegraph* (27 May 2013).

Over the following few years, Ian Craig and his neighbours captured rhino throughout the Laikipia district and brought them to Ngare Sergoi, growing the reserve in area and in conservation importance. Anna Merz handled the international donor side of things, bringing in funding from, inter alia, the American Association of Zoo Keepers and other North American nature conservation lobbies.[14] By 1995, all of Lewa Downs had become a rhino and charismatic game sanctuary. Stripped of most of what had made it a cattle farm in the past, it now resembled the archetypical East African savanna so sought after by foreign tourists. A British charity, Lewa Wildlife Conservancy U.K., was founded in 1997 to manage the conservancy, which by then was made up of the three Craig family properties put into trust by his father in 1987, four properties owned by The Beale Farm Ltd and five properties owned by The Lewa Downs Estates Ltd.[15] The charity had foreign branches to help with fundraising, research and advocacy. The properties are no longer registered under the Craig name or owned by a charitable entity, but rather a series of companies and family trusts controlled by the Craig family, so it is unclear if land tenure under 'trusteeship' makes a significant difference on the ground.[16]

As funds began to come in, Lewa Wildlife Conservancy – which superseded Ngare Sergoi – began increasing its populations of black and white rhino, Grévy's zebra and other animals, not to mention the rest of the 'big five' species most useful for tourism operations. Lewa advertises ten hyper-luxury lodges on the conservancy itself or else on neighbouring Borana Conservancy, owned by the British/Kenyan Dyer family who came into the region as 'soldier-settlers' in the 1920s.[17] Some lodges, such as Borana, state that a portion of the hotel fees are a 'contribution to conservation'. Borana itself contributes 24 per cent of its nightly rate of USD 1,152 to this purpose.[18] Elewana Kifaru House, on Lewa, charges guests a USD 134 'conservancy fee' on top of the nightly rate of USD 1,513 per person.[19]

On the other hand, one could also rent Arijiju house nearby – which advertises no conservation goals apart from being located on a conservancy – for

---

14 Van den Akker, *Monument of Nature*, p. 144.

15 UKCHA #03470186: 'Memorandum of Association of the Lewa Wildlife Conservancy U.K.' – 1 November 1997.

16 Graham R. Fox, personal communication with Bernard C. Moore (email, 15 December 2022).

17 'The Early History of Borana', *Borana Conservancy Kenya*. https://www.boranaconservancy.com/history

18 '2022 Room Rates', *Borana Lodge* https://wetu.com/Resources//20259/borana_lodge_2022_rates.pdf

19 'Rates and Seasons, 2023', *Elewana Collection: Kifaru House* https://www.elewanacollection.com/kifaru-house-lewa/rates-and-seasons

USD 9,000 per night. And it is unclear what percentage of the USD 1,594 nightly fee for tent camping at Lewa Wilderness House (or USD 2,242 for a room) – a property owned and managed by Ian Craig's brother, William – goes towards conservation endeavours.[20] Contributions to conservation – while potentially meaningful – can also be quite vague and misleading. According to Graham Fox, who conducted anthropological research in Laikipia North:

> When lodges say that they direct a portion of revenues for conservation, this can mean a variety of things. It might mean using the security resources from the conservancy to help mitigate human – wildlife conflict in neighbouring African areas. It can mean allowing school groups or researchers to visit the land to learn and appreciate the value of conservation. It can also be as secondary as to include simply hiring people, as it is said that employing people in conservation-related enterprises (i.e., tourism) demonstrates the value of conservation to local communities.[21]

FIGURE 77 Ian Craig (left) speaks with United States Secretary of State John Kerry during the latter's visit to Kenya in May 2015. USAID Kenya remains one of the Northern Rangelands Trust's (NRT) largest benefactors
PHOTO: US STATE DEPARTMENT

20 William Craig is also the founder and CEO of Wilderness Air, an airplane charter company that offers trips to guests travelling to Lewa and NRT. https://www.airwilderness.com/our-team/

21 Graham R. Fox, personal communication with Bernard C. Moore (email, 15 December 2022).

Lewa Conservancy and its ancillary partners have directly targeted the hyper-exclusive luxury tourism market in East Africa, which is the source of most of its local revenue in Kenya. But at the end, it is a pitifully small proportion of the conservancy's global revenue. Lodging does not really matter much to Lewa's ledgers. During the 2019/2020 financial year, Lewa earned more than USD 8 million in revenue, only USD 731,000 of which came from lodging activities – most (USD 5.1 million) came through global tax-deductible donations.[22] A significant portion of the remainder came from other 'conservation fees', which included, inter alia, the fees, profits and donations from translocations and animal exchange programmes. Lewa's international charitable branches in the UK, USA, Canada and Europe helped to bring in massive donations from overseas benefactors, including zoos. Zoo Zürich, for example, donated more than USD 2.7 million to Lewa between 1998 and 2020, with a formal partnership being struck in 2019. The Lewa Savannah exhibit in Zürich is now home to fifteen species, including white rhino.[23] Lewa Wildlife

FIGURE 78 Lewa's savanna exhibit at the Zürich Zoo
PHOTO COURTESY OF ALBINFO: WIKIMEDIA COMMONS

22 Lewa Wildlife Conservancy, *Impact Report, 2019–2020*, p. 56.

23 'Willkommen in Lewa!', *Zoo Zürich* (3 June 2020).

Conservancy is also supported by financial donations from the San Diego Zoo in the USA, Al-Ain Zoo in Abu Dhabi and others, including animal exchange programmes.

Much of the financial support from zoos and other institutions is mobilised through Lewa's satellite NGO, Lewa Wildlife Conservancy USA. Between 2017 and 2021, the US branch raised more than USD 16.8 million for Lewa, transferring USD 3.3 million in fundraising income to its Kenyan parent organisation in 2021 alone.[24] By the end of 2021, Lewa's US investment endowment – into which a significant portion of these donations are placed – reached nearly USD 7.5 million. This is still petty compared to one of Lewa's largest financial benefactors, the US-based The Nature Conservancy, which by the end of fiscal year 2020 held more than USD 8.8 billion in assets, including more than USD 1.6 billion in endowment funds and nearly USD 4.5 billion in land assets and conservation easements.[25]

Lewa has been very successful in attracting donations and investment by framing its work as if all conservation in Kenya depends upon it. According to anthropologist Marlous van den Akker, who conducted research in central Kenya:

> A former employee clarified to me that Lewa Wildlife Conservancy organizes fundraising events in the United States and Britain, which rest entirely on the idea that Lewa is the sole successful private conservation initiative in the entire country, and that without it Kenya's wildlife would be doomed. He explained to me that Lewa needs huge amounts of money to cover their costs, and that in order to raise such amounts it continuously exaggerates its importance and influence. … A conservationist who requested anonymity, as he did not want to antagonize Lewa, said that the conservancy's main point of departure is making itself look bigger by making others look smaller. He described Ian Craig as a 'Rambo conservationist', i.e., someone who is only interested in sensation and spectacle. He said that the things that really matter for conservation, such as butterflies and beetles, were of no interest to Ian Craig, who only concerned himself with fancy cars, private airplanes and large charismatic mammals.[26]

24 US Department of the Treasury, Internal Revenue Service: 'Return of Organization Exempt from Income Tax: Lewa Wildlife Conservancy, USA' (Form 990) – December 2021.

25 US Department of the Treasury, Internal Revenue Service: 'Return of Organization Exempt from Income Tax: Nature Conservancy' (Form 990) – June 2021.

26 Van den Akker, *Monument of Nature*, pp. 146–147.

Lewa has positioned itself as a favoured recipient of private donations from the United States – via The Nature Conservancy and others – or else of governmental support from Danida (Danish government), AICS (Italy), USAID and the United States Department of the Interior. Its success in becoming donor-dependent led to allegations that the organisation had become a reckless spender of its flush cash reserves, and that tenders offered to builders, suppliers and contractors were grossly inflated, driving up general costs for other commercial and communal farmers.[27] Nevertheless, in 2013, the former colonial cattle ranch Lewa Wildlife Conservancy was officially added to an extended Mount Kenya UNESCO World Heritage Site, granting private nature conservation in the region global significance, further securing land tenure in the region.[28] Conservation activities were changing the fundamental economy of central Kenya and beyond.

•••

From the earliest days of conservation activity on Ian Craig's properties, he likely understood that success in local conservation – whether on ecological or economic grounds – depended on reshaping the rural economy in Laikipia County, the Mount Kenya region, and beyond. Craig had little interest in completely eliminating commercial or subsistence livestock production, but he wanted a gradual transition away from it as the primary source of livelihood. Coexistence with conservation could give way to conservation dominating the regional economy, and he found himself at the helm of these transitions. Together with other major landowners in the region, including the novelist Kuki Gallmann, he founded the Laikipia Wildlife Forum (LWF) in 1992 as a collective mouthpiece for (mostly White) private ranchers seeking to expand conservation tourism and wildlife utilisation.[29] According to authors Mbaria and Ogada, the first main agenda of the LWF was to petition for ranchers in Laikipia to be allocated quotas of zebra to cull for the sale of their skins overseas: 'this was an experimental partnership between the KWS [Kenya Wildlife Services] and the landowners aimed at earning regular profits from the significant wildlife populations on their holdings'.[30] The culls proved to be 'unsustainable', and

27 Ibid, p. 146.

28 Ibid, pp. 166–172. World Heritage Committee: Decision 37 COM 8B.9, 'Extensions of properties already inscribed on the World Heritage List: Mount Kenya National Park/Natural Forest' (37th Session, Phnom Penh, 2013). https://whc.unesco.org/en/decisions/5125

29 G.R. Fox, 'The 2017 Shooting of Kuki Gallmann and the Politics of Conservation in Northern Kenya', *African Studies Review*, 61, 2 (2018), pp. 210–236.

30 Mbaria and Ogada, *The Big Conservation Lie*, p. 41.

they were cancelled around 2001; the LWF moved away from seeing wildlife as a source of profits when dead to looking at it alive as a draw for tourism. Other forms of development – even the construction of high-voltage powerlines to connect poor communities to the electrical grid – were vehemently opposed and blocked by the LWF.[31]

Although the LWF has never explicitly opposed the development of pastoral agricultural pursuits by Black Kenyans, for it the development of the livestock sector could come only within wildlife management plans and an expansion of conservation infrastructure, such as game fencing. According to Graham Fox, the LWF provided the land-owning establishment in Laikipia region a mouthpiece through which to assert their interests without 'showing their faces' and attracting scrutiny: 'Though LWF does not have a governance mandate, it [granted] White landowners renewed political power by enabling them to negotiate with the government and their neighbours collectively'.[32] One may rightly praise LWF for being a major force in the growth of the conservation and tourism sector within central Kenyan communities, but it is not really a 'community conservation organisation', in that the 'communities' it has long represented are not subsistence communities but rather moneyed landowners.[33] In order to phrase the Kenyan conservation success story as a success of 'communities', African communities needed to play a role larger than simply as employees of White-owned safari lodges. To accomplish this, Ian Craig and his partners looked northwards.

It is important to remember that there are several legal and institutional structures available to those who pursue conservation initiatives in Kenya. The 'national park' (or other state/county-run protected area) is the most traditional and most extreme form of conservation initiatives, akin to the MacKenzian Orthodoxy referred to in chapter 1. Owned by the national government, managed by the Kenya Wildlife Services, and protected by parliamentary legislation, the long-term maintenance of a national park for pure tourism and research is maintained until legislation is changed; its use is strictly regulated. 'Reserves', on the other hand, are community lands owned by local county governments and managed in partnership with the KWS. They are gazetted, though the KWS must operate in consultation with local communities. Finally, the 'Conservancy' can be either communal land or privately owned land, managed by corporate bodies or communal institutions. Crucially, mixed- and multiple-land-use arrangements are allowed, and the

31 Ibid. p. 42.
32 Fox, 'The 2017 Shooting of Kuki Gallmann', p. 222.
33 Ibid. p. 223.

central government has limited influence on conservancy matters because participation is agreed upon between the landowners and the corporate body.[34] Like other postcolonial African nations, the trend in Kenya has been towards expansion of private and communal conservancies, perhaps at the expense of publicly owned national parks and reserves.[35]

This evolved through the formation of the Northern Rangelands Trust (NRT), increasing referred to as Kenya's 'new government in the north'.[36] Once most White ranchers had joined the conservation bandwagon through membership of the LWF, Ian Craig sought to get participation of communal lands and their residents in his conservation initiatives. There was interest from the Il Ngwesi and Namunyak communities in forming a conservancy, and they sought Lewa's advice about tapping into the tourism and conservation markets. Ian Craig and his backers realised that a partnership was more desirable in the long term. So, in 2004, the Northern Rangelands Trust (NRT) was founded as an 'umbrella organisation' for conservancies in Kenya, with a distinct focus on communal conservancies to the north of Lewa, extending from Laikipia County nearly as far as the border with Ethiopia, with additional western and coastal conservancies.[37] By 2023, NRT had forty three member conservancies, four affiliate private commercial conservancies (Lewa, Borana, Ol Pejeta and Loisaba), and a number of affiliate national parks and national reserves. The scale of NRT is impressive, stretching across 6.2 million hectares of semi-arid rangeland, populated by approximately 465,000 people. According to the current NRT CEO, Tom Lalampaa, 'Thanks to the support of our donor partners, in particular USAID, NRT can assist communities living in these unique landscapes in the protection of their land, wildlife, livestock and culture for themselves and the next generation'.[38]

Despite being only an 'affiliate member' of NRT, Lewa has been undeniably the most important player in NRT's operations and plans for the future. NRT's offices are shared with Lewa, and from the start Ian Craig was appointed as their Chief of Conservation and Development, allowing him and his disciples disproportionate influence in structuring the NRT as a communal wing of

34 See Kenya Wildlife Conservancies Association, 'Conservancies, National Parks, and Reserves' (2021). https://saruni.com/wp-content/uploads/2021/04/KWCA_Differences-between-Kenyas-Conservancies-Parks-Reserves.pdf

35 L. Kalvelage et al., 'Territorialising Conservation: Community-Based Approaches in Kenya and Namibia', *Conservation & Society*, 19, no. 4 (2021), pp. 282-293.

36 See B. Bersaglio and F. Cleaver, 'Green Grab by *Bricolage*: The Institutional Workings of Community Conservancies in Kenya', *Conservation and Society*, 16, 4 (2018), pp. 467–480.

37 See Northern Rangelands Trust, 'About NRT' https://www.nrt-kenya.org/about (Accessed 1 May 2024).

38 Quoted in Northern Rangelands Trust, 'Celebrating Community Led Conservation' https://www.nrt-kenya.org/the-story-of-nrt (Accessed 30 June 2023).

broader goals to develop central and northern Kenya as a wildlife conservation economy (he is now Chief Officer of Partnership, Growth and Innovation).[39] The initial goals of the NRT were to build a system through which communal conservancies could draw on Lewa's expertise and international connections to stimulate investment in and income from wildlife tourism. This investment would be contingent upon community land-use plans that would ensure that livestock production would not come into conflict with wildlife utilisation. Community hunting and poaching would be curtailed – under the NRT, wild animals were valuable only when alive, and livestock was permitted only when it did not interfere with wild animals or tourists.

In the context of localised poverty and drought, nature conservation is a difficult proposition, because its purported benefits are pushed to the future at the cost of short-term sacrifices.[40] Local buy-in would be necessary if the NRT was to grow. Employment as hotel staff and guides was one step towards this goal, but northern Kenya is a pastoral society, so the management targeted cattle. In 2006, the NRT founded the Livestock to Market Programme, a partnership between Lewa, Borana and Ol Pejeta private conservancies and the for-profit branch of the Northern Rangelands Trust, NRT-Trading.[41]

The programme is now called GrazingWORKS, and its fundamental goal, according to its former director, is to 'provide incentives for pastoralists to manage cattle for ready markets and over time reduce herd sizes to avoid losses due to drought [and to] gain conservation leverage by linking market access to conservation outcomes'.[42] If a conservancy consistently follows the agreements of NRT membership, NRT-Trading will allow GrazingWORKS to hold a local cattle market within the community, where the initiative will buy cattle from pastoralists at slightly above local market rates for animals of the same size. On receiving payment for the animals, the pastoralist then gives approximately USD 9 in fees to the conservancy and NRT-Trading gives approximately USD 20.

The purchased cattle are then transferred from the northern African-occupied areas to the private settler conservancies in the more fertile grasslands in the Laikipia area, where grazing is rented to 'finish' (that is, fatten) the animals to prepare for Nairobi.[43] Borana Conservancy, located directly to

---

39 Van den Akker, *Monument of Nature*, p. 149. Northern Rangelands Trust, 'Executive Team', https://www.nrt-kenya.org/executive-team

40 Fox, 'The 2017 Shooting of Kuki Gallmann', p. 221.

41 See P. Ekodere, NRT Livestock Director, 'NRT Livestock to Market', NRT-T Presentation at ECHO Symposium in Nanyuki (6 March 2018).

42 Ibid.

43 J. Dempsey and P. Bigger, 'Intimate Mediations of For-Profit Conservation Finance: Waste, Improvement, and Accumulation', *Antipode*, 51, 2 (2019), pp. 517–538.

the west of Lewa on the southern fringes of the NRT, is one of the destinations for the NRT cattle programme. Owner Michael Dyer sees it as a win-win, in that the pastoralists receive a higher price in Nairobi for a fatter beast, and the abattoirs get a more uniform selection of animals depending on what the meat-dealer needs, and Borana keeps twenty per cent of the sale price as a 'grazing management fee'.[44]

Even more of the GrazingWORKS cattle are brought to Ol Pejeta Conservancy, the farthest-south affiliate of NRT, which measures about 36,000 hectares in size. Between the 1940s and 1960s, Ol Pejeta was a fairly successful cattle ranch owned by the prominent British settler Hugh Cholmondeley, 3rd Baron Delamere; it was later sold to the extravagant Saudi businessman and arms dealer, Adnan Khashoggi, during the 1970s.[45] Today, the land is owned by Flora and Fauna International, whose most important patron is Prince William, Duke of Cambridge.[46] Meanwhile, the business operations are run by Ol Pejeta Ranching Ltd, a subsidiary of the Ol Pejeta Conservancy Ltd, described as a not-for-profit company.

Khashoggi's former mansion has been turned into an expensive hotel, and the conservancy has built a number of campsites. Freehold landowners adjacent to Ol Pejeta have capitalised on the presence of game, transforming their private property into the Mount Kenya Wildlife Estate, a hyper-luxury private, homeowners' association targeting foreign retirees and high-earning digital nomads to migrate to the region, along with their cooks, housekeepers and 'ayahs'/nannies.[47] Homeowners have a private gate to the conservancy.

Through their purchase of GrazingWORKS' cattle, Ol Pejeta argues that 'wildlife/livestock integration can not only work, but it is actually beneficial to the grasslands'. They also employ 120 cattle herders to maintain the cattle as they are fattened for market.[48] Ol Pejeta claims that this revenue is used to maintain their conservation initiatives, which include their black rhinos.

---

44 'All for Conservation: How Community Focused Conservation is Saving the Rhino in Kenya', *Wild Ark* https://wildark.org/journals/protecting-habitat-by-supporting-people-boranas-conservation-success-story/ (Accessed 30 June 2023).

45 It also hosted the founding meeting of the 'Safari Club', a covert alliance of the CIA and Middle Eastern intelligence agencies to run clandestine operations in central and southern Africa during the 1970s, See https://en.wikipedia.org/wiki/Safari_Club

46 Formerly called the Society for the Preservation of the Wild Fauna of the Empire. 'About Us: History', *Ol Pejeta Conservancy* https://www.olpejetaconservancy.org/about-us/our-story/history/ (Accessed 1 June 2022).

47 'Home Purchase FAQs', *Mount Kenya Wildlife Estate at Ol Pejeta* https://mountkenyawe.com/home-purchase-faqs/ (Accessed 30 June 2023).

48 'Conservation', *Ol Pejeta Beef* https://www.olpejetabeef.com/conservation (Accessed 1 June 2022). Ol Pejeta claimed to fatten and market 1,500 cattle each year for GrazingWORKS.

FIGURE 79 NRT livestock grazing on Ol Pejeta together with charismatic megafauna
PHOTO COURTESY OF OL PEJETA BEEF, USED WITH PERMISSION

NRT-Trading and GrazingWORKS operate at a juncture between charity and corporation, epitomised, say researchers Dempsey and Bigger, through the term 'for-profit conservation finance'. This implies that donors structure their contributions as a hybrid of donations, loans and investments.[49] For example, in 2015, NRT-Trading began discussions with NatureVest – the financial investment branch of the US-based NGO behemoth The Nature Conservancy – to help facilitate investment in their GrazingWORKS programme. NatureVest markets itself as putting conservation NGOs in touch with 'impact investors' – since 2014, they have mediated more than USD 1.3 billion in impact investment, including wetlands conservation and carbon sequestration.[50] NatureVest assisted NRT-Trading in obtaining a USD 7 million contribution from an unnamed wealthy donor, half of which would be treated as a 1 per cent interest long-term loan.[51] Part of for-profit conservation finance involves the 'donation' being conceived of as a riskier or else lower rate-of-return investment than could reasonably be obtained through alternative investment venues, and the effect of this is that even charitable or non-profit conservation entities must still structure their operations as businesses in order to obtain funding.

49 Dempsey and Bigger, 'Intimate Mediations'.
50 See NatureVest, *2021 Impact Report* (Arlington, The Nature Conservancy, 2021).
51 Dempsey and Bigger, 'Intimate Mediations'.

GrazingWORKS found itself at a difficult crossroads, however. The ultimate goal of the entire livestock-to-market programme was not really to make money, but to make felt to Kenyan pastoralists the benefits of working within the NRT conservation schemes. After all, only cattle from conservancies that met conservation goals would be taken in by the programme. At the same time, the GrazingWORKS programme was supposed to bring in enough revenue from cattle sales to service the long-term loan. This would be an outstandingly difficult task, because even the massive Ol Pejeta abattoir could not match the scale of Nairobi meat markets. Furthermore, according to NRT-Trading staff, up to twenty per cent of the cattle purchased from pastoralists were smaller than what was anticipated by the farms which were set to fatten the animals. In order to meet what beef markets dictated, more grazing was required to turn a profit (compared to initial estimates), especially when transport, auction and slaughter costs were included.[52] Despite not being very profitable, perhaps the disciplinary element – in a Foucauldian sense – of GrazingWORKS (and its similar craft programme for women, BeadWORKS) could still be successful, in that the organisation could condition the community to conservancy operations and effect behavioural changes to enable pastoralism to coexist with wildlife. This, according to Dempsey and Bigger, signals at least a tentative departure from traditional 'fortress conservation':

> Conservationists cannot put up the fences and kick out the pastoralists: they not only need to avoid the negative attention that comes with conservation dispossessions, but they also view pastoralism as compatible with wildlife, with some modifications. That is, there is more going on here than these processes of commodification, measurement and fungibility.[53]

To avoid accusations of fortress conservation, the NRT had to deliberately build a loss-making enterprise within its conservation operations. Because the vast majority of the 465,000 residents of NRT conservancies would never benefit directly from conservation and tourism employment, the NRT needed to find ways to compensate pastoralists for wildlife competing with the cattle for grass. GrazingWORKS was supposed to accomplish this, and the White-owned conservancies to the south got involved for this reason. The success of the profitable tourism operations on Lewa, Borana, Loisaba and Ol Pejeta depends on the NRT communal conservancies in many ways. In addition to supplementing

52 Ibid, pp. 530–531.

53 Ibid, p. 533.

the genetic pool of valuable game species across the lands, these vast northern territories provide additional lands on which high-paying tourists can ride in a Land Cruiser wearing a pith helmet chasing after game.[54]

This was a precarious balance, which collapsed in 2017 when a severe drought struck much of central and northern Kenya, causing many pastoralists to move their livestock to areas ordinarily reserved for wildlife. In addition, pastoralists across Kenya were selling animals to abattoirs in greater numbers than before, causing beef prices to drop in Nairobi. The business model of GrazingWORKS – to buy quotas of livestock at above market rates – was not sustainable in the context of drought.[55] Furthermore, many of the private conservancies that had been fattening up the NRT's cattle were already regarded with deep suspicion by pastoralists on group resettlement ranches in Laikipia County, because the cattle programme had long proved that the owners' reluctance to share grazing was not really about the wildlife needing all the grass.[56]

As a result, farm invasions ensued. Pastoralists from group ranches and northern communal areas moved their livestock across boundaries into wildlife areas and private conservancies in search of grazing. In the face of drought, poverty and lack of grazing and land, people can become desperate and violent. The most publicised incident of the farm invasions was the April 2017 shooting of the White novelist and landowner/conservationist Kuki Gallmann, which drew concern from the global conservation community and villainised pastoralism in the region, writ large.[57]

The events in 2017 accelerated what was already a trend within the NRT and NRT-affiliated organisations: the militarisation of conservation. The NRT employs more than 1,000 armed 'conservancy scouts' across member communities, most of whom are trained by the NRT at the Kenya Wildlife Service's Law Enforcement Academy. Of these, the NRT's seventy 'rapid response rangers' are equipped with paramilitary-grade weaponry, to respond to poaching threats and other security concerns. Some of this latter group are trained by and work in cooperation with the Kenyan company 51 Degrees Ltd, owned by Ian Craig's son Batian and headquartered on Lewa Wildlife Conservancy. The operations run by 51 Degrees are led by Captain Barry Cork, formerly of the Royal Gurkha Rifles, and other ex-British military officers. They are increasingly being

54 As is shown in chapter 13, GrazingWORKS was also tied to NRT's carbon sequestration schemes.

55 Ibid, p. 555.

56 Fox, 'Images, Uncertainties and Success Stories', p. 168. There were also views by local Black pastoralists that group ranches (owned by their communities) were being specifically excluded from GrazingWORKS programmes.

57 See Fox, 'The 2017 Shooting of Kuki Gallmann', pp. 230–231.

supplemented with paramilitary satellite and tracking software, such as is produced by artificial intelligence companies EarthRanger and KineticSix.[58]

NRT rangers are generally better paid and equipped than their Kenyan state counterparts. Although they are supposed to supplement state efforts as police reservists, NRT and conservancy managers hold disproportionate sway in directing these individuals towards security endeavours that the conservationists deem pressing.[59] Better pay has not necessarily led to a more disciplined security force, however. NRT-equipped rangers dispatched to pastoralist violence have been accused of supporting herders from their own communities at the expense of those from others, leading to allegations of many extrajudicial killings.[60] This was recently the subject of a large exposé by the Oakland Institute (a human rights think tank), which claimed that since the NRT's founding, seventy six people were killed in interethnic clashes in Biliqo Bulesa Conservancy, 'with support from the NRT'.[61] In response, as one of the NRT's largest funders, The Nature Conservancy commissioned a due diligence report into the incidents, which categorically denied most of the Oakland Institute's claims. It admitted, though, given that the security and training of rangers is provided by 51 Degrees and headquartered on Lewa, it was understandable that there would be a 'perception that NRT remains dominated by the Craig family'.[62]

By the middle of 2017, rains would fall again and calm the situation somewhat. But it was not just a temporary lack of rainfall that had precipitated these conflicts. The vast majority of African pastoralists ultimately have insufficient access to land in central and northern Kenya, a situation that is exacerbated as more land is reserved for wildlife.[63] The GrazingWORKS programme of buying livestock above market value – while appearing progressive in international

58 See 'Exercise ASKARI Spear', *Gurkha Brigade Association* (30 April 2020). https://www.gurkhabde.com/exercise-askari-spear-2020/ (Accessed 1 June 2023).

59 C. Schetter, K. Mkutu and M. Müller-Koné, 'Frontier NGO s: Conservancies, Control and Violence in Northern Kenya', *World Development* 151 (2022), pp. 1–12.

60 Ibid.

61 *Stealth Game: 'Community' Conservancies Devastate Land and Lives in Northern Kenya* (The Oakland Institute, 2021), p. 5

62 K. Sena, *Due Diligence Report Concerning the Report 'Stealth Game' Published by the Oakland Institute* (The Nature Conservancy, 2022), p. 23. NRT leadership argued that while they considered moving its headquarters off Lewa to another conservancy, to do so could 'aggravate tribal conflicts'. Nevertheless, just before the report was published, the NRT board approved a motion to strengthen relations with Lewa through, inter alia, increased security investments. See 'Proposed Strengthening of NRT-Lewa Partnership', *NRT Board Paper* (February 2021).

63 Some local politicians, like Matthew Lempurkel (MP for Laikipia North), spoke with vitriol, urging some pastoralists to engage in farm invasions for this reason.

conservation networks – ultimately failed to address the root problem with private conservation initiatives in Kenya, which is that these initiatives build on – and perhaps even depend upon – colonial legacies of land dispossession.

• • •

Ian Craig once remarked that he appreciated Sera Conservancy in Kenya (an NRT member) because 'in a nutshell, it's a great habitat. It is remote and away from civilisation, and, most importantly, the land belongs to just one community, so it's easier to negotiate with them. This also means that security is easier to guarantee'.[64] Thus, Craig facilitated turning Sera Conservancy into a rhino sanctuary. Southern Namibia epitomises this purported goal of Craig's to an extreme. When asked what attracted him to southern Namibia, he replied that what was interesting and important about the region was its remoteness. 'I don't want to use the term *blank slate* to describe those lands, but it's something like that', Craig said. 'It is a remote and unchallenged part of wild Africa with little competition for land use options. When agriculture and other competing land claims are less of a factor, then conservation rises up the ladder'.[65]

Ian Craig agreed to pour a significant amount of his personal money into purchasing farmland in Namibia, in part because it appears that he saw in the region what many likely wished existed in Kenya: a *terra nullius* ripe for 'conservationist' intervention. It is undeniable that Craig and his disciples, followers and protégés have been quite successful in developing conservation as a meaningful share of the economy in central and northern Kenya – with Lewa clearly as the star-child of this conservation family. But it has been far from easy, and the model remains precarious. It is in their interest that conservation be constantly described as under threat from internal and external forces, partly because it stimulates donations from outside by tapping into the fear of species extinction. And it perhaps helps that one of the largest security contractors is owned by the Craig family. Nevertheless, these threats, broadly defined, can be very real. Poaching is a profitable industry for those who can smuggle ivory or rhino horn out of the country, and when drought comes, wildlife can lose out to competition with domesticated livestock. Ultimately, the vast majority of African pastoralists in Laikipia County and the NRT lands will never find secure employment within the tourism or conservation sectors, so a major portion of 'conservation' operations are concerned with finding ways to get local buy-in from pastoralists through programmes like GrazingWORKS or

64 See S. Butler, 'Portfolio: Ian Craig', *Saruni* https://saruni.com/portfolio/ian-craig/

65 Ian Craig, interview with Bernard C. Moore (telephone, 18 October 2022).

ensure – via increased fencing and armed security – that those who do not buy in cannot negatively affect conservation operations.

Absolutely *none* of this is necessary in southern Namibia. Craig, Morkel, Gilbertson, Barthorp, Pawel and their teams can all benefit from the long, winding and sometimes heartbreaking historical events that shaped Namibian land tenure (described in Part Two of this book). Namibia has a secure system of private land tenure, and in southern Karasburg District, the nearest proclaimed African communal areas are more than 50 kilometres away, and even these are sparsely populated and poor. Grand profit-oriented conservation strategies in southern Namibia do not need much local buy-in; they do not need much community outreach. Unlike one of Craig's favourite Kenyan conservancies, Sera, Oana's lands are not even owned by a *community*. These lands are private and owned by only a few individuals and/or companies, of which Craig himself is a major shareholder. KumKum and Pelladrift were probably never really purchased for rhinos alone. It was more likely about acquiring a lot of cheap land in a region where agriculture was struggling, where there were few communal lands nearby, and where private land tenure was secure. The post-apartheid realities of southern Namibia may have made it the best locale to put into place the broader strategies that have been developed in East Africa.

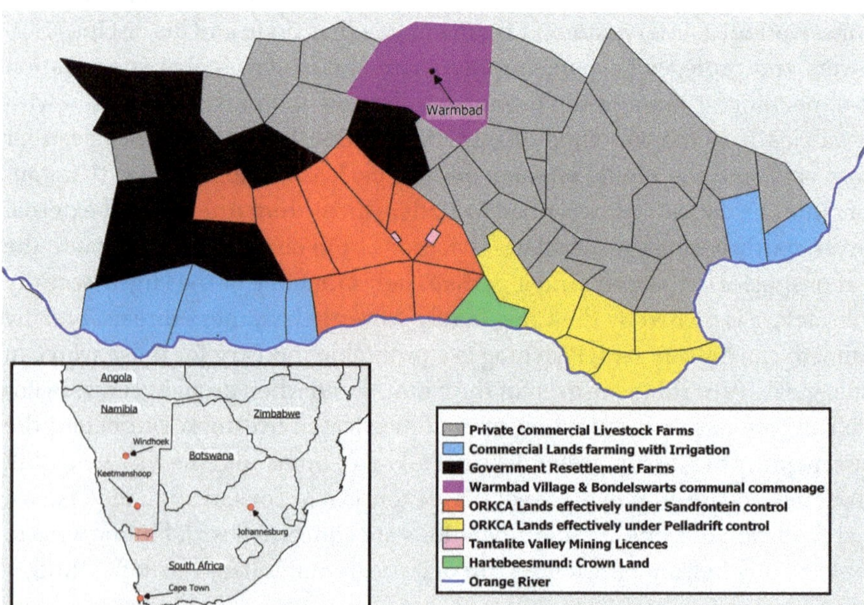

FIGURE 80 Modern-day land tenure in southern Karasburg district
CARTOGRAPHY BY B.C. MOORE

Although local buy-in by the Nama is not really needed, in fairness we should note that Oana did try to make a charitable wing of its conservation operations. The Oana Community Development Project (OCDP) was founded in 2018 by Ian Craig, his wife Jane, and Pete and Estelle Morkel, with the goal 'to provide for and assist in the development and upliftment of the community of Oana, through support to education, healthcare, provisions of clean water, job creation and enterprise through conservation and all aspects ancillary thereto'.[66] The charity is fairly small, though it has facilitated a few projects. In late 2020, the charity received EUR 2,960 from the German Embassy in Windhoek to allow the OCDP to develop a 'soup kitchen project' in Warmbad.[67] Along similar lines, the OCDP partnered with The Dove Trust, a Christian charity in the UK, to do small-scale fundraising in Britain, as on crowdfunding websites like JustGiving.com. During the 2020–21 financial year, The Dove Trust managed to raise nearly GBP 15,000 for Oana,[68] rising to more than GBP 25,000 the following year.[69] One would assume that these donations were for the Warmbad soup kitchen project or other community initiatives, though this has not been made clear.

Indeed, some of Oana/OCDP's charity work appears much less charitable. One of the fundraising schemes coordinated through The Dove Trust was titled 'Protect Oana's Leopards' – ostensibly a conservation goal that could in theory need charitable donations. The JustGiving advert stated that donations would 'allow the company to keep their valuable wildlife trackers, Lisias and Simon, employed'.[70] To the layperson's eyes, this appears to be quite an unusual accounting practice. The trackers were employed by the for-profit entities KumKum Adventure Tourism (Pty) Ltd and Pelladrift (Edms) Bpk, which actually ran the conservation and pay-to-volunteer operations. Thus, the charity was not their employer. The strategy nevertheless worked. The OCDP received

66 It is not clear what is meant by 'the community of Oana'. BIPA File 21/2018/0347: Certificate of Incorporation of a Company Not Having a Share Capital (CM-3) – 10 April 2018.

67 'German Embassy Helps Soup Kitchen Project in the South', *German Embassy Windhoek* (24 November 2020).

68 Charity Commission for England and Wales (hereafter, CCEW), Charity #1001356: 'The Dove Trust: Report of the Trustees and Unaudited Financial Statements for the Year Ended 5 April 2021'.

69 CCEW #1001356: 'The Dove Trust: Report of the Trustees and Unaudited Financial Statements for the Year Ended 5 April 2022'. This donation was technically not given to OCDP but to ORKCA (see chapter 12): BIPA File 21/2021/0913: 'Articles of Association of a Company not Having Share Capital Not Adopting Schedule 1' (CM-44B) – 16 June 2021.

70 'Protect Oana's Leopards', *JustGiving* https://www.justgiving.com/campaign/SupportOurRangers [Accessed 1 June 2022).

GBP 4,332 of their stated GBP 4,000 goal; tax-deductible donations to one charitably entity allowed another for-profit company to pay its employees.

The emotive nature of wildlife protection makes it easy for associated business operations framed as charitable endeavours to solicit donations. Indeed, appeals to emotion and altruism have long been understood as crucial to conservation volunteering, and both fundraisers and conservation companies are aware of this.[71] Whether it is African employees paid by donations or gap-year youths paying to volunteer for the company, both involve cheap or free labour to transform land and landscape into something conducive for the image of conservation. Lisias and Simon appear to no longer be employed by Oana, and they have been removed from the website biographies.

Very little money actually passes from the OCDP charity to the local communities in Warmbad or Karasburg, apart from the short-lived soup kitchen, it appears. The 'Ethos' section of the OCDP website states that the charity operates

> within the paradigms of Nama cultural traditions and laws to enable 100% community buy-in, equitable benefit sharing and cohesive management. [As such, the charity] is committed to conserving and fostering resilience in wild spaces for the benefit of indigenous communities and wildlife in southern Namibia. We create spaces for the restoration and growth of the great Namaqualand's socio-ecological system.[72]

We are not entirely sure what any of this means, especially given that there are no representatives of the Karasburg/Warmbad Nama community on the board of the charity at all.[73] Perhaps 'the community of Oana' as described in the charity's articles of association does not actually include local Nama at all. Regardless, it appears that the charitable branch of Oana may not be operating as a bona fide charity but rather as a marketing avenue to portray the grander and/or profit-oriented goals of the company as community-involved or community-driven.

---

71 The social science literature on volunteering and 'voluntourism' is extensive, and it is not our intention to survey this completely. As a start, see: M. Mostafanezhad, *Volunteer Tourism: Popular Humanitarianism in Neoliberal Times* (Burlington, Ashgate, 2014), pp. 40–42. S. Broad and J. Jenkins, 'Gibbons in their Midst? Conservation Volunteers' Motivations at the Gibbon Rehabilitation Project, Phuket, Thailand', in K.D. Lyons and S. Wearing (eds.) *Journeys of Discovery in Volunteer Tourism: International Case Study Perspectives* (Wallingford, CABI, 2008), pp. 72–85.

72 Oana Community Foundation, 'Our Ethos' https://oanacommunityfoundation.com/ethos (Accessed 1 June 2022).

73 BIPA File 21/2018/0347: 'Contents of Register of Directors, Auditors and Officers' (CM-29) – 5 April 2018.

When asked in October 2022 if he planned to establish a similar programme to GrazingWORKS in southern Namibia, Ian Craig responded by saying that 'the short answer is yes', but he was unsure what this would look like.[74] Pete Morkel was a little more straightforward. On the question of whether African pastoralism and wildlife conservation could go hand in hand, as was described in Kenya, he said: 'Well, if Ian Craig's Maasai were here, then maybe that would make sense for KumKum, but that's not the situation here'.[75] As in Kenya, Ian Craig and his Oana team viewed conservation as the rising tide that lifts all boats, but without large numbers of pastoralists in the vicinity from whom buy-in was needed, community outreach was viewed as less important.[76] Apart from a bit of money or meat donated to a soup kitchen in Warmbad, community benefits were simply just assumed as the natural outcome of conservation endeavours.

The situation on neighbouring Sandfontein was not much different. In late 2021, when we asked Agenbach's successor, the local manager Adriaan Mulder what kind of community relations and outreach Sandfontein had done since Sean Gilbertson bought the properties fifteen years earlier, he said that the lodge had plans to 'work with locals in Warmbad', but that they had not yet figured out anything concrete.[77] More than a year later, Sean Gilbertson finally clarified that the lodge had plans to donate twenty Raspberry Pi computers to the primary school at Warmbad – which would cost around ZAR 46,000, or roughly the equivalent of one single night of full booking at the lodge.[78]

Unlike the charitable foundations and outreach programmes in Kenya that actually have to obtain local buy-in, in southern Namibia it is perfectly possible for all these charitable and community development operations to be

74 Ian Craig, interview with Bernard C. Moore (telephone, 18 October 2022).

75 Pete Morkel, interview with Bernard C. Moore (telephone, 19 October 2022).

76 In early 2024, certain Oana/ORKCA board members held symbolic meetings with Nama leaders (though not the river-folk). Despite a professed desire to 'foster genuine partnerships', it appears that very little concrete action has been taken or decisions made concerning the crucial issues at stake, namely, land and livelihoods. See ORKCA Instagram profile post (29 March 2024).

77 Adriaan Mulder, interview with Bernard C. Moore and Luregn Lenggenhager (Homsrivier farm, 12 November 2021).

78 Sean T. Gilbertson to Bernard C. Moore (7 January 2023), letter in the authors' possession. Gilbertson stated that Sandfontein had sponsored 'some years ago' the construction of a runway near Warmbad, but 'for reasons not clear to us', the community leaders ceased construction at ninety per cent completion. Gilbertson felt that if the airfield were located on townlands, then perhaps customs clearance for his guests could occur there rather than at Keetmanshoop, and a 'local entrepreneur' could set up business selling fuel. These community development goals are obviously deeply connected to the smooth running of his lodge rather than a local-led initiative.

hollow shells. In many ways, the OCDP *is* a hollow shell simply intended to boost the image of Oana as a whole. These properties were purchased in part because of Ian Craig's presumed faith in their existence as *terra nullius* – without large numbers of African residents nearby, they would not need to have local buy-in for conservation. Where the NRT needed one thousand armed and trained game guards to secure their conservation projects in Kenya, remoteness alone would accomplish the same in Namibia. Secure land tenure, along with Craig's and Gilbertson's deep pockets, would maintain this remoteness, and for the equally deep-pocketed tourists coming to Sandfontein and other future projects, the area's distance from centres of power would be a draw. In Adriaan Mulder's words, 'remoteness is what people seek'.[79]

Of course, this land may be remote, but it is not remotely like what Ian Craig, Sean Gilbertson or their employees and disciples believe it to be. As prior chapters have shown, for a long time this land was very successful commercial farmland and hunting grounds, and for centuries before that it was the ancestral land of the Bondelswarts, sustaining their pastoral way of life. It never has been and never will be *terra nullius* or a 'blank slate'. To this day, the Bassons, their relatives and other river-folk continue to farm and graze along the banks of the Orange River, constantly trying to figure out how to reconnect their current restricted areas with the broader Bondelswarts communal lands. For Willem Basson, rewilding involves the original inhabitants of these lands. In this sense, he has read the whole of Wikar's travelogue, seeing the Indigenous residents that fill the pages. Yet it seems that Morkel and his team can see only the wildlife that H.J. Wikar described more than two centuries ago.

The presence of the Bassons perhaps reveals that, despite the land-tenure opportunities that colonial genocide and apartheid removals offered White settlers, these lands are not fully depopulated. Even though this region does not have the density of pastoralists that Laikipia County has, conservationist projects here will struggle to succeed without local buy-in. Yet, in this post-apartheid landscape, these 'conservationists' rarely come to the table to meet the river-folk face-to-face.

We return to the struggles of the Bassons later, but first we must consider that the movement of animals in the days of Wikar meant crossing the Orange River. Therefore, any 'rewilding' project that purports to recreate this landscape must ultimately build it on both sides of the river.

---

79 Adriaan Mulder, interview with Bernard C. Moore and Luregn Lenggenhager (Homsrivier farm, 12 November 2021).

CHAPTER 12

Passports for Animals

'What we're really doing is providing *space*', said Ian Craig about his vision of conservation in southern Namibia. 'It's necessary to give wildlife the space to move where they want, even if that's over the Orange River'.[1] Part of what attracted Ian Craig to southern Namibia in the first place was the fact that large quantities of cheap private land could be acquired, allowing him and his team to 'rewild' the landscape as though it were a *tabula rasa*. In partnership with Sean Gilbertson, the Oana team on KumKum was able to claim more than 150,000 hectares of Namibian property as part of their broad rewilding project. Sean was happy to let Craig take the wheel for their collective conservation operations,[2] though as we will see in chapter 14, Gilbertson would continue to hold control over his properties and how they were used.

In September 2021, Ian Craig's team at Oana – along with various lawyers, professionals and conservationists in Namibia and abroad – registered a Namibian organisation called the Orange River-Karoo Conservation Area (ORKCA), 'to establish a community/private sector conservation model with the aim of expanding the ecological connectivity between the different land units to be acquired by the company'.[3] ORKCA would become the non-profit management agency coordinating the 'conservation' efforts of its constituent members, which at the time of founding included only the properties owned by Sean Gilbertson's Sandfontein Lodge and Nature Reserve (Pty) Ltd and Ian Craig's Pelladrift (Edms) Bpk. ORKCA would also be tasked with investigating expanding the landholdings of the partners, signing custodianship agreements with other neighbouring properties (such as Pete Morkel's farms), and envisioning new models to operate and finance a large-scale conservation project in southern Africa.

As it relates to most of the properties that ORKCA claims under its custodianship umbrella, the elephant in the room is that an international border sits right in their backyard. As the crow flies, the South African village Witbank, with more than seventy-five households, is only nine kilometres away from the

1 Ian Craig, interview with Bernard C. Moore (telephone, 18 October 2022).
2 Sean T. Gilbertson to Bernard C. Moore (7 January 2023), letter in the authors' possession.
3 BIPA File 21/2021/0913: Memorandum of Association of a Company Not Having Share Capital (CM-4) – 15 August 2021. BIPA File 21/2021/0913: Application for Certificate to Commence Business (CM-46) – 3 September 2021.

FIGURE 81 The road descends to the Orange River as one approaches the village of Witbank, on the South African side. Homsrivier Farm is visible in the distance on the Namibian side.
PHOTO: B.C. MOORE, 2022

old mission station on Gilbertson's Homsrivier Farm. Pella town, with more than eight hundred households, sits only seven kilometres from Ian Craig's farm, Pelladrift. Klein Pella date farm, allegedly the largest in the southern hemisphere, is only ten kilometres from Pete Morkel's house on Pelgrimsrust. Finally, the eighth-largest zinc mine in the world, Gamsberg (owned by Vedanta Resources), lies less than fifty kilometres from ORKCA's headquarters on KumKum, with some prospecting sites even closer. Based on their definition of rewilding, ORKCA's conservation has to consider the omnipresence of South Africa in the ecology and economy of southern Namibia, and ORKCA has leaned into this, expressly describing its vision as one of 'transfrontier conservation'.[4]

This chapter examines the ways in which ORKCA views its relationship with South Africa and how its future, in part, depends on cooperation with South African conservation organisations. It elaborates on the intellectual heritage of 'rewilding' as an ecological concept and of transfrontier conversation as a programme to create protected areas in southern Africa and beyond. We examine

4 'Mission', *Orange River-Karoo Conservation Area* https://orkca.org/about/

the Peace Parks Foundation (PPF), noting how its relationship with the centres of apartheid power and the heads of the largest polluting companies on Earth conflicts with its purported goal of protecting nature through planning and building Transfrontier Conservation Areas (TFCAs) across southern Africa. Finally, we show that ORKCA's view – that, in Gilbertson's words, there is 'no need for artificial human constructs in a natural world'[5] – reveals how private nature conservationists conceive of a borderless world as being borderless for wildlife only. Ultimately, as this chapter shows, the private transfrontier conservation zone that ORKCA is planning coincides with the historical reality that Black Namibians and South Africans have lived on and traversed this 'frontier' for generations.

•••

January in Witbank is a most unpleasant experience. In a good year, the village and its neighbouring farm plots are lucky to get even a paltry one hundred

FIGURE 82 Remnants of irrigation furrows built in the 1940s by 'Coloured' ex-soldiers at Witbank, destroyed by the massive Orange River floods of 1988
PHOTO: B.C. MOORE, 2022

5 Sean T. Gilbertson to Bernard C. Moore (7 January 2023), letter in the authors' possession.

millimetres of rainfall, though most of that does not come until February. When we greeted Catherine Cloete, a community leader in Witbank village in her late seventies, we were grateful to get to sit in the shade of the veranda in front of her small house: it was only 10:30am and the thermometer was already reading 43°C. As we heard her story about the history of the village and its people – and eventually followed it up with additional archival research – it became clear to us why more than seventy five families chose to make this seemingly inhospitable place their home.

'My *oupa-groeitjie* [great-grandfather] came here from Bondelswarts country, in Namibia, back when none of these farms were fenced. They were fleeing the war with the colonisers'.[6] She served us the coldest glasses of warm water we ever tasted, since there was no way that ice could have survived in anyone's cooler-box in this heat. In a place like Witbank, even warm water is refreshing. 'Some people went to the Catholic church at Pella to live permanently, but my ancestors came here because they had a lot of goats, and there was open land for us and the river nearby'.

As mentioned in chapter 4, not long after the Bondelswart Nama refugees crossed the Orange River, fleeing the genocidal violence of the Germans, these unsurveyed state lands along the South African side of the river were prepared for sale to private (White) buyers. In 1912, the farms Witbank and Garganab, as well as the two large islands in the Orange River (Marten and Krapohl Islands), were sold to the Oudtshoorn Land Development Co. Ltd., a land speculation syndicate controlled by Charles Weidner.[7] Prior to this, many of these former refugees from Namibia had been able to obtain grazing licences on these properties, but the German government urged its South African counterparts to stop this practice because it felt that former combatants were living on properties bordering the river.[8] Weidner was able to bring together these river-folk (whether identifying as Nama, Damara or Coloured) as a tenant labour force on his properties, which stretched from Goodhouse to Garganab (east of Witbank). Catherine Cloete's great-grandparents were among this group. Others tried constructing small irrigation gardens in out-of-the-way places, or on mission stations at Pella or Onseepkans.[9]

6 Catherine Cloete, interview with Bernard C. Moore (Witbank, 21 January 2022).

7 KAB ACLT 25 File 3588 (vol. 1): Secretary for Lands, Notice re: Sale of Orange River Farms and Islands – 26 April 1912.

8 NASA NTS 7658 File 5/382: Civil Commissioner, Namaqualand to Secretary for Lands 'Cancellation of Exclusive Grazing Rights' – 3 August 1911.

9 See Lenggenhager, 'Onseepkans'.

'It was always a difficult situation for us here at Witbank, especially around the Second World War when I was born', Cloete said. 'The government didn't want us Coloureds here. They wanted to get rid of us and send us to Pella or to the Richtersveld, where there were more Coloureds already'.[10] During the 1940s, there were conflicting voices on whether Witbank, Garganab and the islands – which the government had bought from Weidner during the Great Depression – should be used as a nucleus for a Christian settlement for Coloured farmers or as properties to sell to White settlers.[11] The South African Department of Irrigation had already leased some of the properties to some Coloured ex-soldiers who had returned from serving in Europe, who included some of Cloete's relatives. These former soldiers had manually constructed irrigation furrows near the river as well as a rudimentary diversion weir to bring water into a small canal near the furrows.[12] Indeed, even as late as 1949, irrigation officials from Upington had drawn up detailed plans to settle more than five hundred Coloured families on the stretch of land between Garganab and Goodhouse, who would engage in lucerne, cotton, vegetable and sultana grape production.[13] By the end of the year, however, the Minister for Lands in Pretoria – a National Party appointee – rejected the Coloured scheme, declaring that all Orange River farms apart from Pella should go to Whites.[14] Officials were instructed to evict all non-Whites from Witbank.

Catherine Cloete's family was evicted in 1950 and sent to Pella. By 1957, the evictions were more or less complete, with all 'Coloureds' (broadly defined) being sent to other non-White settlements, particularly Pella, Steinkopf, Eksteenskuil and the Richtersveld.[15] Ironically, very little was being done to develop properties for Whites either, and Witbank was merely leased out to

10 Catherine Cloete, interview with Bernard C. Moore (Witbank, 21 January 2022).

11 For a detailed survey of these considerations, see Moore, 'Swimming Upstream', pp. 126–131, and Lenggenhager, 'Onseepkans', pp. 157–161.

12 KAB ACLT 5 File 3588 (vol. 6): A.J. Ferreira and S. Damon 'Tour to Witbank: Proposed Coloured Settlement along the Orange River' – 11–13 November 1944.

13 KAB ACLT 27 File 3588 (vol. 7): Sekretaris van Lande, 'Memorandum i/s Marten-Krapohl Eilande Besproeiingskema' – 23 September 1949.

14 KAB ACLT 269 File 3588/8 (vol. 1): Asst. Provinsiale Verteenwoordiger, Departement van Lande to Superintendent van Kleurlingnedersettings 'Marten-Krapohl Eilande Besproeiingskema' – 2 November 1949.

15 KAB ACLT 5 File 3588 (vol. 8): Inspekteur van Lande 'Jaarverslag: Afdeling Namakwaland en Calvinia' – 22 May 1957. KAB ACLT 5 File 3588 (vol. 8): N. Esterhuyse (NG Kerk Pofadder) to Streeksverteenwoordiger, Departement van Lande 'i/s Opsigterskap: Witbank en Garganab' – 12 June 1958.

White livestock farmers as temporary grazing over the 1950s and 1960s.[16] The delays were primarily over the building of the Gariep Dam farther upriver – the same construction project that delayed the Odendaal Plan's Coloured settlement on the Namibian side, described in chapter 7.

Nevertheless, the newly formed South African Department of Coloured Affairs and its minister sought to have Witbank and neighbouring farms given back to the department just five years after the evictions were complete, in the hope that both sides of the Orange could be connected in a transriverine Coloured settlement.[17] The government approved this in principle, reversing the 1949 decision that farms be given to Whites alone, but no transfer would take place until all dam works on the Orange River were complete.

Coloureds and river-folk understood that the longer they waited for the government to make up its mind, the greater the chance that the farms would just be given back to Whites and that all the irrigation furrows they had previously dug would be unusable. Catherine Cloete's family left Pella in 1968, returning to Witbank '*sonder toestemming*' [without permission].[18] Other river-folk followed suit, and within a few years the farm was being used for subsistence pastoralism, with a small amount of irrigation and planting on the river banks and on Marten and Krapohl islands. The Witbank community set up a primary school with a hostel – affiliated with the Coloured branch of the NG Kerk – which catered for children in the area and for those whose parents worked at the zinc mine near Aggeneys or on Niemöller's massive karakul farms. During these years, Witbank people were river-folk just like their Namibian counterparts, crossing the Orange River to work the harvests at Chiebella's Haakiesdoorn irrigation works or as piecework mine labourers at Tantalite Valley. On Holy Days, many people from Witbank would cross the river to Homsrivier Mission Station to attend Mass and shop at the store there. As Witbank community elder Anna Magerman put it, 'it didn't matter if the hymnal was for a Catholic church or the NG Kerk, or if it was in Namibia or South Africa, we sang all the songs'.[19]

16 KAB ACLT 5 File 3588 (vol. 8): Streekverteenwoordiger, Departement van Lande to Sekretaresse, NG Kerk Pofadder 'Huur van Staatsgrond op die Plaas Witbank' – 20 June 1957. Nevertheless, many 'Coloureds' still felt that even though the state technically owned the land, the Whites had taken lands from their community. Jan Julie, interview with Luregn Lenggenhager (Pofadder, 24 November 2018).

17 TBK KUS 492 File 4/1/3/B2: Sekretaris vir Kleurlingsake 'Memorandum: Witbank-Garganab' – 28 August 1962.

18 Catherine Cloete, interview with Bernard C. Moore (Witbank, 21 January 2022).

19 Anna Magerman, interview with Bernard C. Moore (Witbank, 21 January 2022).

Throughout this period, however, the Witbank residents did not have secure tenure over these farms. They were not communal lands or lands gazetted under the Coloured Rural Areas Act of 1963. They were merely state lands on which they were technically 'squatting', to which the government was turning a blind eye. By then, the South African government had more or less scrapped all plans for large-scale irrigation downriver from the Neus-Augrabies scheme for Whites, which irrigation officials believed would use so much water that little could be guaranteed for Witbank at all.[20] Catherine Cloete described these years as ones of destitution [*armoede*]. During the drought of the 1980s, so much livestock died that residents had to catch small fish from the Orange River to survive.[21] Coloured people could not rely on the government to provide anything close to what it had provided for 'poor Whites' decades earlier.

Despite being left on their own, after South African independence in 1994 the Witbank community enlisted the help of the Springbok branch of the Surplus People's Project, which helped them raise funds to buy Witbank, Garganab and Hartebeestrivier farms, and Marten and Krapohl islands. These now belong to the Witbank Community Trust, and though the small village faces serious poverty and lack of services,[22] this land is theirs to do with as they please. Additional government trust farms have been allocated to previously disadvantaged communities in the stretch of farms between Pella and Goodhouse, the latter now being a gazetted communal area. Life at Witbank or perhaps anywhere along the Lower Orange River is far from easy or idyllic, especially in the heat of January. Nevertheless, it was a hard-won victory for these river-folk – descendants of Namibian refugees of genocide who became 'Coloureds' in South Africa – to gain security and surety over their own lands. This is something that their brethren on the Namibian side, including the Basson family, have not yet been able to attain.

The river-folk of Witbank have been able to watch from a distance as Homsrivier transformed from a Roman Catholic mission station to an outpost of Sean Gilbertson's luxury tourism operations, and through communication with the Bassons and other Namibian river-folk at religious and cultural festivals, they have seen some of the effects that this has had on African livelihoods.

20 TBK KUS 2/139 File 12/8/B2 (vol. 2): Direkteur-Generaal, Departement van Plaaslike Bestuur, Behuising, en Landbou 'Memorandum: Steinkopf: Ekonomiese Eenhede' – 6 September 1985.

21 Catherine Cloete, interview with Bernard C. Moore (Witbank, 21 January 2022).

22 See Khâi-Ma Municipality, *Annual Report 2021/2022*. Pofadder is responsible for some service provision for Witbank, but the lack of regional funds and vehicles has meant that, as of the end of 2022, domestic waste collection in Witbank has not occurred since 2018, because Pofadder has not sent any trucks.

At the same time, Witbank residents are also keenly aware that conservation does not know national boundaries, as the |Ai-|Ais/Richtersveld Transfrontier Park to the west shows. ORKCA, however, is not about *community* conservation – it is about '*rewilding*'. In order to contextualise both the trepidation of Witbank residents and the outlook of ORKCA as an organisation, it is worth pausing to closely investigate the deceptively simple concepts of 'rewilding' and 'transfrontier conservation'.

• • •

Like many topics and buzzwords within conservation biology, rewilding represents both a social and a scholarly movement to address and rectify the damage that humans have wreaked upon the planet. Rewilding's social origins can be traced back to the turmoil that the American environmental movement was facing in the late 1970s and throughout the 1980s, when many hyper-visible environmental organisations – like the Arizona-based *Earth First!* movement – were moving away from leftist and progressive politics towards libertarianism and nativism. *Earth First!*, founded by Dave Foreman in 1980, dominated media coverage about the environmental movement, in part because of its visible protest of everything that had an environmental impact, from logging to dam construction.[23]

Unlike most social movements at the time, *Earth First!* critiqued capitalism primarily for its damage to the natural environment rather than to society. These 'ecocentrists', claims historian Keith M. Woodhouse, believed that there were no fundamental differences between the human and non-human, yet at the same time, they failed to acknowledge that the concepts of 'wilderness' and 'nature' are in many ways a human creation. While the movement's members were not necessarily honouring capitalism, their critique of it was limited to its environmental record.[24]

By the late 1980s, the organisation splintered, with some of the more right-wing members joining fascist movements and their left-wing counterparts merging environmental causes into a general critique of capitalism. Some,

---

23 *Earth First!* was famous for hammering steel spikes into trees within logging concessions, which would at best destroy chainsaws and machinery and at worst kill the lumberjack himself. D. Foreman and B. Haywood (eds.), *Ecodefense: A Field Guide to Monkeywrenching* (Chico, Abzug Press, 1993).

24 K.M. Woodhouse, *The Ecocentrists: A History of Radical Environmentalism* (New York, Columbia, 2018). This worldview facilitated the emergence of deeply nativist and Malthusian elements within the organisation, such that Foreman and other members made extreme statements about famine and HIV/AIDS as a means to 'bring the human population back to ecological sanity'.

like Dave Foreman himself, sought to enter into the mainstream, bringing an 'ecocentric' ethos with him. In 1991, Foreman teamed up the biology professor Michael Soulé, lobbyist David Brower (of the Sierra Club) and the influential capitalist Douglas Thompkins (owner of Esprit and The North Face) to found the Wildlands Network.[25] This organisation was a distinct shift away from the confrontational strategies of *Earth First!* towards, instead, finding a way to increase the size and quantity of protected and uninhabited areas within North America. It maintained some of the long-held ideological tropes concerning 'pristine' wilderness, but these were now integrated within a market-based practice of acquiring private lands, conservation easements and custodianship contracts in order to create de jure or de facto protected areas.

Rewilding had moved from being a social movement of directly confronting humanity's relationship with nature to one that sought to integrate environmental protection within a human-dominated age, in which private capital would increasingly play a role. To attract the private capital, however, there needed to be a scientific basis to the social and financial infrastructure of the Wildlands Network. In 1998, Michael Soulé and his colleague Reed Noss put pen to paper to articulate the science behind the Wildlands Network, probably representing the first use of the term 'rewilding' in a scientific context. In the conservation magazine, *Wild Earth*, they described rewilding as a focus on the three C's of conservation: *cores*, *connected corridors* and *carnivores*.[26] Large amounts of strictly protected land had to be acquired to become *core* conservation areas, isolated from human habitation and regular engagement. These core lands should be connected through a network of migration *corridors*, on either purchased or leased lands, or else through conservation easements. Finally, these core areas would become nodal points for the reintroduction of 'keystone species', generally large *carnivores*, which would create and regulate resilient ecosystems from the top down.

This belief in top-down regulation depends on two related controversial hypotheses: trophic cascade theory and taxon substitution. *Trophic cascade theory*, as articulated by many rewilders, holds that certain keystone species – generally large predators or mega-herbivores – can and do exert a top-down influence upon the entire ecosystem below them.[27] When ecological systems

25 Ibid, p. 258.

26 M. Soulé and R. Noss, 'Rewilding and Biodiversity: Complementary Goals for Continental Conservation', *Wild Earth* (Fall 1998), pp. 18–28.

27 J.-C. Svenning et al., 'Science for a Wilder Anthropocene: Synthesis and Future Directions in Trophic Rewilding Research', *Proceedings of the National Academy of Sciences*, 113, 4 (2016), pp. 898–906.

are simplified into various trophic levels, the addition or removal of a given level can shape others below it. When a level is missing – whether through natural extinction or manmade extirpation – rewilders believe that species can be reintroduced by bringing in the same species from another region or else via *taxon substitution* – the large-scale transportation of 'similar' species from other continents to fill an empty trophic level.[28] Conceptually, rewilders believe that bringing in carnivores from elsewhere can hold herbivores in check, which rejuvenates grazing.[29]

Even prior to rewilding being articulated as a scientific and social movement, biologists and ecologists were sceptical about trophic cascade theory and taxon substitution, arguing that although top-down regulation is not impossible or unheard of, it is relatively rare, and that food-webs are more complicated than 'trophic levels' allow.[30] Indeed, based on research conducted fifty years earlier, the veteran carnivore ecologist, Paul Errington, remarked: 'I regard the outstanding source of error in appraisals of predator-prey relationships as a confusion of the *fact* of predation with the *effect* of predation'.[31] Errington held that whereas predators dominated the mind and psyche of the naturalist, they did not dominate the landscape. The controlling factors for prey numbers were shelter and food, not predation. In most situations, predators merely skimmed off the excess from prey populations. Errington estimated extraordinarily low ratios of predators to prey, normally in the range of 1 predator to 1,800 prey animals.[32] Even many ecologists and practitioners sympathetic to the ethical and philosophical goals of rewilding have come out against the movement's obsession with trophic cascade theory.[33]

---

28 The most radical proposals concerning taxon substitution is advocated by 'Pleistocene rewilding', which seeks to use African elephants to mimic North American mammoths; sabretooth tigers would be replaced by lion and cheetah. C.J. Donlan et al., 'Pleistocene Rewilding: An Optimistic Agenda for Twenty-First Century Conservation', *The American Naturalist*, 168, 5 (2006), pp. 673–674. J.T. du Toit, 'Pleistocene Rewilding: An Enlightening Thought Experiment', in N. Pettorelli, S.M. Durant and J.T. du Toit (eds.) *Rewilding* (Cambridge, British Ecological Society, 2019), pp. 55–72.

29 P. Jepson and C. Blythe, *Rewilding: The Radical New Science of Ecological Recovery* (Cambridge, MIT Press, 2022), pp. 92–95.

30 G.A. Polis and D.R. Strong, 'Food Web Complexity and Community Population Dynamics', *The American Naturalist*, 147, 5 (1996), pp. 813–846.

31 P.L. Errington, *Of Predation and Life* (Ames, Iowa State University Press, 1967), pp. 235.

32 Ibid, p. 250.

33 M.W. Hayward et al., 'Top-Down Control of Ecosystems and the Case for Rewilding: Does It All Add Up?' in N. Pettorelli, S.M. Durant and J.T. du Toit (eds.) *Rewilding* (Cambridge, British Ecological Society, 2019), pp. 325–354. Hayward and colleagues argue that most defences of trophic cascades are 'blighted by methodological weaknesses, selective use of data to support arguments, and too much over-generalisation'.

Some advocates of rewilding admit that they are really proposing a philosophical rather than scientific approach to nature management and ecology.[34] However, what begins as a 'thought experiment' to test hypotheses of top-down ecological regulation and taxon substitution can shift to represent a general ethos about the ideal ways in which rural areas should exist in the modern world.[35] Jørgensen argues that rewilding as a social movement talks about the concept in troubling ways: 'Taken as a whole, rewilding discourse seeks to erase human history and involvement with the land and flora and fauna. Such an attempted split between nature and culture may prove unproductive and even harmful'.[36] As a social movement surrounding broader conservation goals, proponents of rewilding have blurred the historical baselines to which one should rewild and overall societal goals. Most organisations that espouse the principles and language of rewilding increasingly refer to it, in Jørgenson's words, as 'giving nature space to be wild'. This romanticism, argues Kim Ward, often means that the search for 'pristine' nature unaffected by human habitation risks exclusionary land uses and Indigenous erasure.[37] Indeed, building on William Cronon's critique of humanity's unreflective view of 'wilderness',[38] Jørgensen argues that rewilders seek to recreate a wilderness without people,

> oblivious to the problematic nature of the wilderness construct … Rewilding as currently practiced disavows human history and finds value only in historical ecologies prior to human habitation. The rewilding concept has been deployed in a myriad of ways to exclude humans in time and space from nature.[39]

Although its strategies for achieving non-human wilderness might have changed from its roots in the *Earth First!* group, the anti-humanist sentiment

34 S.M. Durant, N. Pettorelli and J.T. du Toit, 'The Future of Rewilding: Fostering Nature and People in a Changing World', in *Rewilding* (Cambridge, British Ecological Society, 2019), pp. 413–425.

35 D. Johns, 'History of Rewilding: Ideas and Practice', in N. Pettorelli, S.M. Durant and J.T. du Toit (eds.) *Rewilding* (Cambridge, British Ecological Society, 2019), pp. 12–33.

36 D. Jørgensen, 'Rethinking Rewilding', *Geoforum*, 65 (2015), pp. 482–488.

37 K. Ward, 'For Wilderness or Wildness? Decolonising Rewilding', in N. Pettorelli, S.M. Durant and J.T. du Toit (eds.) *Rewilding* (Cambridge, British Ecological Society, 2019), pp. 36–54.

38 W. Cronon, 'The Trouble with Wilderness, or, Getting back to the Wrong Nature', *Environmental History*, 1, 1 (1996), p. 25.

39 Jørgensen, 'Rethinking Rewilding', p. 487.

within some contemporary rewilding discourses has not changed much in essence.

Rewilding's early agendas dealt with projects and programmes in North America and Europe, but by the twenty-first century Africa was being targeted for similar programmes. Rewilding advocates praised the commercialisation of game-farming on private lands in southern Africa – mostly owned by Whites – arguing that, through hunting, 'private landowners have rewilded landscapes, supported an abundance of wildlife, and brought a level of prosperity to rural areas'.[40] Through private ownership of game, these White landowners were increasing the herd sizes of elephants, rhino, buffalo and other megafauna, which had been, argued ecologists Jepson and Blythe, 'virtually wiped out during the 1980s'.[41] Many rewilding projects operating in (southern) Africa play into tropes of heroic White conservationists seeking to save the 'last wilderness' on Earth, often with the guiding hand of well-connected local Whites and wealthy benefactors.[42] Outside of private wildlife estates owned by White ex-colonials, rewilders also look fondly upon transfrontier conservation areas, a form of nature preservation pioneered by the Peace Parks Foundation (PPF) since the 1990s, and which uses the three C's model of conservation in a similar, albeit slightly less exclusionary, way.[43] The PPF also targeted the lower Orange River for transfrontier conservation in the 1990s, and ORKCA today draws from their experience.

•••

By the 1950s, Anton Rupert had established himself as one of the most successful South African capitalists. As mentioned in chapter 7, Rupert had created a network of interconnected Stellenbosch-registered businesses – the Rembrandt Group – with wine and liquor production (especially brandy) at the centre. The Rembrandt-controlled conglomerate Distillers Korporasie Bpk – known now as Distell – controlled the lion's share of southern African

40 Jepson and Blythe, *Rewilding*, p. 159.

41 Ibid. Of course, most of the game was exterminated at the hands of the South African Defence Forces, not the anti-apartheid movement. See, L. Lenggenhager, *Ruling Nature, Controlling People*. Furthermore, as shown in chapter 5, the roots of these commercialisation policies date back to the 1950s, the heyday of apartheid, as a subsidy for White farmers rather than as a conservation strategy.

42 For a particularly egregious case of this, see G. Fowlds and G. Spence, *Rewilding Africa: Restoring the Wilderness on a War-Ravaged Continent* (Johannesburg, Jonathan Ball, 2022).

43 Durant, Pettorelli and Du Toit, 'The Future of Rewilding', pp. 413–425.

wine and liquor distribution, eventually buying up distilleries in Scotland and irrigation land along the Orange River in Namibia. The Rembrandt Group held a controlling interest in British cigarette manufacturers, and its Stellenbosch companies controlled distribution across southern Africa. Afrikaner capital was expanding out of South Africa, capturing large segments of European and global markets.[44] Back at home in Stellenbosch, Rembrandt diversified into property-holding, land speculation, printing, hotels and even mining. In 1972, in order to hedge against potential anti-apartheid sanctions, the Rupert family spun off their European holdings into the Switzerland-domiciled luxury goods company, Richemont, which eventually came under the directorship of Anton's son, Johann Rupert.[45] Rembrandt eventually reformulated its South African operations into Remgro Ltd, also controlled by the Rupert family.

In 1960, Anton Rupert founded the South African Nature Foundation (SANF) as a means to promote (White) South African culture and natural heritage on a world scale. Through Rupert's business connections with the German-born Dutch royal, Prince Bernhard von Lippe-Biesterfeld, who served as president of the Worldwide Fund for Nature (later renamed the World Wildlife Fund [WWF]), the SANF was recognised as the South African national branch of the WWF. Through the WWF, South African businessmen and politicians could meet their counterparts abroad, clothing the meetings in the language of nature conservation while there was clearly much more going on.[46] Many key officials within the WWF, such as Director-General Charles de Haes, bounced back and forth between employment for one of Anton Rupert's Richemont companies and those within Prince Bernhard's network.

In 1997, Anton Rupert and Prince Bernhard established the Peace Parks Foundation.[47] The PPF sought to depart from fortress conservation models, maintaining that there needed to be a 'people-and-parks' approach whereby

---

44 See Du Toit, *The Stellenbosch Mafia.*

45 The financial services and investment side of Rembrandt/Richemont/Remgro – Reinet Investments S.A. – was spun off in 2008 and registered as a securitisation vehicle in the tax haven of Luxembourg. See https://www.reinet.com/

46 M. Spierenburg and H. Wels, 'Conservative Philanthropists, Royalty, and Business Elites in Nature Conservation in Southern Africa', *Antipode*, 42, 3 (2010), p. 658.

47 Nelson Mandela was also a founding member, who agreed in part because Prince Bernhard convinced the Dutch Postcode Lottery to make a very large donation to Mandela's children's charity fund. The timing of the PPF's birth corresponds with the exposé of the Kumleben Report, detailing Prince Bernhard's and the SANF's relationship with the SADF and rhino-horn smuggling. The PPF was likely an effort to rehabilitate their image. See L. Lenggenhager, *Ruling Nature, Controlling People*, pp. 133–134. M.E Kumleben, 'Commission of Inquiry into the Alleged Smuggling of, and Illegal Trade in Ivory and Rhino Horn in South Africa' (Report delivered to the RSA State President, January 1996), pp. 61–71.

conservation could drive economic development. In many ways, this merged conservationist goals and capitalist accumulation; many in the PPF saw their work as pioneering conservation via consumerism. The PPF's main plans involve the creation of transfrontier conservation areas (TFCAS) straddling the boundaries of southern African nations, presumably promoting peace between the countries because collaboration with nature conservation is necessary. The TFCAS were marketed to promote employment through tourism, and most operations are public – private partnerships (with private, generally South African, businesses holding the wheel).[48]

Through its global network of fundraising offices, the PPF raises funds to build and operate these TFCAS, areas that the PPF claims should represent 'the confluence of several mutually-reinforcing interests, mainly those of biodiversity conservation, economic development, cultural integrity and regional peace and security'.[49] This seemingly holistic approach became highly influential and earned the support of politicians and regional governments, including the heads of state of South Africa, Namibia, Botswana, Angola and Zimbabwe. Governmental support has been bolstered by successful fundraising, making the PPF one of South Africa's wealthiest NGOS. Prince Bernhard founded 'Club-21', an exclusive list of donors, who honoured the organisation with a cheque for USD 1 million. The club contains wealthy individuals, foundations and companies, including many connected to the Rupert family business network: Prince Bernhard, Cartier, De Beers, Exxaro coal, Hansjörg Wyss, Reinet Foundation, Remgro, Richemont, the Rupert Family Foundation, Total oil and many others.[50] Other prominent donors include companies clearly hoping to improve their public image, including Anglo American, Distell, Microsoft, Nedbank Private Wealth, UBS and others.[51]

Particularly in their early phases, many of the transfrontier conservation areas were described as comparably 'harmless' greenwashing and marketing instruments. However, observers began to see that although the PPF did not propose conservation areas that were completely devoid of humans – as their more extreme rewilding colleagues did – African occupation of lands marked for conservation needed to be compatible with the profit-driven model of these public – private partnerships. For example, the PPF's Limpopo National

48 Spierenburg and Wels, 'Conservative Philanthropists', p. 662.

49 A. Hammill and C. Besançon, 'Measuring Peace Park Performance: Definitions and Experiences', in S.H. Ali (ed.) *Peace Parks, Conservation and Conflict Resolution* (Cambridge, MIT Press, 2007), p. 25.

50 Peace Parks Foundation, *Annual Review 2021*, p. 71.

51 Peace Parks Foundation, 'Donors and Partners', https://www.peaceparks.org/about/donors-and-partners/

Park in Mozambique, which was slated to join a TFCA, resulted in the forced and/or coerced removals of more than 7,000 residents from villages that would fall within the park, in part because the removals would, as one study put it, 'render the area more attractive to private tourism operators'.[52] Beyond removals and evictions, the Peace Parks Foundation's TFCAs have occasionally become sites of 'green violence' and 'green militarisation', where concession companies promote the transfer of technologies to buttress the war against poaching, which takes various forms, from rhetorical condemnation of locals living nearby to blunt instruments like shoot-to-kill policies.[53]

The PPF and other large international NGOs epitomise a shift over the past few decades in how conservation operates in southern Africa. Throughout most of the twentieth century, conservation areas – such as national parks and wildlife preserves – were the domain of the state. Individual landowners engaged in commercial trophy-hunting, but private investment and international interest in conservation was limited because of apartheid-era sanctions. With the end of apartheid, this barrier to investment disappeared. Moreover, the two post-apartheid governments found that the budgets they had inherited from the old regime were designed to meet the needs of the White population only. A newly equal citizenry means equal spending, and certain cuts had to be made. While both governments continued to maintain nature conservation governmental divisions – holding on tightly to their national parks – the field opened the door for more private investment.

The confluence of capitalism and conservation does not only arise when capitalists themselves purchase areas deemed for 'conservation' use. South African NGOs like the Peace Parks Foundation and the African Parks Foundation operate in a manner which opens government lands and national parks to private investors and interested parties. When the PPF organises and lobbies for the creation of a TFCA spanning multiple countries' national parks, the promise of tourism investment decentralises decision-making power to a broader cohort of 'stakeholders', including the PPF itself, and potentially to the capitalist investors who build hotels and lodges on park concessions.[54] A more extreme case is the African Parks Foundation, which negotiates with African governments to fully take over long-term management of certain national

52 M. Spierenburg, C. Steenkamp and H. Wels, 'Enclosing the Local for the Global Commons: Community Land Rights in the Great Limpopo Transfrontier Conservation Area', *Conservation and Society*, 6, 1 (2008), pp. 87–97.

53 B. Büscher and M. Ramutsindela, 'Green Violence: Rhino Poaching and the War to Save Southern Africa's Peace Parks', *African Affairs*, 115, 458 (2016), pp. 1–22.

54 B. Büscher, *Transforming the Frontier: Peace Parks and the Politics of Neoliberal Conservation in Southern Africa* (Durham, Duke University Press, 2013).

parks, rebuilding their governance and management structures on a public – private partnership model to facilitate international investment in tourism. It currently has twenty two national parks in its portfolio, from Zambia to Angola, DRC, Chad and South Sudan.[55]

Whereas large amounts of the funding for the PPF and Africa Parks Foundation come from governments or explicitly charitable initiatives, like the European Union or the Dutch Postcode Lottery, larger figures are donated by explicitly profit-driven companies and their corporate foundations. The explosion of corporate foundation funding channelled to environmental NGOs across the globe has led to a softening in how these NGOs handle capitalism as a mode of production that is deeply destructive to the environment.[56] Foundations tied to capitalist wealth increasingly steer NGOs towards researching 'market-based' or 'profit-based' solutions to the problem of climate change. Furthermore, individual and corporate donations are often a means to divert excess wealth in order to receive tax breaks while simultaneously advancing corporate ideological agendas. As an example, the African Parks Foundation received a massive donation from Walmart's Rob Walton Foundation. This was not intended for its operations directly, but rather for its USD 57.3 million endowment fund, which is in turn invested into the Jersey-based private equity fund, CVC Capital Partners, which spreads it across numerous industry investments, from glass manufacturing in Florida to oil drilling in Norway and Algeria, to chemical production in India.[57] As for the Peace Parks Foundation, its funds are primarily reinvested in the Rupert family's own global portfolio of businesses, which include Remgro, Richemont, British American Tobacco, Reinet Investments, etc.[58]

It is hard to imagine that this financial relationship with deeply polluting industries does not shape the conservation behaviour and goals of NGOS that receive dividends from them. Some large international NGOs have placed restrictions on how they direct their endowment funds and investment portfolios, in order to focus on 'green' financial projects. The Nature Conservancy, for example, forbids its team from investing any of its funds in companies that

55 O. van Beemen, *Im Namen der Tiere: Wie eine NGO große Teile Afrikas beherrscht* (Münich, C.H. Beck, 2024).

56 M.M. Betsill et al. 'Philanthropic Foundations as Agents of Environmental Governance: A Research Agenda', *Environmental Politics*, 31, 4 (2022), pp. 684–705.

57 See African Parks, *Annual Report 2022*, pp. 110–115. Stichting African Parks Foundation [Netherlands], *Annual Report 2022*, pp. 4–5. United States Security and Exchange Commission, 'CVC Capital Partners VIII Associates L.P.' (Form D) – 5 October 2020. CVC, 'Our Portfolio' https://www.cvc.com/strategies/our-portfolio/ (Accessed 30 June 2023)

58 This portfolio also includes FirstRand and RMB Holdings, which are partly owned by Rupert family companies, as well as Stonehage Fleming and Nedbank Private Wealth. See, Peace Parks Foundation, *Annual Review, 2021*, pp. 92–93.

generate five per cent or more of their annual revenue from extractive activities, or that inflict an estimated damage to the environment via greenhouse gas emissions and/or biodiversity destruction that exceeds five per cent of revenue.[59] Honourable as these prohibitions are, it is hard to in turn understand The Nature Conservancy's partnership with and consideration of BHP, Dow Chemical Company, Shell and Walmart as 'companies investing in nature'.[60]

From the start of the twenty-first century, there was an increasing alignment of interests between those interested in rewilding and those interested in transfrontier conservation, especially the PPF. Part of this was ideological. In 2001, the PPF launched its 'Rewilding Africa' project, whereby the PPF funded the translocation of tens of thousands of game animals 'from areas of overpopulation, to areas of decimation'.[61] In turn, the PPF argued, the reintroduction of these species in large quantities would restore 'biodiversity' broadly.[62] In many ways, the PPF's efforts to secure, connect and repopulate TFCAs resembled trophic cascade theory and the three C's model of conservation. Furthermore, because rewilders' critique of capitalism was fairly mild, they had little complaint about where the money came from or whether the gears for rewilding were 'manual' or 'automatic'.[63] Conservation in southern Africa was not about transforming human behaviour or economic structures; it was instead focused on acquiring property, securing easements and corridors and paying to fill these areas with wildlife and tourists.[64]

•••

Land tenure and property rights ultimately determine the shape that conservation operations will take. Up to this date, none of the Peace Park Foundation's TFCAs in southern Africa have private lands serving as their core areas. Almost without exception, the land is either state owned or gazetted communal lands

59 The Nature Conservancy, Office of Investments, 'TNC's Customized Restrictions' https://www.nature.org/en-us/about-us/who-we-are/how-we-work/finance-investing/office-of-investments/ (Accessed 1 May 2024).

60 The Nature Conservancy, 'Companies Investing in Nature', https://www.nature.org/en-us/about-us/who-we-are/how-we-work/working-with-companies/companies-investing-in-nature1/ (Accessed 1 May 2024)

61 Peace Parks Foundation, 'Rewilding Africa' https://www.peaceparks.org/rewilding-africa/ (Accessed 30 June 2023)

62 See Peace Parks Foundation, 'A Wild Dog Revival in Limpopo National Park' (3 March 2023). https://www.peaceparks.org/a-wild-dog-revival-in-limpopo-national-park/ (Accessed 30 June 2023).

63 See Jepson and Blythe, *Rewilding*, p. 159.

64 For a similar analysis in the American West, see Farrell, *Billionaire Wilderness*, pp. 106, 156.

governed by government-recognised traditional authorities.[65] This does not necessarily mean that private landowners are inherently against TFCAs, but rather that – as was the case for landowners near the proposed Greater Mapungubwe TFCA – they preferred to convert their properties into photographic tourism or hunting farms in order to profit from the nearby conservation zone without having to abide by its rules.[66] Negotiating with each individual landowner to expand TCFAs via a custodianship agreement is burdensome, and purchasing lands is expensive. This was the case with the Kgalagadi TFCA, which stretches across national parks and communal lands in South Africa and Botswana. The Namibian side of the boundary was excluded from the discussion, in part because the Namibian side is White-owned private farmland.[67]

Quite a lot of this reasoning is emblematic of how ownership of wildlife functions in Namibian and South African law. As mentioned in chapter 5, apart from a few species of protected game (such as black rhinoceros), landowners in Namibia also own the wildlife within their property's fenced boundaries. Namibian legislation confers ownership rights upon the landowner himself as a landowner, not as an individual; i.e., individuals who do not own farms therefore do not own wildlife. Throughout the apartheid era in Namibia, because Whites were the only landowners, they were the only people who owned game animals. Wildlife residing on African communal areas belonged to the state, which had the final say over its utilisation.[68] After Namibian independence, the new government changed policies to allow communal areas to have ownership rights over game by reformulating their land tenure into a communal conservancy, governed by additional environmental use restrictions.[69] This transferred wildlife ownership not to individual conservancy members, but rather to the conservancy itself. This decentralised ownership away from the government in Windhoek but centralised it in the hands of the

---

65 L. Lenggenhager and M. Ramutsindela, 'Property Killed a Peace Park Dream: The Entanglement of Property, Politics and Conservation along the Gariep', *Land Use Policy*, 105 (2021), pp. 1–9.

66 Ibid, p. 3.

67 M. Ramutsindela, *Transfrontier Conservation in Africa: At the Confluence of Capital, Politics, and Nature* (Wallingford, CABI, 2007), p. 74. However, there are ongoing efforts, funded by The Nature Conservancy, to formulate a plan to purchase Namibian farmland in the vicinity of Twee Rivieren and Mata Mata to expand Kgalagadi. See SADC, 'Securing Land for a Namibian National Park in the Kalahari' (10 August 2021) https://tfcaportal.org/securing-land-namibian-national-park-kalahari (Accessed 1 May 2023).

68 See B.C. Moore, 'The Night Harvests: From Royal Game to Property Rights in Namibian Wildlife Policy', Paper presented at the conference: Realising Nature(s) with Protected Areas in Namibia and Germany: Pasts, Presents, and Futures (Cologne, June 2024).

69 Lenggenhager, *Ruling Nature, Controlling People*, pp. 183–185.

conservancy committee. As it relates to the PPF planning a TFCA, it is only necessary, then, for it to negotiate with the central government and conservancy officials rather than with each individual resident of the communal areas, because property rights over land and game are not individualised.

Of the Peace Parks Foundation's originally planned transfrontier conservation areas (see Figure 83), the only one that failed completely – the Gariep TFCA – sought to use private lands to form its core conservation area. In 1997, the PPF hired Hugo Bezuidenhout of South African National Parks (SANParks) to conduct a survey of the Lower Orange River to identify key natural features and vegetation and ultimately narrow down where the TFCA should be situated. He divided the river into Region A (Augrabies Falls National Park to Onseepkans) and Region B (west of Pella as far as Goodhouse).[70] Bezuidenhout drove a 4 × 4 vehicle throughout the two regions, observing the intensive grape, date and lucerne irrigation along many parcels on the South African side of the river, as well as mining areas closer to Aggeneys and Goodhouse. He did not cross the river to investigate anything on the Namibian side, and it is unclear if the PPF ever sent anyone across the border at any point. Bezuidenhout remarked that Region B to the west (that is, ORKCA's future zone) was less desirable for conservation on biodiversity grounds. More importantly, he believed that the existence of community trust lands and villages like Witbank ultimately posed the greatest difficulty to securing the lands for conservation.[71] He therefore urged the Peace Parks Foundation to focus on Region A in the east, ideally finding ways to connect the TFCA to Augrabies Falls National Park.

The PPF took his report and funded several University of Cape Town dissertation projects to gather data to inform future plans; none of the studies involved crossing to the Namibian side. Christine L. Jardine interviewed various stakeholders and found that buying out the farmers would be exceptionally expensive, even for the PPF. This was in part because there were a number of grape producers using intensive irrigation.[72] The livestock farms were cheaper, but Jardine recommended delaying their purchase because 'in addition to facing a low threat, their integrated conservation value is low'.[73] The whole Gariep TFCA was abandoned by the end of the 1990s, because it was not

70 H. Bezuidenhout, 'Confidential Report on the Landscape/Vegetation of the Possible Transfrontier National Park in the Onseepkans Area' (Peace Parks Foundation, August 1997).

71 Ibid, p. 6.

72 C.L. Jardine, 'The Proposed Gariep Transfrontier Conservation Area: Conservation Overlay Evaluation of the South African Section' (MPhil thesis, University of Cape Town, 1998), p. 9.

73 Ibid, p. 73.

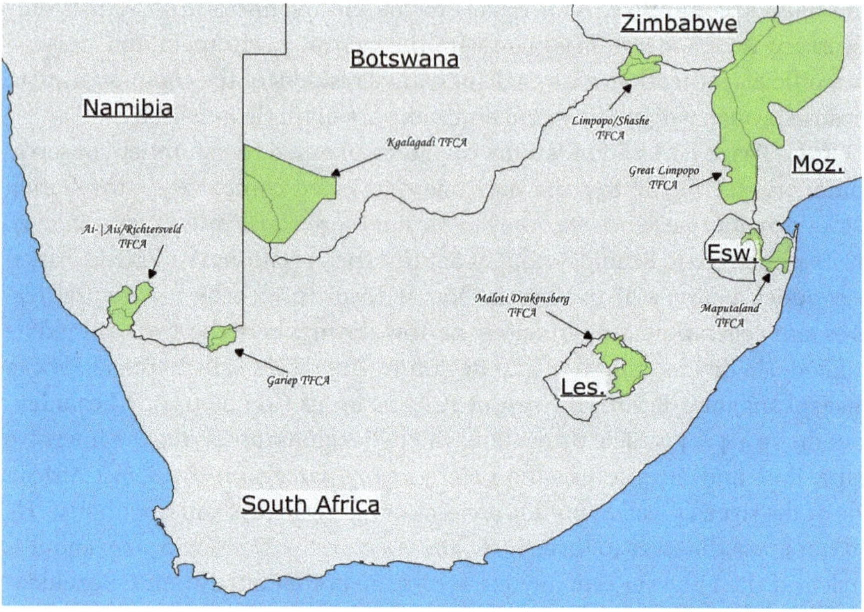

FIGURE 83 First plans for the Peace Parks Foundation's transfrontier conservation areas, 1997
MAP: CARTOGRAPHY: B.C. MOORE, BASED ON C. JARDINE, 1997.

deemed financially prudent to purchase thirteen Namibian and sixteen South African farms in remote areas, especially when the farmers themselves were reluctant to sell.

By 2005, however, plans were back on the table to create a larger, albeit more open-ended TFCA, stretching from Augrabies to the Atlantic Ocean, straddling both sides of the Orange River as far south as Pofadder and as far north as Keetmanshoop.[74] The Lower Orange River TFCA was imagined as a tourism promotion partnership between the RSA Department of Water Affairs and Forestry, SANParks and the PPF.[75] The plans were further motivated by the declaration by the RSA government that tourism benefits reaped from hosting the 2010

74 This includes all the farms owned by Sandfontein and Pelladrift companies, which are part of ORKCA.

75 RSA Department of Water and Sanitation, 'Lower Orange River Transfrontier Conservation Area Planning: Background Information Document' (August 2007).

World Cup would be spread across the SADC region.[76] This new initiative did not envisage any purchase of private farmlands at all, or even changing the land use of the farms explicitly towards conservation, per se. Rather, the PPF and other planners sought to convince the farmers to remove most of the boundary and camp fencing from their properties, stop controlling predators and reduce grazing intensity. The PPF reasoned that these changes would gradually increase tourism for landowners in the region. The 2008 final plan allowed for the higher-value economic land uses – such as date and grape irrigation along the river, and lead and zinc mining at Aggeneys farther inland – to continue more or less unimpeded, with stock farmers bearing much of the conservation burden. Furthermore, the PPF's investigators did not consult most of the actual residents of the proposed TFCA: namely, farm workers, communal area residents, urban dwellers, squatters and others.[77]

Some landowners were thrilled by the proposals. Daberas Farm, for example, near Augrabies on the South African side, holds no irrigation potential, and the mountainous landscape makes intensive stock-farming difficult. After working most of his career outside of Namakwaland, the owner had returned to the area, developing a small tourism business and constructing a lodge on Daberas. 'I came here not to farm, but because I love the landscape and the beauty of this area', he reflected.[78] As the owner of one of the few lodges existing between Augrabies and Onseepkans, he would be well positioned to earn a good income from tourism, and the restrictions on fencing use would not affect him much because his land was not really a productive farm anymore.

In contrast, White livestock farmers on the South African side of the river were very organised in their condemnation of what the PPF's plans meant for meat production and rural livelihoods. 'For you behind your desk looking at maps, it may look like a good plan', wrote farm owner Erna Claassens. 'But it's not at all for us livestock farmers who are affected by it'.[79] Transfrontier conservation evokes sharp emotions from those close to it, such that more than a decade after these (ultimately failed) conservation proposals, Claassens was still alert, mistakenly accusing us of being spies for some international cabal of renegade conservationists. Claassens argued, with mathematical accuracy, that the PPF's decision not to buy up land actually missed the point. The 'value'

76 |Ai-|Ais/Richtersveld Transfrontier Park Joint Management Board, 'Lower Orange River Management Plan: Unlocking the Ecotourism of the River' (October 2008), p. 4.

77 Lenggenhager and Ramutsindela, 'Property Killed a Peace Park Dream', p. 7.

78 Interview with Luregn Lenggenhager (Daberas farm, 8 February 2020).

79 ['*Vir u agter u lessenaar en op kaarte mag dit 'n goeie plan lyk, maar glad nie vir ons veeboere wat daardeur geraak word nie*']. Erna Claassens, personal communication with Luregn Lenggenhager (email, 19 February 2020).

of sheep-farming land is not the ground upon which the sheep walk and graze but rather the millions of rands' worth of infrastructural improvements, like the fencing and water points that criss-cross the properties.[80] Most of the tourism benefits would be reaped by those who own river frontage and irrigated grape farms – or else farm owners who got in early – yet the financial and logistical burdens would be carried by the stock farmers.[81] She concluded her critique in exaggerated language familiar to commercial and communal livestock farmers, language that ultimately doomed the TFCA project: 'This global conservation imperialism is plunging countries into a food crisis, all the while livestock farmers have been proving over and over again that meat production can continue in harmony with nature'.[82]

Claassens was the most vocal critic, but there were many reasons why the Lower Orange River TFCA never came to fruition – from the RSA-centric focus of the scheme, to ignoring the communal residents, to placing all the burden on the livestock farmers. In addition, the institution of private property – especially community trust lands – consistently complicated how to plan an overarching transfrontier conservation area. This would be the last time the PPF would look to this stretch of the Orange River for future conservation plans, but it would not be the last time that transfrontier conservation was proposed for the Orange River. ORKCA was entering the picture.

•••

When Ian Craig fronted the money in 2010 to pay commercial hunter Willem F. Mans for the purchase of KumKum, he and Pete Morkel were on excellent terms, and Morkel became the majority shareholder of Pelladrift (Edms) Bpk and the lands that would become Oana. This did not last for long, it appears. Morkel said:

> I didn't get along with those young kids which Craig brought in … I didn't really care that they were hosting these gap-year adventure travellers from Europe, but they were mismanaging the place in so many other ways. Eddie Barthorp really alienated a lot of the nearby farmers. After

80 E. Claassens, 'Rupert se Droom Nagmerrie vir Boere', *Landbou Weekblad* (21 November 2008). Of course, much of this capital investment was highly subsidised by the apartheid state, as shown in chapter 5.

81 Erna Claassens, personal communication with Luregn Lenggenhager (email, 21 February 2020).

82 Claassens, 'Rupert se Droom'.

> all these years, he still can't even understand *hoe gaan dit* in Afrikaans. It's crazy.[83]

When the droughts of 2015 came to southern Namibia, Morkel stated that they needed to cull more game in order to protect the grazing, which Craig apparently did not want to do. It is possible that Morkel was the one who invited Willem Mans to hold one last trophy hunt on KumKum in 2015, for which hunters from the Dallas Safari Club were brought to Oana's properties and to Sandfontein.[84] Either way, by the time we spoke with the personnel at KumKum, they were clear that Pete Morkel had *no* involvement with Oana/ORKCA, apart from owning lands nearby.[85]

Through his partnership with Morkel, Ian Craig was introduced to Peter F. Koep, whom Craig hired as his lawyer. Koep also became his business partner. In 1982, Koep founded a corporate law firm in Swakopmund, which has long focused on mining and energy and has represented some of the largest uranium companies in the Swakopmund and Arandis areas. Koep travelled extensively through southern Namibia at the turn of independence, when he was hired to be Steen Severin's lawyer and conveyancer (see chapter 8). 'Tantalite Valley was a bit hopeless in those days', reflected Koep on those years. 'I respected Severin, but, you know, he was getting old, and I could tell that Tantalite Valley was more of a retirement project than some bull business interest. You should see his house there, though. It's massive!'[86] Koep's connections to the area continued in the years that followed, in that he represented the Canadian uranium mining company, Xemplar, which operated on Sean Gilbertson's Sandfontein (see chapter 9). He was also a director and shareholder of Xemplar.

Ian Craig was clear that his intention was not just to own a bit of land in Namibia to undertake some small-scale conservation projects. He had bigger plans, and he wanted Peter Koep on board. 'I travelled with Ian to Kenya to see what he had done there, and I was really impressed', said Koep. 'It's on another scale compared to what we've got here in Namibia. They're far ahead of us in conservation'.[87] Together with Ian Craig – and in the early days, Morkel – Peter Koep jointly founded Craig's company KumKum Adventure Tourism (Pty) Ltd

83 Pete Morkel, interview with Bernard C. Moore (telephone, 19 October 2022).

84 Dallas Safari Club, '4-Day Plains Game Hunt for Two Hunters in Namibia – Includes Trophy', *'Traditions 2015' Auction Catalogue* (14 January 2015). https://bidding.biggame.org/4-Day-Plains-Game-Hunt-for-Two-Hunters-OR-One-Hunter-and-One-Non-Hunter-in-Namibia-Includes-Trophy_i21066060

85 Andreia Pawel, interview with Bernard C. Moore (telephone, 10 October 2022).

86 Peter F. Koep, interview with Bernard C. Moore (telephone, 11 January 2023).

87 Ibid.

(trade name Oana), and the section-21 non-profit called Oana Community Development Project. Around 2020, Koep stepped in for Morkel, becoming officially the 51 per cent 'owner' of Pelladrift (Edms) Bpk and all the lands that the company owned.[88] Then, in September 2021, he became a founding board member of Orange River-Karoo Conservation Area (ORKCA), a section-21 non-profit company that would manage conservation initiatives across the lands owned by Morkel, Craig and Gilbertson, with the aim of expanding across the Orange River into South Africa.

ORKCA was explicitly founded as a rewilding organisation. Via a memorandum of association between Sean Gilbertson on Sandfontein and Ian Craig on KumKum, each of their properties would become a *core* conservation zone. Custodianship agreements had been and would be signed with other neighbouring properties, such as Pete Morkel's lands, as well as perhaps some commercial livestock and irrigation farmers in Karasburg District.[89] This would provide *connected corridors* enabling and facilitating the movement of game and wild animals across the region. Finally, within the core areas, ORKCA would reintroduce large *carnivores* and other predators, such as cheetah, spotted hyaena and wild dog.[90] Additional reintroductions would include hippopotamus in the Orange River, and megaherbivores such as rhinoceros and Angolan giraffe.

According to Ian Craig, the big-picture goals of ORKCA involve 'opening up fences to make one giant landscape, which will increase game numbers and provide financial windfall to the communities through increased employment'.[91] ORKCA is drawing from Ian Craig's long experience in building private conservation in Kenya, but in this region there are some significant differences. Perhaps, if there were not a large population of African pastoralists in central Kenya and beyond, Craig would have pursued a strategy closer to 'rewilding' there, but the NRT and Lewa are not operating on the three C's model of

88 It was alleged that Peter Koep (whether individually or via Pelladrift) is in the process of purchasing the eastern portions of Ramansdrift farm for NAD 12 million. William Gennen McDowall, interview with Bernard C. Moore (telephone, 1 February 2023). Peter Koep declared that he was not the majority owner, but rather that the shares were held by a trust with him as a trustee. This does not reflect the state of the BIPA file at time of research, however. Peter F. Koep, personal communication with Bernard C. Moore & Luregn Lenggenhager (email, 29 October 2024).

89 Andreia Pawel, interview with Bernard C. Moore (telephone, 10 October 2022).

90 Orange River-Karoo Conservation Area, 'ORKCA Objectives' https://orkca.org/objectives/

91 Ian Craig, interview with Bernard C. Moore (telephone, 18 October 2022).

conservation. There are simply too many humans living nearby and using the land to create the 'pristine wilderness' sought by ORKCA and other 'rewilders'.[92]

ORKCA describes its mission as 'to protect, restore and rewild the Orange River-Karoo landscape. ORKCA catalyses change by bringing together fragmented land to create a contiguous, transfrontier conservation area for the benefit of local communities and wildlife'.[93] This transfrontier element involves finding ways to either acquire or manage land on both sides of the Orange River. Working towards this goal, ORKCA partnered with the South Africa-based Wilderness Foundation Africa, to investigate the logistics of making ORKCA a transfrontier project. With funding from the Stellenbosch-based Millennium Trust, the Wilderness Foundation launched its Lower Gariep Conservation Corridor project in 2022.

'I generally try to avoid language like "Transboundary" or "Transfrontier" to describe what we're doing', said Matthew Norval, Wilderness Foundation Africa's Chief Operations Officer for conservation (and ORKCA liaison). 'It triggers emotions regarding Namibia's sovereignty. I prefer to say that we're trying to build a shared vision on both sides of the Orange River'.[94] Pete Morkel and Norval have worked together on a number of conservation projects, and it was Morkel who recommended that the Millennium Trust fund the project. Norval and his team took a 4 × 4 drive along the South African side of the Orange River in May 2022, speaking with (mostly White) stock farmers, irrigation growers, mining companies, and other interested parties regarding conservation projects like this.

'I was pleasantly surprised at people's openness to conservation, especially the irrigation farmers', Norval said. 'Ideally, we'd like to include all kinds of lands within the conservation zone (irrigated, non-irrigated, pasture, mountains), and we've got an idea that some of these irrigation companies might get on board financially, as it will make them look good in the long run'. When we asked if Norval and his team had read up on the previous plans by the PPF, he remarked that he was surprised that there had been something in the 1990s. 'Our core concept today is eighty per cent the same as what the Peace Parks Foundation was doing, but they were a bit too ambitious, a bit old-school. They

92 Even so, ORKCA's board member, Sean Gerrity, is the darling of the rewilding movement; through his private land acquisitions in Montana for the American Prairie Reserve, cattle are replaced with bison.

93 Orange River-Karoo Conservation Area, 'Mission' https://orkca.org/about/

94 Matthew Norval, interview with Bernard C. Moore (telephone, 12 January 2023).

FIGURE 84 Witbank village, viewed from the south bank of the Orange River
PHOTO: B.C. MOORE, 2022

focused too much on trying to buy land rather than entering into stewardship agreements, which is our strategy at the moment'.[95]

The expansion of Gamsberg zinc mine at Aggeneys meant that Vedanta had to buy up and allocate some land for conservation in order to get its mining licence renewed. This potentially makes some land available to the Wilderness Foundation/ORKCA, which could become another *core* area for the broader rewilding scheme. As it relates to *corridors*, while Norval understands this to be quite important in the long-term goals, 'it is not our immediate objective. On the South African side, the removal of fences is unrealistic at this time. Maybe one day we could build some small corridors, like a miniature game crossing at Pelladrift, but this would be in the future'.[96] We then asked what would actually change in the short and medium term, if mining, irrigation and stock-farming were to continue as usual. Was this just business as usual being clothed in conservation? Norval was unable or unwilling to concretely answer our question, apart from saying that, no, it would not be business as usual, but he was unable to communicate what would change for individual landowners apart from less hunting of predators. He articulated their goals in the lofty, exaggerated language of conservation, that this part of the Northern Cape was one of the 'last pristine wildernesses in South Africa' and that they had to do what they could to conserve what was not being actively mined or farmed. 'Opportunities aren't great anymore, and you've got to take what you've got now'.[97]

95 Ibid.
96 Ibid.
97 Ibid.

Norval and his team did not spend much time visiting non-White communities like Witbank or Pella. 'How to deal with settlements like them is the million dollar question', said Norval. 'We've got the option to either engage with them or work around them. We haven't decided which route we'll take. At the moment, we've got enough keeping us busy, so we'll probably come to the table in the future'. His ORKCA colleagues on the Namibian side have generally dealt with the question of 'settlements' and 'communities' (dog-whistle terms for Africans) by praising the potential employment opportunities that will come out of conservation. Perhaps rightly so, Norval quickly dismissed this illusion.

> Look, tourism potentials for this kind of conservation project in this region are really quite low. I've tried to emphasise to my colleagues who talk about employment that there really won't be many jobs created. At least year-round permanent jobs. We deliberately shut down any possibility for even a fact-finding investigation into tourism.

He referred to Sean Gilbertson's Sandfontein Lodge – possibly the area's largest employer in the tourism sector – stating that even it employed only a few people, varying by season.

•••

It is still early days for ORKCA's transfrontier plans, and so assessing them is ultimately examining a moving target. That being said, the presence of private lands on both sides of the Orange River boundary complicates transfrontier conservation and rewilding in unusual and occasionally unpredictable ways. On the Namibian side, a slow, gradual process of land accumulation into the hands of just a few individuals, via the Sandfontein and Pelladrift companies, facilitated a uniform strategy towards private conservation. Sean Gilbertson's lodge would attract the high-value tourism clientele, while Ian Craig's conservation disciples would initiate a rewilding programme. On the Namibian side, the institution of private property is conducive to this process. On the South African side, things are slightly different, and private property has complicated things. Property ownership there is more diverse. Lands are held by a variety of fairly successful commercial sheep farmers, irrigation growers and a number of highly successful mines. Furthermore, a significant amount of the private properties are owned and governed by community trusts, such as Witbank. Negotiating with these communities is something that must be done on a democratic basis, because government has less influence over these communities compared to formal communal areas. The institution of private property on the South African side has been a hindrance from the days of the PPF up to today.

FIGURE 85 Aerial view of the irrigation settlement, Onseepkans, on the South African side of the River, 2023. ORKCA's transfrontier rewilding programme would need to contend with the reality that vast swathes of both sides of the river are not Terra Nullius
PHOTO COURTESY OF ROSITA SMEENK AND MAURITS MATERS, USED WITH PERMISSION

Since Sean Gilbertson took over Sandfontein more than fifteen years ago, he and his business partners have spent a lot of money importing game from outside of southern Namibia to display on their properties. The most visible of these are the black rhinoceros, owned by the Namibian government through the Black Rhino Custodianship Programme. Whereas the rhino do not belong to Sandfontein, much of the rest of the game on the property does, from the kudu to the giraffe to the cheetah to the springbok. When we asked Gilbertson's manager why they do not remove the game fencing on Homsrivier to make it easier for the wildlife to reach the Orange River (which seems in line with rewilding's goals), Adriaan Mulder's response was simply that the animals were private property. If they crossed to South Africa, Sandfontein's assets would diminish.[98] Pursuing transfrontier conservation in the context of capitalist property relations is difficult. Private nature conservation is a curated kind of nature for a reason; allowing game to migrate, especially across the border to another country, means surrendering control over property rights.

98 Adriaan Mulder, interview with Bernard C. Moore and Luregn Lenggenhager (Homsrivier farm, 12 November 2021).

Nearly all the ORKCA staff, board members and former board members we have spoken to view the Orange River as both a site of immense possibilities (both for conservation and profits) and as a site of immense danger (both to conservation and profits). Much of this trepidation regarding the Orange River centres around the fact that the Bassons, other river-folk and residents of Witbank and Pella live there. Matthew Norval described the Lower Orange River as 'a lawless place' and that with their conservation initiatives, security along the river would finally come.[99] Pete Morkel told us that many believe that allowing the Bassons or other river-folk to stay near the Orange could pose a serious veterinary problem for Namibia.[100] Sean Gilbertson echoed accusations of the Bassons being smugglers.[101] For rewilders, the Orange River is an insecure place because there are people there; in this context the goal of transfrontier conservation is to control and limit the movement of humans while facilitating the international movement of wildlife.

Rewilding is compatible with private property under three circumstances: (1) when there are very, very few private owners; (2) the owners are of the same mind concerning conservation goals; (3) the owners are ideally very wealthy. Yet at the same time, rewilding in many ways *depends* on the institution of private property, and it is private landowners, foundations and patrons who have sustained the movement from its earliest days. In southern Africa, where most game can be private property, rewilding can be big business for a well-connected tourist lodge. It also can be a pleasure project for wealthy patrons. In recent years, however, rewilders have targeted a new avenue to fund their operations and acquire more land: carbon sequestration. According to ecologists Jepson and Blythe:

> In order to develop rewilding into a climate solution, practitioners need to become savvy in the emerging field of carbon finance … If rewilding projects can demonstrate measurable reduction in carbon emissions and/or carbon capture, they can sell these as credits and generate an income stream for rewilding projects.[102]

As the following chapter shows, carbon finance is on the agenda for the ORKCA and many of the projects that have inspired it.

---

99 Matthew Norval, interview with Bernard C. Moore (telephone, 12 January 2023).

100 Pete Morkel, interview with Bernard C. Moore and Luregn Lenggenhager (Pelgrimsrust farm, 10 November 2021).

101 Sean T. Gilbertson to Bernard C. Moore (7 January 2023), letter in the authors' possession.

102 Jepson and Blythe, *Rewilding*, p. 176.

CHAPTER 13

Carbonated Lands

'Old Ian's been scratching around in all directions, hasn't he?' Pete Morkel laughed when he learned that the Orange River-Karoo Conservation Area (ORKCA) was planning a carbon offset credit scheme. 'Look, I find this carbon market to be a bit bullshit, but hey, if well-meaning rich people want to give us their money to do conservation in exchange for carbon credits, then why not?'[1] Morkel implied that in the early days after Ian Craig had purchased his Namibian farmlands, Craig was under the impression that the team would just find a couple of rich benefactors to fund the entire scheme. The voluntary carbon offset market was possibly a new way to tap into benefactors. Rather than begging for or soliciting donations from English royalty or billionaire philanthropists, ORKCA could offer something to 'sell' to business and corporations in the global north looking to 'green' their enterprises and offset their carbon emissions.

'You see, carbon sequestration is not like planting wheat and harvesting it to sell', said Ian Craig. 'While it is a sort of conservation dividend, it's not really about getting a pay-cheque. These carbon credits are an objective measure of land rejuvenation and offsetting of pollution.'[2] Craig echoed many established voices within the carbon finance industry. In many ways, carbon finance has fundamentally changed how 'conservation' operations work. Even more broadly, it has reshaped the relationship between rural, peripheral areas and the global centres of power. This chapter examines the institutional and scientific background of carbon credits, whether voluntary or governmental, and considers the relationship between emissions producers in the Global North and envisioned 'carbon sinks' in the Global South, showing how locales in the Global South continue to be, as some researchers argue, 'rural feedstock for the industrialised North'.[3] However, instead of producing wheat, meat or skins to export to Europe or North America, the 'dividend' – as Ian Craig phrased it – is proof that a given quantity of carbon dioxide equivalent has been removed from the atmosphere.

1 Pete Morkel, interview with Bernard C. Moore (telephone, 19 October 2022).

2 Ian Craig, interview with Bernard C. Moore (telephone, 18 October 2022).

3 R.P. Lejano, W. Shan Kan and C.C. Chau, 'The Hidden Disequities of Carbon Trading: Carbon Emissions, Air Toxics, and Environmental Justice', *Frontiers in Environmental Science*, 8 (2020), pp. 1–6.

DOI:10.1163/9789004712454_014

FIGURE 86 Prosopis glandulosa (honey mesquite) growth on southern Homsrivier Farm, surrounded by sands from the Orange River floods
PHOTO: B.C. MOORE, 2021

After exploring the institutional background of carbon offset credits, our narrative moves again to East Africa, particularly Ian Craig's operations on the Northern Rangelands Trust, which controversially pioneered one of the first soil organic carbon (SOC) sequestration schemes. We explore the social, financial and technological elements of the Northern Kenya Grassland Carbon Project, considering what these schemes mean for the future of conservation financing. Finally, we return to southern Namibia, where Ian Craig and his staff on ORKCA envision a carbon scheme to finance their operations along the Orange River, one that potentially conflicts with their idealistic goals of 'rewilding' the landscape. On the whole, this chapter reveals troubling trends emerging within conservation finance, which ultimately rewards those projects that do not utilise their lands and structure their definitions of conservation to be most favourable to emissions-producing companies in the Global North.

•••

The modern carbon finance market emerged out of the 1997 Kyoto Protocol of the United Nations Framework Convention on Climate Change, which sought to use market-based incentives to commit national governments to reduce emissions, on a global stage. So-called Kyoto Units were allocated to each participating nation, granting them the emissions rights at each country's 1990 levels. These units could be bought and sold among signatory parties via a newly formed emissions trading system, such that if, say, Germany successfully reduced its emissions for a period of time, some of its Kyoto Units could be sold to a country which exceeded its quota.

In addition to emissions trading, the Kyoto Protocol built the nucleus for the modern-day carbon offset credit market – the Clean Development Mechanism – which allowed signatory parties to invest in greenhouse gas reduction projects in developing nations in order to offset a nation exceeding their Kyoto Unit quota. It was envisioned that the financial pressures placed on developed nations that exceeded their quota would incentivise them to reduce emissions. Likewise, it was also believed that potential profits from the sale of offset credits would incentivise developing nations to invest in cleaner energy and green development schemes.[4]

The Clean Development Mechanism was the most important aspect of the Kyoto Protocol's great compromise, which would taint emissions-reduction schemes in the years to come. Ultimately, the most polluting states in the Global North did not really want to stop polluting. Nevertheless, it was believed that emissions could be reduced on a global scale if the Global South got involved through the sale of carbon offset credits, which represented the certified removal or avoidance of one tonne of CO_2 equivalent (hereafter, tCO_2-e) from the atmosphere. As is the case today, if they were to receive Kyoto funds, approved offset projects needed to prove *additionality*, and anticipate and avoid *leakage*. All carbon offset or sequestration projects revolve around these elusively simple terms. First, it must be shown that whatever intervention is being made will reduce carbon emissions to below what would ordinarily be the case if no project were supported.[5] Second, it must be shown that the emissions that are being locally expunged or reduced are not simply being moved elsewhere. If a coal furnace in one place is replaced with a solar plant, and the coal furnace is merely reconstructed elsewhere for the same purpose, this would be an example of *leakage*.

Carbon markets depend on complicated calculations of greenhouse gas equivalence. Even though there are scores of gases emitted into the atmosphere, aggregate emissions and removals are to be expressed only in tonnes of CO_2 equivalent (tCO_2-e), calculated based on 'global warming potential values' specific to each greenhouse gas.[6] These calculations have been challenged and revised on a number of occasions, and this was only part of the reason for the volatility of the emissions trading system in its early years.[7] On a

4 UNFCCC, *Achievements of the Clean Development Mechanism, 2001–2018: Harnessing Incentive for Climate Action* (2018).

5 D. MacKenzie, 'Making Things the Same: Gases, Emission Rights and the Politics of Carbon Markets', *Accounting, Organizations, and Society*, 34 (2009), p. 443.

6 See details and revision in UNFCCC, 'Report of the Conference of the Parties on its Nineteenth Session, held in Warsaw [FCCC/CP/2013/10/Add.3]' (11–23 November 2013), pp. 11–13.

7 M. Paterson, 'Governing Mobilities, Mobilising Carbon', *Mobilities*, 9, 4 (2014), pp. 570–584.

more fundamental level, carbon finance requires legibility and measurability: one must be able to calculate what is emitted and what is sequestered and express this in monetary terms. Not only is much of this research funded by or supported by companies that contribute disproportionately to national and global greenhouse gas emissions, it has also given rise to new 'ecoinformatics' companies that seek to profit from either the sale of offset credits or a more sophisticated measurement of them.[8] Some may idealistically think that these market-based pressures would incentivise behavioural change on the part of industries that pollute, but the evidence points less to the success of the Kyoto model and more to the costs of carbon offsets being placed increasingly upon workers and consumers rather than on profit margins.[9]

By the turn of the twenty-first century, there was increased mobilisation for a voluntary carbon market that would be at least partly disconnected from the Kyoto model. Whereas the United Nations system was primarily targeted at governments, a voluntary system would reflect private efforts of companies themselves to 'green' their operations in response to consumer and civil society pressures. Companies understood that consumers and investors increasingly wanted their dollars not to be spent on operations that contributed towards global warming and climate change. Like the great Kyoto compromise described above, the voluntary carbon offset market was ultimately a means for corporations to claim that they were quickly cleaning up their act while not necessarily taking swift action to reduce emissions. Where true emissions reductions require changing technologies, changing lifestyles and changing priorities, reducing both production and profits, carbon offset schemes allow for companies to potentially 'reduce' emissions by 100 per cent in a fairly short period of time.[10] 'Net Zero' or 'Carbon Neutral' companies may not need to drastically reduce their emissions at all, as long as they buy sufficient credits.

In order for companies to purchase carbon offset credits on a voluntary basis, a parallel system needed to be put in place whereby conservation organisations and companies building projects to sequester or offset carbon could have their operations independently verified as scientifically grounded. Carbon finance itself necessitates a global equivalence of carbon offset programmes, such that one tCO_2-e sequestered or offset in one country can be verified as true and equivalent to projects in other countries. The largest player in the certification

8 See S. Sullivan, 'Banking Nature? The Spectacular Financialisation of Environmental Conservation', *Antipode*, 45, 1 (2013), p. 205.

9 R. Felli, 'On Climate Rent,' *Historical Materialism*, 22, 3–4 (2014), pp. 251–280.

10 H. Lovell, H. Bulkeley and D. Liverman, 'Carbon Offsetting: Sustaining Consumption?' *Environment and Planning A*, 41 (2009), pp. 2357–2379.

of carbon projects is the Washington-based organisation, Verra. Founded in 2007 as a joint project by the London-based hedge fund Cheyne Capital Management and the American financial advisory firm Climate Wedge Ltd, Verra was intended to be an impartial review agency for the scientific methodologies by which voluntary carbon schemes proved their additionality and managed their projects.[11] Verra's 'Verified Carbon Standard' – which laid out the terms by which carbon schemes could be approved – was developed in consultation and cooperation with heavy polluters like Rio Tinto, Glencore, British Petroleum, Cargill, Bank of America, Goldman Sachs and others.

Fifteen years later, Verra was certifying and overseeing more than two thousand carbon offset projects, facilitating the sale of verified credits on the international market for over seventy five per cent of all carbon offset credits globally.[12] Despite being a not-for-profit NGO, Verra has become a massive organisation on which the entire carbon finance market depends. After all, without verification that *additionality* is present, that *leakage* has been avoided and *permanence* has been maintained, the entire system of trust that specific quantities of tCO_2-e indeed have been offset collapses. Verra has situated itself as an impartial broker mediating the exchanges between the company seeking to 'green' its operations, the international project looking to sequester carbon, and the scientists and inspection agencies who write the methodologies and monitor carbon capture.[13]

Although the vast majority of global emissions originate from developed economies in Europe and North America, as well as industrialised East Asia, the vast majority of Verra's carbon offset and sequestration projects are situated in the Global South. It has certified more projects in Africa than Europe and North America combined. This may not necessarily matter as it relates to atmospheric CO_2, but it does matter as it relates to strategies for development and where a national economy or national environment fits into the global economy. Even before carbon offset credits were sold by projects in the developing world, many scholars understood that the Global South was the source of 'ghost acres', lands coerced to support the expanding diets and expanding

---

11 Climate Wedge Ltd, 'Company Overview' http://www.climatewedge.com/company.html (Accessed 30 June 2023).

12 See T. Fischer and H. Knuth, 'Grün getarnt', *Die Zeit* (19 January 2023).

13 Although Verra's offset programmes are from many sectors – renewable energy, waste disposal, emissions capture and others – they are most known for their relationship with REDD+ schemes (reducing emissions from deforestation in developing countries), which protect and replant forests. Verra also receives a small cut of all credits sold under their auspices.

consumer behaviours of citizens of the Global North.[14] Indeed, critical social scientists have observed parallels, whereby ghost acres in the Global South are gradually being reformulated. Instead of being cash-crop plantations they are becoming locales where carbon emissions from the Global North can be easily and cheaply offset, cleansing the conscience of polluters and easing societal pressures upon them.[15]

•••

In the first five years of the voluntary carbon market, there was an explosion in private donations and investment into carbon sequestration and offset schemes across the globe. It followed, then, that conservation projects increasingly targeted carbon finance as a means to support their initiatives. This was ultimately part of a broader shift whereby conservation initiatives structured their operations and foci to meet the investment and social corporate responsibility goals of for-profit firms. The massive organisation, The Nature Conservancy, particularly its in-house investment wing, NatureVest, was at the forefront of this transformation.[16]

As early as 2009, The Nature Conservancy entered into discussions with Ian Craig at Kenya's Lewa Wildlife Conservancy to explore the possibilities of building a carbon project within the (then recently formed) Northern Rangelands Trust (NRT).[17] Lewa and the NRT were in many ways the poster children of the possibility to wed private and community nature conservation operations. Both the Kenyan parties and the international agencies viewed carbon as a possible way to further fund these initiatives. However, Ian Craig and the NRT did not really have forests to conserve – they had only grass and soil. If they were to sell carbon offset credits, they would need to reconceptualise and establish new ways of measuring carbon stocks and carbon sequestration.

They called Mark E. Ritchie, Professor at Syracuse University in the United States. After receiving his PhD in Environmental Science from the University of

14 See G. Borgström, *The Hungry Planet: The Modern World at the Edge of Famine* (New York, Macmillan, 1972 [1965]), pp. 74–76. Björn-Ola Linnér, 'The World Household: Georg Borgström and the Postwar Population-Resource Crisis' (PhD thesis, Linköping University, 1998), pp. 205–206.

15 Lejano et al., 'The Hidden Disequities of Carbon Trading, pp. 1–6.

16 See Dempsey and Bigger, 'Intimate Mediations of For-Profit Conservation Finance'.

17 See 'The Northern Kenya Rangeland Carbon Project: Partners', *Northern Rangelands Trust* https://www.nrt-kenya.org/carbon-project-partners

Michigan in 1987, where he wrote a dissertation on ground squirrels,[18] Ritchie gradually shifted focus and became established as an expert in plant-herbivore interactions and the effect of grazing behaviour on soil fertility. In addition to his research and teaching, Ritchie would take up contract work regarding landscape regeneration and offer his research services to conservation operations in the American West and occasionally East Africa. Over time, his interest and expertise centred on how one can measure changes in the composition of grassland soils, particularly soil organic carbon (SOC) levels.

'I first grew seriously interested in soil carbon when I was working on a regenerative grazing project near the site of an old, abandoned gold mine in Nevada', Ritchie said. 'My colleague asked me if there was concrete data on the relationship between changing grazing patterns and SOC capture rates. After digging in my computer a little bit, I saw that there definitely was'.[19] In April 2010, Ritchie founded a Delaware-domiciled limited liability company, Soils for the Future, which provides technical services and project management to companies, landowners and other entities seeking to improve their soil. Most of their operations are concerned with measurement of SOC stocks, as well as supporting carbon offset credit programmes which deal with soil.[20]

At this stage, there was no accepted methodology for assessing and verifying SOC storage projects for the production of carbon offset credits, and Mark Ritchie was working on building one. The Nature Conservancy and Fauna and Flora International – Lewa and the NRT's largest supporters – hired Ritchie to design what eventually became known as the VM0032 methodology, which was submitted to Verra in late 2010 and was eventually fully approved after modifications in 2015.[21] Ritchie's methodology would be one of the first to provide for large-scale soil organic carbon capture projects, and most future SOC schemes, including NRT's, would be based on it.

It is worthwhile to delve briefly into the methodology required to ascertain SOC storage in grasslands, because this will enable us to contextualise the socioeconomic foundations of soil carbon capture for conservation projects in semi-arid and arid areas, like Kenya and Namibia. VM0032 projects are intended to be those that manipulate grazing strategies, timing and grouping of domesticated livestock 'in ways that sequester soil carbon and/or reduce

18 M.E. Ritchie, 'Natural Selection of Optimal Foraging Behavior in Columbian Ground Squirrels' (PhD thesis, University of Michigan, 1987).

19 Mark E. Ritchie, interview with Bernard C. Moore (telephone, 13 December 2022).

20 See 'Solutions through Soil', *Soils for the Future* https://www.soilsfuture.com/welcome

21 Verra Verified Carbon Standard database (hereafter, VCS) no. VM0032: 'Methodology for the Adoption of Sustainable Grasslands through Adjustment of Fire and Grazing' (version 1.0, 16 July 2015).

methane emissions'.[22] Ritchie designed the methodology to be applied to projects taking place on grasslands that have 'historically experienced carbon loss', such that one could frame additionality on the assumption that without the SOC project, carbon loss and land degradation would continue. Before one can measure the growth of soil organic carbon stocks, one needs to establish a baseline from which future growths would deviate. Here is where soil becomes quite difficult.

'You could theoretically take soil sample cores from maybe 200–500 points along the project site, and then come back after a sufficient period of time to take new measurements from the same locations to measure the growth', Ritchie explained.

> The problem with the measured strategy is that while SOC capture still occurs in more arid areas, the rate of change is so slow that one would have to return after ten years or so to get statistically sufficient measurements. For landowners or companies sponsoring these projects, this just isn't cost-effective.[23]

Ritchie's methodology instead relied on a mathematical modelling approach, which would model and simulate the initial soil core samples based on a comparable case study.[24] In short, if a comparable published data set existed that claimed an average carbon sequestration per hectare, then future SOC projects that controlled land use in similar ways could simply measure the baseline using soil core samples and then extrapolate the gains from the model. One still needs to take core samples occasionally to continue to show that the project adheres to the other published study, but the advantage of the modelled approach is that one does not need to constantly measure samples in a laboratory to claim the sale of carbon offset credits.[25] Finally, the methodology required monitoring and recording leakage, such that if livestock were moved out of the target area, this would be reflected in the model.

With the methodology submitted for review in 2010 – and expected to pass with some small modifications – the Northern Rangelands Trust officially launched what would become known as the Northern Kenya Grasslands Carbon Project (NKGCP), in coordination with its main funding bodies, The Nature

22 VCS no. VM0032, sec. 2.1, pp. 4–5.

23 Mark E. Ritchie, interview with Bernard C. Moore (telephone, 13 December 2022).

24 VCS no. VM0032: sec 2.3, p. 9.

25 Mark E. Ritchie, interview with Bernard C. Moore (telephone, 13 December 2022).

Conservancy and the governmental agencies USAID and Danida.[26] It brought Mark Ritchie and his team to Kenya in late 2012 to take 233 soil core samples from the fourteen participating NRT communal conservancies, to establish the baseline on which the changes in SOC capture could be compared. It was envisioned that the NKGCP would go hand in hand with the GrazingWORKS livestock-to-market programme (see chapter 11), which was established in 2010 with USD 3.5 million in 'impact investment' funding from NatureVest.[27]

The NKGCP spanned a territory of more than 2 million hectares, and it was home to more than 112,000 primarily Samburu and Maasai residents. It applied for Verra certification in 2015 with the goal of rehabilitating rangeland and sequestering SOC through shifting pastoralist herding strategies from 'continuous, unplanned grazing' to 'planned rotational grazing' across the project area.[28] The project would prove additionality based on the old-fashioned 'tragedy of the commons' view of communal grazing as inherently destructive, 'depleting soils of organic matter'.[29] In many ways, SOC projects in the Global South are indirectly pressured to condemn traditional grazing systems in order to conform to modernist development logic: a logic which works in favour of carbon projects seeking such additionality. Justifying additionality can therefore go hand-in-hand with resistance by local pastoralists to such a scheme, as the latter's hesitancy to change their grazing strategies can in turn be reported back to funding bodies as evidence of antiquated traditional worldviews in need of intervention. The NKGCP promoters planned to use funds from credit sales to institute a rotational grazing system whereby livestock would be constantly herded, never grazing on one particular pasture more than once per wet and dry season, increasing annual SOC sequestration by approximately 1.85 million tCO_2-e across the constituencies.[30] On its final approval in 2020, the NKGCP became the largest soil organic carbon scheme in the Global South.[31]

---

26 VCS Project 1468: 'Proof of Concept Document' – 21 July 2015. The NRT refers to it in their promotional materials as 'rangelands project' rather than 'grasslands project', but the latter is the legal name for the project.

27 IAWB: 'Livestock to Markets', *NatureVest.* https://web.archive.org/web/20150621062742/http://www.naturevesttnc.org/our-projects/lands/livestock-to-markets/

28 VCS Project 1468: 'Northern Kenya Grasslands Carbon Project: Project Description' – 15 June 2015.

29 VCS Project 1468: 'Proof of Concept Document' – 21 July 2015.

30 VCS Project 1468: 'Northern Kenya Grasslands Carbon Project: Project Description' – 15 June 2015.

31 NRT-Kenya, 'Frequently Asked Questions about the Northern Kenya Grassland Carbon Project', *Northern Rangelands Trust* https://www.nrt-kenya.org/carbon-project-faq (Accessed 1 June 2023)

In SOC projects involving livestock, measuring or preventing leakage is a central concern, especially when droughts are taken into account. Not only would droughts potentially lower the possible SOC sequestration gains, but they could also result in livestock moving off the boundaries of the carbon project in search of grazing, potentially nullifying the project's adherence to the methodology. The NKGCP was remarkably vague in its methods of preventing leakage, implying that the value of the project itself would render leakage an impossibility.

> (1) Rapidly build grass banks that can be used as forage during droughts. (2) The accumulation of soil carbon will lead to greater infiltration of rainfall, which will make communities progressively less vulnerable to climate variability and drought as the project proceeds. (3) Further build the livestock-to-market program instituted by NRT to provide cash compensation for de-stocking in advance of intense droughts. (4) Through education associated with monitoring social benefits of the project, continue to convey the community and personal revenue advantages linked to project success (and thus participation). (5) The project will progressively entrain sustainable livestock management processes as a new "tradition" of planned rotational grazing.[32]

Mark Ritchie framed their strategies as maintaining livestock numbers and grazing pressures steady across the whole project area but increasing movement from area to area.[33] Of course, the fundamental ecological and ethical problem of the NKGCP is that if livestock numbers must be maintained steady across the project area for a given period of time – necessary to prove both additionality and avoidance of leakage – then African pastoralists whose communal conservancies signed on to the programme are *effectively banned* from increasing their wealth in livestock.

This is where the GrazingWORKS livestock-to-market programme became so crucial: it needed a means to de-stock the conservancies if carbon credits were to be certified and sold. Additional livestock would violate the terms of the methodology and reduce the SOC sequestration rate, yet at the same time, community buy-in would be lost if all accumulation of wealth in cattle were banned. GrazingWORKS had been founded with broader conservation goals regarding community buy-in – that is, permission to sell cattle within

32 VCS Project 1468: 'Northern Kenya Grasslands Carbon Project: Project Description', sec. 2.3, pp. 17–18.

33 Mark E. Ritchie, interview with Bernard C. Moore (telephone, 13 December 2022).

the programme necessitated acceding to certain wildlife conservation practices. But with an additional USD 7 million impact investment from J.P. Morgan Chase (via NatureVest) in 2015, the project shifted focus and scale.

During the earlier years of USAID and Danida grant funding alone, GrazingWORKS was purchasing only about 1,000 cattle per year from NRT conservancies to fatten on Ol Pejeta, Borana and other conservancies. After the NatureVest funding came in, the programme immediately ramped up production to purchase up to 10,000 cattle per year.[34] Some critics of GrazingWORKS have argued that the programme was intended to de-stock communities to make way for wildlife,[35] but it is more likely that GrazingWORKS provided the infrastructure to avoid leakage within an intended SOC sequestration project. By constantly buying increasing numbers of young cattle from African pastoralists, livestock numbers could remain fairly stable, and the average age and weight of animals on the communal areas – where SOC was being sequestered – would likely decrease, potentially reducing grazing burdens and increasing SOC. However, larger and larger quantities of heavier cattle were now grazing outside the project area, on the private commercial conservancies, likely negating some of the benefits of whatever sequestration would occur on the NKGCP lands.

The GrazingWORKS system also set in place the disciplinary mechanisms by which African pastoralist grazing behaviour could be monitored and controlled by conservationists like the NRT, Lewa, Soils for the Future and others. Not only would both GrazingWORKS and the NKGCP use investor funds to hire separate cattle herds, livestock wardens and game rangers – individuals whose livelihoods depended upon adherence to these new models of grazing behaviour – the programmes instituted new technological shifts in surveillance. Mark Ritchie remarked that his team used satellite imagery to monitor long-term vegetation changes and short-term grazing patterns. They could track pastures that had already been grazed, direct livestock herders employed by the programme and police those grazing the same sections twice.[36]

Kenyan law grants conservancies ownership over carbon rights, so it was necessary for the NRT to obtain signed 'Carbon Consent Declarations' from

---

34 IAWB: 'Investing in Conservation and Communities: How Impact Investment Generates Benefits for People and Nature in Kenya', *NatureVest* (May 2015). https://web.archive.org/web/20150701174407/http://www.naturevesttnc.org/wp-content/uploads/2015/05/LTM_Infographic.pdf

35 See Fox, 'The 2017 Shooting of Kuki Gallmann', pp. 225–226.

36 Mark E. Ritchie, interview with Bernard C. Moore (telephone, 13 December 2022). This included using NDVI infrared satellite imagery, as well as NASA's MODIS 16-day composite vegetation indices.

each conservancy before it could enter into negotiations to market credits on the conservancies' behalf. This took a troublingly long time to accomplish – they were not all signed until 2021, despite initial surveying on NRT lands beginning in 2012 and the NKGCP itself being approved in 2015.[37] In 2017, the NRT entered into a sales contract with the company NativeEnergy, one of the largest developers and retailers of carbon finance projects and offset credits in the United States. NativeEnergy became the sole marketer of credits from the Northern Kenya Grasslands Carbon Project, of which 3.2 million tCO_2- e for the period 2013–2016 were certified by Verra and its auditors in 2020.[38]

By the middle of 2022, according to Ian Craig, all 3.2 million SOC offset credits were sold for USD 12 per credit, bringing in a gross revenue of USD 38.4 million.[39] The purchasers of the NKGCP credits are well-known international technology, travel and banking companies.[40] In February 2022, the first payments to NKGCP conservancies were made, with each conservancy receiving KES 38,880,000 (approximately USD 324,000 at the time), and it was expected that they would receive the same payment in 2023, bringing the total amount disbursed from the 2013–2016 monitoring period to about USD 9 million.[41] While this is not an insignificant amount of money, when spread out across the population of the conservancies – 112,000 people and assuming an average family size of seven – it is only about USD 112 per family per year for all the work, personal sacrifice and surveillance to sequester carbon. This is certainly less impressive than framing the NKGCP revenues in aggregate. Furthermore, if the NRT's numbers are reliable, more than seventy five per cent of the revenue from the sales of carbon credits was lost to middle-men, scientists and agents, including Verra, NativeEnergy and Syracuse University. One may reasonably wonder if this is an acceptable overhead for a charitable project that has won international awards.[42]

37 VCS Project 1468: 'Comments received by Verra during the public comment period 12 January 2023 to 11 February 2023 for the verification of the Northern Kenya Grassland Carbon Project' – 23 January 2023.

38 VCS Project 1468: 'Northern Kenya Grassland Carbon Project VCS Validation Report' (Aster Global Environmental Solutions Inc, 7 April 2020).

39 Ian Craig, interview with Bernard C. Moore (telephone, 18 October 2022).

40 The top purchasers were Netflix (180,000 credits), NatWest Bank (120,000), Meta/Facebook (90,000), Kering (75,000) and British Airways (25,000). Most credit sales from the NKGCP are anonymised, however, or referred to simply as 'NativeEnergy Clients'. See VCS Project 1468: 'VCUs Issuance' https://registry.verra.org/app/search/VCS?programType=ISSUANCE&exactResId=1468

41 Northern Rangelands Trust, *State of Conservancies Report, 2023*, p. 233.

42 Ibid, p. 235.

The NRT admits that thirty per cent of revenue from sales goes to the costs of verification, issuance, ecological monitoring, reporting, and so on, but this is still far from its claim that more than sixty per cent of revenue would go – in cash, kind or service – to the fourteen communities.[43] It is unclear where the rest of the funds are used. Of the money that does get disbursed to conservancies, forty per cent goes towards conservancy operational costs – half of which goes to the rangeland programme itself – and the rest is directed towards 'social development projects chosen by the community'.[44] In its 2023 report, it stated that some carbon revenue went to important projects, such as water infrastructure installations, tourism enterprise development and more than six thousand bursaries for primary and secondary school students.[45] More than USD 150,000 was allocated for drought mitigation efforts, as well.

These funds are significant, and they were meaningful enough that the GrazingWORKS livestock-to-market programme was in the process of being phased out.[46] By 2021, GrazingWORKS had been scaled back to nearly fifteen per cent of its height in 2015–2018.[47] Ian Craig stated that, in the future, the NRT and other conservation operations would focus more heavily on the duo of tourism and carbon as a means to finance their operations. Carbon was the new cash cow, and they were ready and willing to double down on their efforts to sequester carbon. After all, the NRT estimated that the Northern Kenya Grasslands Carbon Project would be able to offset more than fifty million tCO_2-e over the thirty-year lifespan of the project.[48] Many, including Ian Craig, viewed carbon offset credits as the future of conservation finance, and Craig intended to bring this approach with him to his new operations in Namibia.

•••

As Oana gradually morphed into Orange River-Karoo Conservation Area (ORKCA) in 2021, its funding strategy shifted from gap-year environmental education and 'adventure tourism' programmes to one much more intertwined with the financial connections of conservation operations like Lewa, the NRT and The Nature Conservancy. Andreia Pawel noted that by 2022 KumKum Adventure Tourism (Pty) Ltd – the business-side of Oana – was effectively

43 NRT-Kenya, 'Frequently Asked Questions about the Northern Kenya Grassland Carbon Project', *Northern Rangelands Trust* https://www.nrt-kenya.org/carbon-project-faq
44 Ibid.
45 Northern Rangelands Trust, *State of Conservancies Report, 2023*, pp. 266–267.
46 Ian Craig, interview with Bernard C. Moore (telephone, 18 October 2022).
47 Northern Rangelands Trust, *State of Conservancies Report, 2021*, pp. 83–84.
48 Northern Rangelands Trust, *State of Conservancies Report, 2023*, p. 232.

FIGURE 87 Aerial view of the steep and mountainous Pelladrift Farm, within ORKCA's conservation zone, 2023
PHOTO COURTESY OF ROSITA SMEENK AND MAURITS MATERS, USED WITH PERMISSION

dormant, and efforts were being rerouted towards developing ORKCA as a broader initiative.[49] The strategies included increasing land acquisition and custodianship agreements on both sides of the Orange River and finding new revenue streams. Although Ian Craig and Sean Gilbertson likely had deep pockets, they were not going to fund ORKCA for all eternity. Andreia Pawel and Ed Barthorp would remain the young, optimistic faces of ORKCA, but standing behind them was ORKCA's new board chairman, Jonathan Baillie.

Born in Montreal in 1970, Jonathan Baillie earned degrees in Canada, the USA and a PhD in biology from Imperial College London, after which he worked for the National Geographic Society and the Zoological Society of London.[50] In 2016, Baillie married Ian Craig's daughter, Jecca, in a ceremony on Lewa Wildlife Conservancy, attended by more than three hundred people, including Prince William, Duke of Cambridge.[51] Probably through his close

49 Andreia Pawel, interview with Bernard C. Moore (telephone, 10 October 2022).

50 Jonathan Baillie, *LinkedIn*. https://www.linkedin.com/in/jonathan-baillie-82379713/ (Accessed 21 November 2024).

51 S. Jamieson and R. Savage, 'Prince William Joins Champagne Reception Celebrations for Ex-Girlfriend Jecca Craig's Wedding', *The Telegraph* (26 March 2016).

connections with his father-in-law, Baillie would be drawn into Craig's private conservation efforts in Namibia, and like Ian Craig, Jonathan Baillie would increasingly draw on new technology to better measure possible ways to fund conservation.

Since April 2020, Baillie has been President of a Washington-based charitable company called Natural State. Perhaps echoing the call from billionaire industrialist/philanthropist Hansjörg Wyss, who called for preservation of at least thirty per cent of the earth's surface, Natural State seeks to contribute to this 'through large-scale restoration and rewilding', whereby they will 'secure biodiversity, sequester and store carbon, and help achieve the Sustainable Development Goals'.[52] Along these lines, the organisation looks to develop technology – including acoustic sensors and drones – to measure and monitor the impact of conservation activities, in order to produce 'robust and cost-effective site-level data'. This data would, according to Natural State, 'help unlock billions of dollars for nature' in part through pioneering new forms of conservation financing, including 'an Integrated Carbon and Biodiversity Credit, Rewilding Credit, and Single Species/Custodian Credit'.[53]

Natural State have taken the lead to promote ORKCA as a par excellence case of 'rewilding at scale', arguing that their efforts would 'ecologically restore the fragile semi-arid ecosystem, quadruple the number of native grazers and predators, restore wildlife migration patterns, secure fifteen million tCO_2-e and triple employment in the region through eco-tourism'.[54] Similar to the operations of agencies like the Taskforce on Nature-Related Financial Disclosures, which works with major banks and manufacturers to investigate and report on the impact that companies have on the environment, it appears that Natural State has tried to position itself and its supported projects as a means through which corporate funds could be channelled towards ecological regeneration programmes.

It follows, then, that Natural State would need to conduct research and monitoring programmes, develop appropriate technology to quantify corporate ecological damage and prove that corporate dollars are indeed regenerating what has been degraded.[55] Natural State utilises new remote imaging technologies, acoustic measurement devices, and camera traps to build

52 Natural State, 'About Us' https://www.naturalstate.org/about-1 (Accessed 21 November 2024).

53 Ibid.

54 Natural State, 'Restoration and Rewilding at Scale', https://www.naturalstate.org/restoration-rewilding-at-scale (Accessed 21 November 2024).

55 Natural State also keeps an office and staff in Kenya near Lewa.

'predictive' models for soil organic carbon and 'AI models' for birds and terrestrial vertebrates;[56] presumably, the organisation believes these models can evaluate the biodiversity gains when certain changes are made to particular landscapes. In short, conservation organisations are beginning to understand that the more one can improve measurement regimes, the easier it is to 'prove' additionality for carbon offset projects. Implementation of these strategies within ORKCA is being funded by the Rolex Perpetual Planet Grant, via Natural State, which provides the funding to purchase and install the remote-sensing systems that will help identify baselines.[57] New technologies enable conservationists to find new ways to 'prove' that their work adds value, and (as we authors see it) if you cannot measure it, you cannot sell it.

When asked about the applicability of his SOC sequestration methodologies in hyper-arid landscapes like southern Namibia, where rainfall is well under 100 millimetres per annum, Mark Ritchie remarked that he would not try SOC projects in any area with less than 250 millimetres of rainfall. Furthermore, he believed that where there were steep mountainous areas – with gradients exceeding 10 degrees incline, like much of southern Namibia – a good deal of the land has to be effectively removed from the calculations of SOC sequestration, because carbon will not accumulate as much where land is too steep and rocky.[58]

Natural State's reported operations indicate that they do not concur with Ritchie's methodology. ORKCA's decision to hire a project manager whose tasks include carbon and biodiversity monitoring implies that they believe that there is revenue to be had in pursuing this. As ORKCA's job advert reads:

> With Natural State and their best-in-class partners, we will be using aerial and ground level remote sensing data, including aerial scans, LiDAR technology, camera traps, and acoustic sensors, along with satellite data, to measure changes in carbon and biodiversity as the land transitions from a degraded to a restored state. This data will be used to inform AI modeling systems that will enable the analysis of vast datasets to quantify impact over time. The technology will be used to measure different impact parameters, such as above-ground carbon, soil organic carbon, species richness, relative abundance, functional diversity, and financial benefits to local communities, providing the rigor and transparency

56 Natural State, 'Impact' https://www.naturalstate.org/impact (Accessed 21 November 2024).

57 See ORKCA, 'Job Description: Project Coordinator' https://orkca.org/wp-content/uploads/2023/06/Job-Description-ORKCA-Project-Coordinator-.pdf (Accessed 21 November 2024).

58 Mark E. Ritchie, interview with Bernard C. Moore (telephone, 13 December 2022).

> needed to catalyze nature-based solutions markets. All of the resulting data will be extremely useful in modeling for similar properties within the wider ORKCA landscape, as well as being scalable to other arid and semi-arid landscapes.[59]

Perhaps the amounts of tCO2-e sequestered would be less than in the NKGCP, but evidence suggests that the team at Natural State and ORKCA believe that SOC sequestration would nevertheless occur. Furthermore, ORKCA also seeks, through their relationship with Natural State to tap into other kinds of conservation finance, including their 'integrated carbon and biodiversity credit' scheme, whereby the credit purchased would reflect perhaps only a small amount of carbon sequestered but it would purportedly represent broader biodiversity gains.[60]

Natural State's 'biodiversity credit' idea is rooted in subtle shifts in the conservation finance market over the past several years.[61] Carbon finance is based on an understanding of atmospheric universality: one tCO_2-e emitted in one country is equivalent to one tCO_2-e sequestered in another. There are important ongoing scientific debates about how one precisely measures this and calculates equivalency, but this is accepted by many within the carbon finance sector as well as many scientific researchers. Biodiversity, however, is more qualitative; protecting a wetland in Uganda does not necessarily equate to destroying one in Florida.[62] This is increasingly being challenged, however, with some critics alluding to consumer price indices as comparisons. Although the content of the 'basket' of goods that one uses to calculate inflation rate and other economic indicators may vary from country to country, one may nevertheless compare composite baskets.[63] As it relates to biodiversity credits, this may mean valuating a 'basket' of five or more ecological metrics – such as abundance of a specific species, soil health and air quality – and tracking the

59 ORKCA, 'Job Description: Project Coordinator' https://orkca.org/wp-content/uploads/2023/06/Job-Description-ORKCA-Project-Coordinator-.pdf (Accessed 21 November 2024).

60 Natural State, 'Impact' https://www.naturalstate.org/impact (Accessed 21 November 2024). ORKCA board member Sean Gerrity's main project, the American Prairie Reserve in Montana, appears to be moving in a similar direction. See C. Freese, K. Kunkel and M. Sommer, 'American Prairie Reserve: Climate Change Mitigation and Adaptation through Large-Scale Grassland Conservation' (2016).

61 See also, A. Buller, *The Value of a Whale: On the Illusions of Green Capitalism* (Manchester, Manchester University Press, 2022), pp. 250–255.

62 See J. Song, 'An Incomplete Guide to Biodiversity Credits', *Conservation Finance Network* (14 February 2023).

63 'Biodiversity Credits', *Operation Wallacea* https://www.opwall.com/biodiversity-credits/

inclination or degradation of the collection of ecological indicators, quantified in the calculated value of the 'basket' itself. There are many definitions claimed by various organisations, but according to the recent methodology published by the Wallacea Trust – in consultation with major financial institutions – one biodiversity credit is defined as a '1 per cent uplift or avoided loss in the median value of the basket of metrics per hectare'.[64]

In contrast to the universality of carbon offset credits, most biodiversity credit schemes are proprietary and specific to each company, NGO or country pioneering them. This is not very surprising, however, because biodiversity as an ecological concept is far more qualitative in essence; it is not as universal as tCO_2-e. Biodiversity credit companies and projects have leaned into this fact, proposing that their vast array of new data from new technologies could be individualised to the specific hectare of land via distributed ledger (blockchain) technology and reported back to the buyer.[65] This is the position of South African company ValueNature, which plans to use blockchain technology to 'represent various biodiversity indices in a scorecard format'.[66] ORKCA staff are leaning in this direction as well, arguing that non-fungible tokens (NFTS) and 'conservation investment in the blockchain' is in the near future.[67]

This is a troubling trend. Individualising biodiversity programmes and ecological regeneration – such that theoretically a buyer could trace each and every hectare of rejuvenated land – ultimately renders the equivalence of the units null and void. Each tCO_2-e released into the atmosphere is anything but unique, and multinational corporations are not meticulously documenting each hectare of tropical rainforest removed in the Amazon and the Cerrado. Destruction is generalised, pollution is generalised. As biodiversity credits become more specific and individualised on the blockchain – such that no two credits are identical – enforcement and oversight *completely disintegrate*. When Netflix purchased 180,000 carbon offset credits from the Northern Rangelands Trust, the company had no intention of examining each and every credit: it would be impossible. Verifiers of biodiversity credits – if they ever

64 World Economic Forum, *Biodiversity Credits: Unlocking Financial Markets for Nature-Positive Outcomes* (September 2022), p. 8. See also 'Biodiversity Credits', *rePLANET* https://www.replanet.org.uk/what-are-biodiversity-credits/. See also the definition by the Colombian company Terrasos, 'Protocol for Issuing Voluntary Biodiversity Credits: Executive Summary', *Terrasos* (2022).

65 Z. St. George, 'Pricing Nature: Can Biodiversity Credits Propel Global Conservation?' *Yale Environment 360* (6 April 2023) https://e360.yale.edu/features/biodiversity-credits

66 See 'Our Credits', *ValueNature* https://valuenature.earth/#credits (Accessed 30 June 2023).

67 Andreia Pawel, interview with Bernard C. Moore (telephone, 10 October 2022).

become widespread – simply cannot examine inherently unique, qualitative tokens, which makes these kinds of credits more a novelty item than a secure means to offset climate change and ecological destruction.

Throughout our interviews with representatives from ORKCA, Natural State, Oana, and/or other parallel co-traveller organisations, we have been unable to find a single report or single person who has been able to clearly and meaningfully articulate how they are quantifying biodiversity – a category pregnant with individual value judgements – in the context of conservation finance. Furthermore, biodiversity credits appear limited to non-human 'nature', and contain little to no information about human rights, ancestral land, culture or the destruction of the lives of humans who live on or have lived on these lands. The eviction of Indigenous peoples does not necessarily reduce the 'value' of biodiversity credits.[68] On a more ecological level, though, biodiversity as a whole requires weighing positives and negatives of various environmental interventions. Let us briefly examine the relationship between carbon and *Prosopis* as an example.

From the very first days of Oana's operations as an 'adventure tourism' destination to 'rewild' KumKum and the Orange River area, Ian Craig and his employees Andreia Pawel and Ed Barthorp targeted the invasive tree species *Prosopis glandulosa* (honey mesquite). Oana partnered with an English company, Mossy Earth Ltd,[69] to fund the planting on KumKum of a 'tree nursery' for quiver trees (*Aloidendron dichotomum*) – not actually a tree but an aloe, native to the arid regions of southern Namibia and the Northern Cape. Despite quiver trees not sharing an ecosystem with *Prosopis*, Oana and Mossy Earth implied that the invasive *Prosopis* threatened the biodiversity of the region, including the quiver tree; thus, *Prosopis* had to be removed en masse.[70]

Originating from the south-western United States and Central America, *Prosopis* was brought to southern Africa during the mid-twentieth century by commercial farmers – including R.G. Niemöller – who cultivated fields of them to harvest the seed pods.[71] The pods can contain, on average, thirteen per cent protein and thirty per cent sucrose, which makes them attractive as a fodder, which is ground up and mixed with grasses or lucerne. Depending on rainfall

68 This has been proposed by proponents of biodiversity credits, but never implemented or systematically considered. See A. Ducros and P. Steele, *Biocredits to Finance Nature and People: Emerging Lessons* (London, IIED, 2022).

69 UKCHA #10927633: Certificate of Incorporation of a Private Limited Company (22 August 2017).

70 'Building a Tree Nursery: Preparing for Future Restoration and Habitat Restoration', *Mossy Earth* https://www.mossy.earth/projects/building-tree-nursery (Accessed 1 February 2023).

71 Louise Niemöller, interview with Bernard C. Moore (Pofadder, 21 January 2022).

FIGURE 88 Oana staff burning Prosopis on KumKum
STILL FROM OANA PROMOTIONAL FILM, 2020

and irrigation, 50–200 trees per hectare can yield 4,000–10,000 kilograms of pods per hectare.[72] *Prosopis* enriches soil nitrogen to suit its own growth, and for this reason it spreads very well in semi-arid environments, especially along river beds like the Orange River. It is one of the leading species in the 'bush encroachment' revolution occurring in drylands across the world, especially in the United States, with Acacia varieties being the dominant bush-encroaching species in South Africa and Namibia.

Prosopis is indeed invasive. The spread of its seeds occurs through natural means (water and wind, for example), as well as through the digestive tracts of animals. While scholars acknowledge that domesticated livestock can facilitate the spread of the seeds, many research studies in Arizona and Texas have used controlled experimental ranges to study the causes and effects of *Prosopis* growth. These studies have noted that without a large-scale fire regime, *Prosopis* spreads equally well – if not better – in areas *protected* from livestock.[73] This is in part because of the species' adaptability to dryland environments, but also because of the presence of wild animals, rodents and other means of spreading the seeds to other locales. As early as the 1980s, scholars understood that expansion of *Prosopis* did not necessarily mean a 'degradation' of the landscape, though it did mean an economic harm to one single

72 R.W. Bovey, *Mesquite: History, Growth, Biology, Uses and Management* (College Station, Texas A&M University Press, 2016), p. 24.

73 D.M. Browning and S.R. Archer, 'Protection from Livestock Fails to Deter Shrub Proliferation in a Desert Landscape with a History of Heavy Grazing', *Ecological Applications*, 21, 5 (2011), pp. 1629–1642. For a southern African case of this, see B.J. Wigley, W.J. Bond and M.T. Hoffmann, 'Bush Encroachment under Three Contrasting Land-Use Practices in a Mesic South African Savannah', *African Journal of Ecology*, 47 (supp. 1) (2009), pp. 62–70.

industry: commercial cattle ranching.[74] Cattle prefer grass and appreciate the ability to manoeuvre through the landscape unencumbered, unlike goats, donkeys and wild game, which regularly browse on bush and can adapt well in 'encroached' pastures.[75]

It appears that Barthorp, Pawel and the Oana/ORKCA team likely could see only the 'invasive' traits of *Prosopis*. Eradicating this invasive species – which really exists only along the banks of the Orange River – gave the team purpose: how could one rewild a landscape if foreign invasive species were present? Many of their promotional films featured Oana staff destroying *Prosopis* (see Figure 88), and they wrote about it as follows:

> It is spread by livestock and has deep roots which win the race against other plants for nutrients and water, killing off our indigenous flora. It's a nightmare of a plant and needs getting rid of. Together we will continue to eradicate it quadrant by quadrant and poisoning the stump to ensure it does not return. Once the Prosopis is cleared hopefully our bird life will flourish once again.[76]

Most studies of *Prosopis* show that once the bush grows into a tree at canopy stage, not only does it provide shade for wildlife and roosting places for birdlife, as well as fixing soil nitrogen, it also acts as a 'nutrient pump', bringing water and minerals from deeper soils closer to the surface. This is the origin of the hypothesis that woody bush encroachment creates 'islands of fertility' that bring additional benefits to flora and fauna.[77]

More pertinent for ORKCA, however, is that woody bush encroachment often represents the largest gain in soil organic carbon (SOC) sequestration in arid and semi-arid landscapes.[78] Across the board, an abundance of studies have found that eradication of *Prosopis* and bush encroachers – whether by

74 R.A. Virginia and W.M. Jarrell, 'Soil Properties in a Mesquite-Dominated Sonoran Desert Ecosystem', *Soil Science Society of America Journal*, 47, 1 (1983), pp. 138–144.

75 Steven R. Archer, Professor Emeritus: University of Arizona, interview with Bernard C. Moore (telephone, 26 May 2023).

76 IAWB: 'Oana Flora and Fauna: Management' (11 February 2018). https://web.archive.org/web/20180211220335/http://oana-ff.org/management

77 S.R Archer, T.W. Boutton and K.A. Hibbard, 'Trees in Grasslands: Biogeochemical Consequences of Woody Plant Expansion', in E.D. Schultze et al. (eds.) *Global Biogeochemical Cycles in the Climate System* (Amsterdam, Academic Press/Elsevier, 2001), pp. 115–137.

78 R.L. McCulley et al., 'Soil Respiration and Nutrient Cycling in Wooded Communities Developing in Grassland', *Ecology*, 85, 10 (2004), pp. 2804–2817.

chemical, pyric or manual means – results in a decrease in SOC and capture rates.[79] However, for those seeking to profit from increased SOC storage, measuring SOC sequestration via *Prosopis* is much more difficult than for grasslands, because the gains are really only measurable on decadal levels, rather than annually.[80] This means that soil organic carbon sequestration via *Prosopis* is less immediately profitable in the short term.

The ORKCA team's goals of both sequestering fifteen million tonnes of CO_2 equivalent *and* removing invasive species that allegedly threaten biodiversity reveals some of the contradictions of 'integrated carbon and biodiversity credits'. When we presented this quandary to individuals sympathetic to ORKCA and Natural State, few had constructive answers. Most responded that although *Prosopis* does sequester carbon, it is also an invasive species, and that native plants were inherently better for biodiversity. Ultimately, as one prominent conservationist familiar with and supportive of ORKCA and Natural State (and who wished not to be named in this study) remarked: 'even though they are called Natural State, what they really look for is a Desired State'.

•••

This last quotation sums up most 'rewilding' efforts taking place on southern Namibian landscapes. The historical and ecological evidence that we have presented over the course of this book reveals that while climate change and persistent drought have made it harder to farm over the past decades, individuals have found ways to continue making use of the land. ORKCA's teleological belief in inevitable degradation if any farming continues simply does not hold up. From livestock to rhinoceros to carbon to *Prosopis*, under the microscope ORKCA's rewilding resembles less and less a meaningful ecological project. Truly 'rewilding' to the Orange River region's 'natural state' may actually mean

79 T.R. Filley et al., 'Chemical Changes to Nonaggregated Particulate Soil Organic Matter Following Grassland-to-Woodland Transition in a Subtropical Savanna', *Journal of Geophysical Research: Biogeosciences*, 113, G3 (2008), pp. 1–11. Bovey, *Mesquite*, p. 102. Research in southern Africa confirms the association of *Prosopis* encroachment with increased SOC. See A. Mureva et al., 'Soil Organic Carbon Increases in Semi-Arid Regions while it Decreases in Humid Regions Due to Woody-Plant Encroachment of Grasslands in South Africa', *Scientific Reports*, 8,. 15506 (2018). S. Mosweu et al., 'Modification of Soil Properties by Prosopis L. in the Kalahari Desert, South-Western Botswana', *Open Journal of Ecology*, 3, 2 (2013), pp. 145–150. However, the abundance of termite intrusions in the region can lead to surface SOC being brought deeper into the sub-surface soil; see S. Mosweu, 'Environmental Factors and Management Options for Prosopis Invasive Species: A Case Study in Botswana's Kgalagadi District' (PhD thesis, North West University, 2012), pp. 87–90.

80 McCulley et al., 'Soil Respiration and Nutrient Cycling'. Archer et al. 'Trees in Grasslands'.

returning to transhumant pastoralism, as practised by the Bassons and other river-folk for centuries, but perhaps this may not fit into ORKCA's 'desired state'. As the anonymous conservationist above stated, 'I'm not sure that sheep are needed anymore in that part of Namibia. In the big picture of the Earth, domesticated livestock have only been around for a minute'.

These views are held not only by Ian Craig and his disciples on Oana. When we met Adriaan Mulder (Sean Gilbertson's local manager on Sandfontein), he spent nearly half the interview explaining that *Prosopis* seeds had been spread *almost entirely* by Willem Basson's goats, seeming to forget that the *Prosopis* problem extended hundreds of kilometres upstream along the Orange River as far as the Free State. When we reminded him that, in September 2021, a team of mammal ecologists from the University of South Africa used *Prosopis* seeds from Homsrivier to help identify baboon droppings, Mulder quickly changed the subject.[81] One could reason that the invasive prosopis question was more than just a way for young conservationists to rally behind a 'rewilding' scheme – it was possibly also an ostensibly apolitical, scientific means to advocate for evicting Black farmers from conservationists' 'desired state' along the Orange River. In the words of Pete Morkel, 'When I go down to my river banks, I don't want to step in cow shit'.[82]

ORKCA has situated itself on the cutting edge of new technologies that could theoretically open up new avenues of funding for 'conservation' and 'rewilding' projects and programmes. This involves finding new ways to convince those who buy credits that landscape improvement is indeed happening. Whether that could be done with traditional soil core sampling, camera traps and acoustic devices to prove 'biodiversity' gains, or by training AI predictive models, the ORKCA team is on board. If you cannot measure it, then you cannot sell it. But although ORKCA appears new and cutting edge, its views are not. For the Bassons and other river-folk who seek to continue their pastoral activities unmolested, it makes little difference whether it is an apartheid-era capitalist farmer or a post-apartheid 'conservationist' who blocks their way. It is outsiders who have come and sought complete control over more than 150,000 hectares of their ancestral lands, from which they and their forebears have fed their families for centuries. New technologies may be used to assist in

81 Adriaan Mulder, interview with Bernard C. Moore and Luregn Lenggenhager (Homsrivier farm, 12 November 2021). The team represented T. McIntyre's ecology lab at UNISA. Candice Lewis and Tracy Basch, interview with Bernard C. Moore (Homsrivier farm, 19 September 2021).

82 Pete Morkel, interview with Bernard C. Moore and Luregn Lenggenhager (Pelgrimsrust farm, 10 November 2021).

'modernising' subsistence pastoralism, to benefit the pastoralist and the natural environment, but that is not the purpose of these tools for these conservationists. In this way, ORKCA is both dependent on and indirectly honours past histories of colonial and capitalist land dispossessions, and its supporters appear to have very little interest in changing strategy by building meaningful relationships with African pastoralists who would surely welcome this new technology and new perspectives on sustainability.

This book has ultimately been the tracking of a moving target – not only has ORKCA evolved as an organisation while we have been studying it, but conservation finance markets have also undergone great changes over the past year. In January 2023, a nine month investigation by news agencies *Die Zeit*, *The Guardian* and *SourceMaterial* into forest carbon offset credits certified by Verra came to a close. Examining more than one hundred million certified forest credits, the research concluded that Verra had drastically overstated its results.[83] It found that, in many cases, the additionality of specific projects was downright fraudulent, and where additionality was possible, carbon sequestration figures had been inflated by up to four hundred per cent. Much of this had to do with Verra's methodology itself, which constantly overestimated the 'threat' that various landscapes were facing.[84] Verra responded to this investigation by stating that for each project that underperforms, there is another that overperforms, yet they were unable to furnish evidence of this claim. In March 2023, Verra issued a press release declaring that it was substantially revising its methodologies regarding forest credits, though it was unclear how this would affect existing projects.[85]

Similar accusations have been flung at SOC sequestration schemes, primarily at the largest SOC scheme in the world, the Northern Kenya Grasslands Carbon Project. In January 2023, two leading researchers from the indigenous rights NGO, Survival International, wrote to Verra claiming that the project 'is fundamentally flawed at all levels'.[86] Expanding on this analysis in a report titled *Blood Carbon*, the organisation argued that the project's claim for additionality – that traditional grazing inherently leads to degradation – is not

83 P. Greenfield, 'Revealed: More than 90% of Rainforest Carbon Credits by Biggest Certifier are Worthless, Analysis Shows', *The Guardian* (18 January 2023).

84 Tin Fischer and Hannah Knuth, 'Phantom Offsets and Carbon Deceits', *Die Zeit* (19 January 2023).

85 P. Greenfield, 'Biggest Carbon Credit Certifier to Replace its Rainforest Offset Scheme', *The Guardian* (10 March 2023).

86 VCS Project 1468: 'Comments received by Verra during the public comment period 12 January 2023 to 11 February 2023 for the verification of the Northern Kenya Grassland Carbon Project' – 23 January 2023.

supported by meaningful evidence. Further, this flawed understanding of traditional Kenyan pastoralism had led to a baseline that implied more degradation than was present. Survival International claimed that efforts to prevent and quantify leakage are 'little more than guesswork'. Finally, it held that forcing Kenyan pastoralists to practise new grazing strategies damaged traditional governance structures and cultural norms.[87]

On 10 March 2023, Verra officially placed NKGCP on hold, suspending all future credit issuances until a review was completed.[88] Verra implied that the decision was part of its general review of all large carbon offset schemes, likely prompted by the forest carbon credit scandal, but the timing seems to imply that the Survival International report had something to do with Verra's decision to review the scheme.[89] Mark Ritchie condemned Survival International's report as lacking scientific rigour, and stated that new findings of SOC gains are awaiting publication after peer review.[90] In the NRT's 2023 Annual Report, which dedicates many pages to the purported success of the NKGCP, it is not mentioned at all that its carbon project had been put on hold by Verra.[91]

Nevertheless, over the course of 2023, the NRT and its partners were obligated to respond to a series of inquiries from Verra's review board regarding how the NRT and NKGCP communicated the project to Kenyan conservancies and crucial questions relating to implementation. Many of their responses continued to justify the NKGCP on the basis of old 'tragedy of the commons' fallacies.[92] Other responses perplexingly claimed that the pre-NKGCP grazing

87 S. Counsell, *Blood Carbon: How a Carbon Offset Scheme Makes Millions from Indigenous Land in Northern Kenya* (Survival International, March 2023).

88 VCS Project 1468: C. McCarthy, Director of VCS Program Management to J. Worden, Northern Rangelands Trust 'Section 6 Quality Control Review Notification' – 17 March 2023.

89 A. Mukpo, 'Carbon Credits from Award-Winning Kenyan Offset Suspended by Verra', *MongaBay* (21 March 2023).

90 Mark Ritchie did not respond to our repeated follow-up questions concerning the suspension of the NKGCP and Survival International's critique of his methodology. Ritchie and his team have continued to publish about the success of his VM0032 and SNAP methodology for measuring SOC capture, though he does not appear to have directly addressed any of Survival International's critiques about additionality, leakage or societal impact. M.E. Ritchie et al., 'Increased Pastoralist Livestock Mobility is Associated with Large-Scale Rangeland Restoration and Soil Carbon Sequestration', *agriRxiv* (2024). Mark Ritchie has since left his position at Syracuse University.

91 Northern Rangelands Trust, *NRT Annual Report, 2023*. The NKGCP's website also makes no mention of it: https://www.northernkenyacommunitycarbon.org/ (Accessed 1 May 2024).

92 [*'Without the project, livestock management in the region would likely continue to drive a cycle of impoverishment associated with continuous grazing, loss of soil organic matter and forage productive capacity'.*] See Verra, 'Section 6 Quality Control Review: 1468 – Northern Kenya Grassland Carbon Project' [Final Issue] (22 November 2023), p. 14.

intensity levels were at nearly 99 per cent on the Kenyan rangelands, which would imply that virtually any reduction would inherently sequester carbon.[93] It appears that the NRT's responses were sufficient for Verra, because the NKGCP was able to sell off an additional 870,000 credits between December 2023 and February 2024.

The carbon market is volatile and controversial, yet when projects position themselves correctly they are able to reap large financial rewards from it. Although we have quite a bit of data for how Ian Craig's projects in Kenya are operating and seeking to sequester carbon, ORKCA's situation in Namibia is still in flux. At the time of writing, no project had been formally proposed to Verra, and it was unclear if it would take shape as an SOC sequestration project or a biodiversity credit scheme.[94] Regardless, ORKCA is still young, but it is being shaped by individuals and projects that are long established. The final section of this book reviews the question of the Bassons and the river-folk, examining some of the social, cultural and (crucially) *legal* means that they are using to hold onto their ancestral lands. As stated earlier, we are observing a moving target, and our analyses are inherently shaped by the situation at the time of writing.

---

93 Ibid, pp. 29–30.

94 ORKCA still claims that the project will sequester 15 million t-CO_2-e. The ORKCA team also took a trip with Natural State to visit the latter's new research centre near Lewa in Kenya, as well as visit the NRT itself. ORKCA also intends to raise additional revenue through 'joint-venture irrigation farming' of 'high-yield' crops. Perhaps the Orange River as an irrigation zone will continue in the future, though with the landowners branding themselves as 'conservationists'. See Andreia Pawel, interview with Ben Goldsmith, *Rewilding the World Podcast* (23 January 2024).

PART 4

Land, Wealth and Post-Apartheid Nature

CHAPTER 14

Vetustas

Apart from creating an international border on a map, Namibian independence did not change much immediately for the lands on which river-folk lived near the Orange River. As previous chapters show, the new Namibian government did not urgently buy up farmland to expand communal areas or even to cultivate a Black commercial farming class. Although land-reform policies would come, this would be only in the mid-1990s. Nevertheless, the remoteness of the lower Orange River meant that the Bassons and other river-folk could still move along the river somewhat unencumbered for some time, so long as they did not interfere with Chiebella's irrigation works at Haakiesdoorn on the Namibian side or Niemöller's date production on the South African bank. Much of the Namibian lands were owned by absentee landlords like Steen Severin, or else aged and infirm farmers like Gerrit Luttig. Furthermore, crossing the river was not heavily policed yet, so it was possible still to work on the South African side if needs arose.

However, this apparent independence and autarchy was really an illusion because it was a result of isolation, and ignorance on the part of the state and of international capital. In fact, the river-folk's land tenure was no more secure than during the apartheid years. With the arrival of Sean Gilbertson and Willem Agenbach in the twenty-first century, they would be isolated no more. As the Norechab Sand & Tent Club paddled their kayaks down the Orange River in October 2005, Sean saw portions of Severin's and Luttig's properties, which were utterly stunning, beautiful, and offered the possibility of both adventure and luxury. Once he brought in Agenbach as his Namibian business partner and finalised the sale in 2007, Sean's new company quickly sought to make better use of the Orange River for its 'conservation' activities.

Although Namibian independence did not necessarily change the political economy of land tenure for the Bassons and other river-folk, it did change the way they related to the state and to the law: Whites and Blacks were now fully equal citizens. The land question could now be viewed as both a *political* project and a *legal* one. Whereas White landowners invoked a libertarian discourse of property rights and business expediency to secure their interests in post-apartheid Namibia, many Black Namibians increasingly entered the land debate from an oppositional stance – one that emphasised the legacies of historical injustices and advocated for the restoration of ancestral land rights. These two viewpoints – property rights vs. ancestral rights – stand in direct opposition

 | DOI:10.1163/9789004712454_015

FIGURE 89 Graveyard of the Basson family on Girtis farm
PHOTO: B.C. MOORE, 2021

to one another and are in many ways incompatible. Yet both viewpoints are supported within the Constitution of the Republic of Namibia.[1]

This chapter steps back from considering the Orange River-Karoo Conservation Area (ORKCA) and its constituent members as a conservation organisation – the subject of the past several chapters – and instead, examines the ways in which these private nature conservationists have used the law to resist post-apartheid ancestral land claims. This chapter, therefore, understands these conservationists as landowners with title deeds undertaking extraordinary measures to maintain exclusive ownership and use of these landscapes. We also consider the struggles by Willem Basson III, his family and other river-folk to invoke legal precedents concerning ancestral land and restitution in their efforts to maintain access for the Bondelswarts community. Finally, we reflect on how rewilding as advocated by ORKCA is a sort of ancestral land project as well, though one that does not consider the human inhabitants who claim ancestral connections.

•••

1 Article 16(1) grants the right to moveable and immoveable property. Article 23(2) allows the government to enact policies and programmes to redress the 'imbalances in Namibian society' that arose from colonialism and apartheid. The latter gave rise to numerous pieces of land reform legislation, many of which are covered in this chapter.

In 2006, Willem Basson was running out of patience. After sixteen years of national independence and the implementation of government programmes that purported to improve the lives of rural Namibians, nothing had been done at the Orange River. He put pen to paper and wrote to SWAPO politician Jerry Ekandjo, Minister of Lands and Resettlement in Windhoek, and to focus on the plight of his river-folk, he made an example of Pelgrimsrust Farm. He described how Michael J. Hayes had been able to obtain the farm after the Second World War, only to find that Pelgrimsrust was, in Basson's words 'already inhabited by our people, then simply referred to as black people'.[2] When Hayes sold the farms to Gerrit Luttig, it only got worse: 'Mr Luttig was an apartheid fanatic, torturing the people, robbing them of their livestock, and he threatened to shoot them if he finds any of them on his farm'. By then, Luttig was old and absent from the farm, and the grazing had been leased to a White South African named Josef de Witt. Willem Basson, who attentively followed the politics of land in rural Namibia, could not understand how a government that claimed to enact policies to redress the ills of apartheid could ignore the continuance of its structures.

> I personally feel that us who have survived thus far should get our farm back to start afresh on a commercial basis. I, Mr. Basson am with a commercial brand number in the communal land. I am desperately in need of land and find it unfair that Josef de Witt who is a born South African and lives in South Africa possesses my rightful land whilst I am struggling.

Jerry Ekandjo never wrote back to Basson, and Pelgrimsrust would be sold to Pete Morkel a few years later. Nevertheless, this letter reveals ways in which Black Namibians increasingly realised that the land question was not only a political one based on the promises of the anti-apartheid liberation movement; the *law* had to be used to uphold these promises.

By the second half of the 1980s, it was clear to most that South Africa's presence in Namibia was unsustainable, and it was time to adopt United Nations Resolution 435 to grant Namibia political independence. Very few White Namibians were supporters of SWAPO or of complete separation from South Africa, but they generally split into two camps regarding how to control the transition to independence to hold onto their own interests as securely as possible. On the one hand, many Whites from within the National Party endorsed the Privatisation Committee of 1989–1990, which sought to remove White schools, hospitals and social services from state ownership entirely, such that

2 LAC BCF: Willem Basson III to Ministry of Lands and Resettlement – 31 July 2006.

the new SWAPO government would not inherit these assets.[3] They had limited success. White liberals, on the other hand, sought not to remove assets from SWAPO's hands but rather reshape SWAPO's vision of a future Namibia. In 1987, the Namibia Peace Plan Study & Contact Group (NPP-435) was founded to educate Whites within business, agriculture and civil society that there was a way for Whites to have a future in post-apartheid Namibia. This was done in part through transforming the expectations of Whites themselves but also through transforming the goals of the anti-apartheid movement. Their chairperson was the lawyer and eventual ORKCA board member, Peter F. Koep.[4]

Twenty two NPP-435 members travelled to Stockholm in June 1988 to meet leading members of SWAPO in a consultative conference mediated by the Swedish Foreign Ministry. For the Whites, it was a fortuitous and productive gathering. They did not meet the same young Marxists who had left Namibia for exile in the 1960s but rather an ageing, weary group of veteran politicians who were falling out of favour with the Soviet Union and Cuba and looking to return to their native land.[5] According to the main agricultural representative of NPP-435, Anton von Wietersheim, who was the game-farming chairman for the South West Africa Agricultural Union (see chapter 5), 'the main thing we spoke about was how the future economy would look and how we would deal with land'.[6] 'We knew that there would have to be some reforms in the way Namibian land would be handled, but we were able to agree that there would be no nationalisation of farmland in a future Namibia'. Indeed, the right to private land ownership would be enshrined in the Namibian Constitution, and all future land reform programmes would have to operate within a market-economy framework. Although it may not have been the organisation's explicit goal, one could argue that NPP-435 accomplished far more than the National Party ever could have done to maintain White control over land.

Most Namibians – Black, White and Coloured – accepted this Faustian bargain in 1989–1990 as a means to end a bloody war. Peace was achieved, but in order to maintain stability and achieve prosperity more would need to be done, and the new national government understood this. In July 1991, the newly

3 For the extensive records of the Komitee van Privatisering and the Afrikaner organisation Cultura 2000, see NAN SWAS 267 File A.S.40/1/6/1/8 (vols. 1–3). The full archives of Cultura 2000 are held in Windhoek West at the Erfdeel Argief en Kultuursentrum.

4 Peter Koep took over from the founding chairman, Bryan O'Linn, in 1989.

5 For an account of the conference, see C.L. Pettersson, *In the Footsteps of Mr. Anderson: Milestones in Swedish-Namibia Relations* (Parkwood, Ulwazi, 2008), pp. 79–89.

6 Anton von Wietersheim, interview with Bernard C. Moore (Swakopmund, 26 October 2020). The NPP-435 and SWAPO also agreed to unite the militaries after independence and to make English an official language.

formed Ministry of Land Reform held the First National Conference on Land Reform and the Land Question. The conference represented an important step in opening dialogue about what future land reform legislation in Namibia could look like, but it epitomised the disconnect between, on the one hand, views of land reform as a way to deracialise agriculture and, on the other, land reform as synonymous with returning ancestral land. For this reason, the Ministry's conclusions from the conference were vague and general. It concluded that 'there was injustice concerning acquisition of land in the past, and that something practicable must be done'. At the same time, the Ministry viewed expropriation as the most extreme option, reserved for absentee landlords and underutilised or abandoned lands. Concerning ancestral land rights, despite many submissions from across Namibia advocating for increased attention to it, the Ministry would not commit to anything: 'The conference concludes that given the complexities in redressing ancestral land claims, restitution of such claims in full is impossible'.[7]

Over the following few years, however, no new legislation emerged regarding the land question and many Namibians were growing impatient, leading scores of Black farmers from across the country to meet in Mariental in September 1994 to hold the People's Land Conference, organised completely outside the auspices of the Ministry of Land Reform. Most of the resolutions urged government to be quicker in addressing the land question, and the People's Land Conference was also much more direct concerning ancestral land. They criticised the government's 1991 stance, writing that the ancestral land issue

> should not simply be put aside. It should be maintained on the policy agenda concerning land questions ... Land claims [must] be given serious attention, taking into account constitutional principles and policies of affirmative action for the historically disadvantaged, impoverished communities and the moral obligation to promote social justice.[8]

They further advocated regional grassroots commissions to investigate ancestral land claims across the country.

Early the following year, the government finally signed into law the Republic of Namibia's first piece of legislation to address the land question in the country. The Agricultural (Commercial) Land Reform Act of 1995 set into motion the 'willing-buyer, willing-seller', market-based land-reform programme.

7 National Conference on Land Reform and the Land Question, 'Consensus of the Conference' (July 1991), p. 2. See also, E. Tjirera, 'Namibia's Intractable Land Question', *Current History*, 122, no. 844 (2023), pp. 167-171.

8 EAKS JWF 15 File 7: Report of the People's Land Conference (1994), pp. 16–17.

Section 17 of the Act mandated that all agricultural land must be offered to the state for purchase before it could be sold on the open market, at which point a waiver would be provided.[9] The law also set up advisory boards to determine whether the state should buy farms or – in limited cases – expropriate farmland.[10] In addition, as mentioned in chapter 9, the law set out the restricted terms by which foreigners could obtain agricultural land.

This Act set in motion what would eventually be formalised in 2001 as the National Resettlement Programme (NRP), which would target Namibians from specific groups for resettlement on purchased farms. They would include San citizens, soldiers of the former anti-apartheid army, Namibians formerly in exile, disabled citizens, those living in overcrowded conditions, as well as 'displaced, destitute and landless Namibians'.[11] Settlement would be predicated on a willingness to engage in formal agriculture and to occupy land on a leasehold basis – resettlement farms belonged to the government, not the farmer.[12] The NRP would run alongside the parastatal Agribank's Affirmative Action Loan Scheme (AALS), which was founded in 1992 to provide low-interest loans to upstanding farmers from previously disadvantaged communities who were looking to buy commercial farmland on the open market.[13]

Both the NRP and the AALS have been a modest success. However, these tentative gains belie structural problems, which led to their rapid decline over the past decade. The problem with market-based land reform is that unless the government suspects disparities in the price offered to it compared with the open market, post-waiver, it is the farm owners themselves who set the price that the Ministry (or the AALS farmers) must pay. Once land reform became a political project, farm owners deliberately increased sale prices dramatically; if the post-apartheid government wanted to uphold its promises to Black Namibians, it would have to pay the Whites dearly. In the twenty five years after independence, the average farm price rose from NAD 76 to nearly NAD 1,100 per hectare,

9 'Agricultural (Commercial) Land Reform Act of 1995', *Government Gazette of the Republic of Namibia* (no. 1040 of 3 March 1995), pp. 17–18.

10 See Ibid, sections 20–23.

11 Republic of Namibia, *National Resettlement Policy* (Windhoek, Ministry of Lands, Resettlement and Rehabilitation, 2001), pp. 3–5.

12 At this time, the Ministry of Lands and Resettlement took over the existing ad hoc willing-buyer willing-seller programme run by the Ministry of Agriculture from 1993–1999, which had purchased fifty three farms over this period, mostly for San group resettlement projects. For a survey of perspectives on the land question, see R.V. Nghitevelekwa, *Securing Land Rights: Communal Land Reform in Namibia* (Windhoek, University of Namibia Press, 2020).

13 From 1992–2015, the AALS gave Namibians 642 loans totalling more than NAD 750 million, which facilitated the purchase of more than three million hectares of farmland across the country.

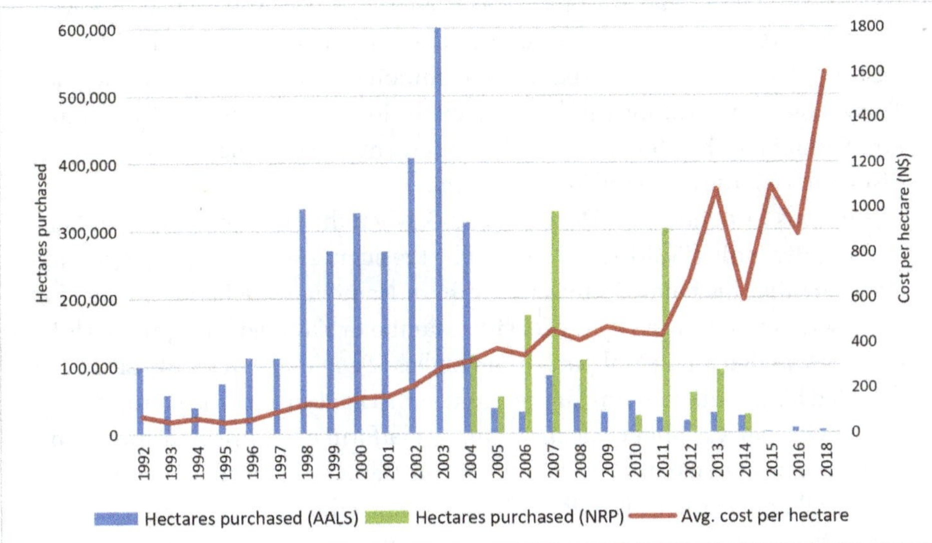

FIGURE 90 Farm purchases through Agribank's Affirmative Action Loan Scheme (1992–2018) and the Namibian Government's National Resettlement Programme (2004–2014), plotted alongside increasing land prices in Namibia.[14]

with many transactions at higher rates.[15] AALS farmers were the first to realise that sale prices did not reflect agricultural capacity, and by 2004 more than one third of Agribank loan accounts were in arrears, causing many to default. Indeed, as Figure 90 shows, swift price increases rendered it impossible for the government or AALS buyers to acquire significant amounts of farmland. The government simply did not have the money.

The number of AALS farmers declined dramatically after 2004, and although the Ministry continued to purchase farmland through the NRP, this was insufficient to redress the longstanding injustices of colonialism and apartheid. Furthermore, the broad focus of Namibian land reform remained centred on agricultural production and bringing Black people into a sector once dominated by Whites. There was little consideration of using resettlement farms for other purposes, such as tourism, research, commercial hunting or conservation. Lease contracts explicitly forbid building hotels or engaging in hunting

14 Data from: R. Sherbourne, *Guide to the Namibian Economy, 2017* (Windhoek, Institute for Public Policy Research, 2017), pp. 447–455. R. Sherbourne, *Guide to the Namibian Economy, 2022* (Windhoek, IPPR, 2022), pp. 572–575.

15 Sherbourne, *Guide to the Namibian Economy, 2017*, pp. 447–453.

beyond 'own-use' (that is, for the dinner table).[16] Resettlement farms were also not allowed to be a means to expand communal areas, and the Ministry ensured that they did not expand the commonages of traditional authorities. This increased frustrations and reinforced the local belief – especially in southern Namibia – that the National Resettlement Programme was completely divorced from the ancestral land question.

Tensions rose sharply in December 2016, when the Deputy Minister of Land Reform, Bernardus Swartbooi, was fired by President Hage Geingob for publicly declaring that his fellow Namas in southern Namibia were being sidelined in the resettlement process by those with Ovambo or Kavango backgrounds.[17] He quickly founded a political party, the Landless People's Movement (LPM), and declared his intention to run for president with a land-first political programme. Though his party is not ethnic in origin, its platform unsurprisingly meant that it sourced most of its support from the Nama, Damara and Herero communities of central and southern Namibia. However, since the main political agenda of LPM is far removed from the daily lives of people living in northern communal areas – who make up more than forty five per cent of the population – the LPM and Swartbooi could only ever be a very vocal opposition movement carrying the majority of the sparsely-populated constituencies south of Windhoek.

The rise of Swartbooi accompanied increased demands from the rural poor and civil society for greater accountability in the land reform programme. Not only was an insufficient quantity of farms being purchased, but there were also allegations that some land reform beneficiaries were in fact high-ranking officials and businesspeople with connections to the anti-apartheid movement.[18] This sparked demands by traditional authorities and civil society for a list of NRP beneficiaries to be published, and for a second conference to reformulate land reform policy.[19] The master beneficiary list was supplied only in October 2018, days before the agreed-upon Second Land Conference. The list revealed that although ethnic preference and elite capture were not completely absent

16 After a certain period of time, resettlement farmers would be permitted to construct a campsite, but lodges are forbidden. Timotheus Ekandjo, interview with Bernard C. Moore (Vergenoeg farm, Maltahöhe District, 1 December 2020).

17 'Bernardus Swartbooi is Fired', *Namibian Sun* (13 December 2016). The government claimed that it was a resignation, albeit under distinct pressure. 'Swartbooi Resigned, not Fired – Kapofi', *New Era* (14 December 2016).

18 S. Likela, 'Top Government Officials Benefit from Resettlement Farms', *The Namibian* (25 September 2018).

19 See H. Melber, 'Colonialism, Land, Ethnicity, and Class: Namibia after the Second National Land Conference', *Africa Spectrum*, 54, 1 (2019), pp. 73–86. See also, P. Kaapama, 'The Ancestral Land Debate prior to and during Namibia's Second National Land Conference: The Redistribution-Recognition Dilemma', *Anthropology Southern Africa*, 42, no. 3 (2019), pp. 232-246.

from the land reform programme, perceptions within the public perhaps overestimated these trends, especially in the ǁKaras Region.[20]

In the months running up to the 2018 Second National Land Conference, civil servants and traditional authority members across the ǁKaras Region gathered in Keetmanshoop to prepare their list of concerns, recommendations and demands. Front and centre were issues of historical injustice and ancestral lands, and the Keetmanshoop congress began the demands with a simple request: 'Government needs to recognise that ancestral land exists'.[21] Further recommendations were more specific: right of access to grave sites on commercial farms; seventy per cent of resettlement farms should be allocated on an ancestral basis to local traditional authorities; independent ancestral land tribunals should be established to review claims. Importantly, the participants understood that part of the problem with advancing an ancestral land agenda lay in the Namibian Constitution itself. Thus, they recommended the 'Constitution to be amended, where required, to meet the needs and demands of Indigenous people'.[22]

When the participants of the Second National Land Conference met in October 2018, it became clear that the conference was poorly organised and reflected an increasing distance between the government in Windhoek and the rural poor elsewhere.[23] Most of the resolutions were about reforms to urban and communal land tenure, but the conference did make a few crucial points relevant to the situation in southern Namibia. First, it was made clear that the 'willing-buyer, willing-seller' programme was not working and needed to be replaced with something new (which it did not define) to accelerate land reform.[24] Second, it called for the establishment of a Presidential Commission of Inquiry on Ancestral Land, which was endorsed by President Geingob in a press release a few days later.[25] At the same time, the conference resolved that

---

20 Ministry of Land Reform, 'Master List of Beneficiaries Resettled since 1990–2018' – 11 October 2018. There were 189 individual beneficiaries of land reform in the ǁKaras Region spread over 97 resettlement farms, adding up to nearly 700,000 ha. The 189 beneficiaries for ǁKaras is quite small compared to the 5,782 beneficiaries across Namibia.

21 ǁKaras Regional Council, 'Regional Consultations in Preparation for the Second Land Conference' (19–20 July 2018), p. 23.

22 Ibid.

23 See Melber, 'Namibia after the Second National Land Conference', and P. Gilolmo Lobo, 'In the Wake of the Namibian Second National Land Conference', *Agrarian South* (29 January 2019).

24 Ministry of Land Reform, 'Resolutions of the Second National Land Conference' – 5 October 2018.

25 Republic of Namibia: Office of the President, 'Post-Mortem and Review of the Resolutions of the Recently Concluded Second National Land Conference' – 10 October 2018.

'alternative restorative measures' needed to be found to address the ancestral land question apart from land reform itself. Most of these alternative proposals revolved around changing colonial names of places, 'ensuring the participation' of these communities in land decisions, as well as – revealingly – 'investigating possible unintended consequences of ancestral land claims and restitution'.[26]

Although some progress had been made since independence in addressing the land question, these gains were limited and had not solved the problems that the rural poor face concerning insecure land tenure. The Second National Land Conference revealed that government was increasingly removed from the struggles which its people faced on the ground, especially in more remote parts of the country far from Windhoek. This was particularly the case regarding ancestral land. Nevertheless, previously disadvantaged Namibian citizens continued to petition for recognition of their ancestral rights, and they increasingly turned to the law to do so.

•••

Amidst the gradual progress of land reform legislation in Windhoek, in 2010 Willem Basson III's situation at the Orange River was rapidly disintegrating.

FIGURE 91 Gate on the D292 road at the entrance to Homsrivier farm
PHOTO: B.C. MOORE, 2021

26 Ministry of Land Reform, 'Resolutions of the Second National Land Conference' – 5 October 2018.

His brother Ambrosius, who was staying with the livestock near Girtis Farm, trekked cattle and goats westward, passing by the old Homsrivier Mission Station to meet the national district road D292. He wanted to take some animals to sell to the abattoir and relocate some to grazing farther north in the Bondelswarts communal areas. He found that Willem Agenbach had built a new gate and locked it, blocking the connections between the lands along the Orange River and the national road heading northwards to Warmbad. Furthermore, Agenbach had apparently extended the game-proof fence well beyond the high-water mark into the river. Additional fences had been constructed by the irrigation farmers upriver at Khaais and Velloorsdrift, which extended into the middle of the riverbed.[27]

Basson requested the assistance of the Legal Assistance Centre (LAC) – a non-profit, human rights law firm in Windhoek – to seek an interdict or spoliation order to stop the locking of gates and the construction of private fences on state lands. His brothers Paulus, Ambrosius and Laurentius were trapped at their kraals with no means to exit with their livestock. Facing a locked gate and an apparently intransigent Agenbach, Willem Basson could not take food rations and medicine to his family and his workers, which – if left to continue – would require the river-folk crossing illegally to Witbank on the South African side; certainly not ideal. The LAC drafted a letter to Agenbach demanding that the gate be opened, or legal action would be taken in the High Court against him for unlawful spoliation.[28] Agenbach reopened the gate temporarily, but the Sandfontein company continued to build more game-fencing across Homsrivier Farm, rendering northbound movement increasingly difficult.

Agenbach apparently did not want Bondelswarts moving between various kraals along the river (see Figure 93 for map). He took the initiative and hired a law firm in Keetmanshoop to serve eviction orders to the Bassons living near Girtis Farm, which had been purchased by Agenbach's partner, Sean Gilbertson, in 2009 and registered under the corporation Marwilben Farming cc.[29] The eviction notice gave the pastoralists ten days to evacuate the property with their livestock, otherwise Marwilben would demand NAD 100,000 in damages. Willem and Paulus Basson again sought the assistance of the Legal Assistance Centre, which challenged the eviction notice, triggering legal procedures over the following four years. Although the LAC took the case based on their

27 LAC BCF: Lesle Jansen, notes from consultation with Willem Basson – 20 July 2010.

28 LAC M/B: Natasha Bassingthwaighte to Lesle Jansen 'Bondelswarts Community: Spoliation Order' – 12 August 2010.

29 RoHC File I1295/13: Attorneys Lentin, Bothma and Van Den Heever to Willem Basson et al. 'Eviction from Farm Girtis no. 109, Karasburg District' – 23 June 2011.

principles of post-apartheid human rights, many of the lawyers involved were also intrigued by the unusual geographical situation that related to the eviction notice and the history of the Bassons.[30]

Once the pre-trial discovery and exchanges of claims and pleas began in 2013, it was clear that there was far more at stake in the Basson case than simply the eviction of small-scale farmers by a larger commercial landowner. In May 2013, Agenbach – representing Marwilben – claimed that the Bassons were 'unlawfully and maliciously' occupying his farmland, and he sought eviction, an increased NAD 400,000 in damages, 20 per cent interest on any delayed payments, as well as 'further and/or alternative relief'.[31] Based on knowledge carried down from the generations of his family staying along the river between Homsrivier and Pelladrift, Basson urged the LAC to challenge Agenbach's claim over the lands on which the river-folk farmed.[32] He believed (correctly) that Girtis and other neighbouring properties did not have any river frontage south of the highest high-water mark of the Orange River floods; the river-folk therefore resided on state lands.

The LAC lawyers did exactly this, digging up German-era maps of Girtis, which indeed showed that the southern boundary did not extend to the Orange River waters. They wrote back to Agenbach's lawyers, requesting them to confirm whether or not Marwilben believed that Girtis extended to the river waters.[33] After nearly an entire year of deliberation, the law firm confirmed that, yes, it did believe that Girtis extended south of the high-water mark to the middle (Thalweg) of the river.[34] The Bassons, via the LAC, immediately denied the allegations of the plaintiffs' claims, arguing that they were not on Girtis in the first place, and the exorbitant 'damages' to grazing they were alleged to have caused could not have been suffered.[35] Furthermore, they initiated a counterclaim, arguing that Agenbach's and Marwilben's decision to erect game fences, lock gates and inhibit the Bondelswarts' ability to utilise lands that their people had occupied since 'time immemorial' was a form of illegal

30 Peter Watson (LAC), interview with Bernard C. Moore (Windhoek, 30 July 2022).

31 RoHC File I1295/13: Marwilben Farming cc, 'Particulars of Claim' – 2 May 2013.

32 Willem Basson III, interview with Bernard C. Moore and Luregn Lenggenhager (Karasburg, 11 November 2021).

33 RoHC File I1295/13: Legal Assistance Centre 'Request for Particulars in Terms of Rule 21' – 30 May 2013.

34 RoHC File I1295/13: Lentin, Bothma and Van Den Heever 'Further Particulars' – 28 May 2014.

35 RoHC File I1295/13: Defendants' Plea – 17 June 2014.

spoliation. They demanded a court interdict against further construction of fences or restriction of livestock movement.[36]

Agenbach's team responded to the counterclaim by denying that the Bondelswarts had ever historically occupied the area, denying that colonialism and apartheid had dispossessed lands, denying ancestral land claims, denying even that the company had constructed fences at all.[37] As the case proceeded towards trial in March 2015, the presiding judge was clear that the two main issues that would be debated were: (1) whether the land on which the Bassons were residing actually formed part of Girtis; and (2) if it did not, whether the Bassons had the right to any form of public servitude through Girtis and other farms.[38] However, strangely, and abruptly, Marwilben Farming cc withdrew the case the day before it was to go to trial.[39]

Some court reporters assumed that Agenbach's legal team must have realised that Girtis indeed did not touch the Orange River, and that whether the riverbanks belonged to Namibia or South Africa, they were not private lands.[40] An alternative possibility, however, is that Sean Gilbertson, long absent in Europe, realised that his company Marwilben was about to go to court, with Willem Agenbach falsely claiming to be a director of it. Agenbach and Gilbertson were indeed business partners on Sandfontein, but Marwilben – which owned Girtis and Norechab – was completely different, and Agenbach had no financial or legal relationship with the company at all. Between 2014 and 2016, Gilbertson's companies were simultaneously involved not just in internal squabbles between directors, but they were also involved in legal action against the river-folk along the Orange River.

In 2017, while in the process of buying back the assets of his recently liquidated company, Sandfontein and Houmsrivier Properties (Pty) Ltd, Gilbertson hired a team of GIS technicians to establish the southern boundary of Girtis and the location of the river-folk's kraals. The team did not visit Girtis, but using Google Earth and other software, they declared that the kraals and huts existed higher than the north bank of the river. The report was, however, replete with uncertainties and errors. The technicians acknowledged that they did not actually have the coordinates of the high-water mark, so they had relied merely on vegetation density to calculate this. Furthermore, the report incorrectly claimed that the boundary between Namibia and South Africa was

36 RoHC File I1295/13: Defendants' Conditional Claim in Reconvention – June 2014.
37 RoHC File I1295/13: Plaintiff's Plea to Conditional Counterclaim – 7 July 2014.
38 RoHC File I1295/13: Pre-Trial Report – 11 March 2015.
39 Peter Watson (LAC), personal communication with Bernard C. Moore (26 April 2023).
40 C. Rickard, 'A Borderline Case that Never Was', *Legal Brief* (21 October 2015).

FIGURE 92 Small-scale farming at the Orange River: The Basson kraal near Girtis farm
PHOTO: B.C. MOORE, 2021

the Thalweg of the Orange River and, strangely, that 'it is not clear if the farm boundary now extends to the middle of the Orange River as per the constitution' – as though a national government increasing its land area would result in giving away land to private individuals gratis.[41]

The question of where exact boundaries lie – whether of farms or of nations – was first asked more than a century ago. As shown in chapters 3 and 4, when South Africa took over Namibia in 1920 and prepared to survey Karasburg District's farms for sale to Whites, the surveyors were clear that the southernmost farms in Namibia would either be delayed from gazettal until the boundary question was settled or sold explicitly without riparian rights. Even some thirty years after the land was surveyed, the lands branch repeatedly had to send letters to farmers near the Orange River reminding them that they had no riparian rights at all.[42] During these years, the administrations on both sides of the river understood that the lack of riparian rights on the north bank was not that Namibia was *denied* Orange River waters. Rather, Namibia *possessed no such water at all.* All waters flowing within the Orange River and all lands that the Orange River regularly touched during high flood periods belonged to South Africa.[43]

The Republic of South Africa recognised Namibian independence in March 1990, repealing all laws on its books that applied to Namibian territory, albeit

41 LAC M/B: Geo Pollution Technologies (Pty) Ltd, 'Farm Girtis 109, Division V, Border Assessment' – March 2017.

42 NAN LAN 710 File 2137/2: Hoofamptenaar van Lande, Windhoek to R.P.C. Ludick, KumKum 'Oewer-Regte Oranjerivier' – 27 February 1949.

43 NAN WAT 91 File WW.77/4: Surveyor-General, Windhoek to Secretary for SWA 'Vioolsdrift Irrigation Settlement, Orange River: Report of Inspection' – 11 December 1934.

with a caveat that this 'shall not be interpreted as recognition of the validity of any provision of the Constitution of Namibia which purports to derogate from the sovereignty of the Republic [of South Africa] over its territory'.[44] Article One of the Namibian Constitution deems the middle of the Orange River to be the international boundary, but this does not appear to be in line with the terms of the RSA Recognition of the Independence of Namibia Act, because this would result in the RSA losing sovereignty over its territory. Although there have been discussions between the two governments about addressing the boundary issue, no official boundary treaty has been signed to replace the 1890 Anglo-German Treaty.[45] Given that Namibia is a member of the African Union – which holds that boundaries at independence must be maintained – there appears to be slippage between the Namibian definition of the international boundary and the South African one.[46] Regardless of which government owns the land between the Thalweg and the high-water mark – if one can ascertain precisely where either is – the land is not private property, but state lands.

On 25 June 2020, at the height of COVID pandemic lockdowns in Namibia, Sean Gilbertson – as the representative of his companies – filed a lawsuit against the entire Basson family. He claimed that they were trespassing on Girtis and Homsrivier 'by virtue of them driving their cattle and other animal herds onto the plaintiffs' farms'.[47] He sought either full eviction of the river-folk and their animals or else an interdict preventing them from trekking their animals northwards or westwards. At the same time, Pete Morkel – owner of Kambreek and Pelgrimsrust, and at the time an affiliate of Oana/ORKCA – began to put pressure on the Bassons and other river-folk who were trekking their animals eastwards towards Ian Craig's farm KumKum and Morkel's Kambreek. Morkel supplied Gilbertson with photos taken near Kambreek, of livestock, kraals and branding irons belonging to the river-folk.[48] Pete Morkel filed a similar

44 'Recognition of the Independence of Namibia Act, 1990', *Government Gazette of the Republic of South Africa* (Vol. 297, No. 12372 of 20 March 1990).

45 For more on how the border has been debated over the years, consult W. Rutishauser, 'The Orange River Boundary and the Ongoing Dispute between Namibia and South Africa', in L. Lenggenhager et al. (eds.) *The Lower !Garib/Orange River: Pasts and Presents of a Southern African Border Region* (Bielefeld, Transcript Verlag, 2023), pp. 211–224.

46 See AHG/Res. 16(I): Assembly of Heads of State and Government, Meeting in Cairo, 17–21 July 1964.

47 RoHC File HC-MD-CIV-ACT-CON-2020/02424: Plaintiffs' Particulars of Claim – 25 June 2020.

48 RoHC File HC-MD-CIV-ACT-CON-2020/02424: Sean Thomas Gilbertson: Discovery Affidavit (first schedule) – 25 June 2021. Based on the discovery appendices, the photos were taken between March and August 2020.

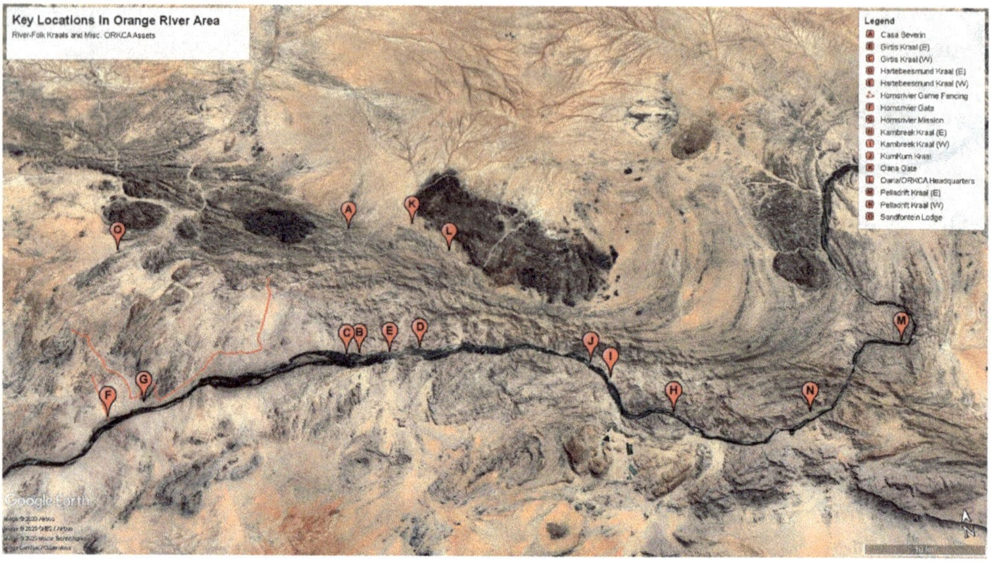

FIGURE 93 Key locations along the lower Orange River, approximated via Google Earth
CARTOGRAPHY: B.C. MOORE

lawsuit against the Bassons in 2021, representing himself correctly as owner of Kambreek but falsely of KumKum and Pelladrift,[49] and like Gilbertson he sought eviction of the river-folk and their animals from KumKum, Kambreek and Pelladrift.[50] In April 2022, Gilbertson and Morkel decided to consolidate their court cases into one massive eviction proceedings, spanning the Lower Orange River from Homsrivier to Pelladrift.[51]

Just as in the case against Agenbach a few years earlier, Willem Basson rejected all of the allegations laid out by Gilbertson and Morkel. In an affidavit confirmed by his brothers and other river-folk, Basson attested that the plaintiffs simply do not own the property between the middle of the river and

49 As mentioned previously, KumKum and Pelladrift are owned by Pelladrift (Edms) Bpk, which was primarily financed by Ian Craig. To maintain status as a Namibian company eligible for land ownership, Pete Morkel was for many years listed as a director and majority shareholder, to be replaced with lawyer Peter F. Koep c. 2020. There was no extraordinary meeting by the company filed to allow Morkel to stand legally for it, and the timeline suggests that Morkel was no longer affiliated at all with Pelladrift (Edms) Bpk when he filed the lawsuit.

50 RoHC File HC-MD-CIV-ACT-CON-2021/01637: Peter van der Byl Morkel, Particulars of Claim – 23 April 2021.

51 RoHC File HC-MD-CIV-ACT-CON-2020/02424: R. Linde (Theunissen, Louw and Partners), 'Rule 41(1) Application: Founding Affidavit' – 14 April 2022.

the highest high-water mark: the independence of Namibia did not simply increase the size of surveyed farmlands near the river. This river frontage was state lands.[52] Because of the broader implications of the international boundary question, the LAC and Basson brought, in a joinder, both the Ministry of Agriculture, Water and Land Reform as well as the Registrar of Deeds as additional defendants in the case. If Morkel and Gilbertson wanted exclusive use of the Orange River by evicting small-scale pastoralists, they would have to take on the Namibian state. This is likely what caused Gilbertson – through his companies – to abandon all territorial claims to the Orange River, which would lead to Morkel doing the same:

> The plaintiffs aver that they do not lay claim to any area beyond the duly demarcated and surveyed borders of their farmlands and to that end accept that they do not have any legitimate claim to the inner banks (the area between the southern border of their farmland and the Orange River) or islands situated in the Orange River.[53]

Regarding the use of the land near the river and the trekking of livestock across private farmlands, Basson and his legal team invoked a rarely considered legal principle in order to advocate for ancestral land rights: *Vetustas*.

•••

In March 2019, the Ancestral Land Commission – as recommended by the Second Land Conference – was officially appointed by President Hage Geingob and tasked to inquire into the historical, economic, legal and political elements of the ancestral land question in Namibia.[54] Throughout the year, the team gathered oral testimonies from traditional authorities and information from historical documents and public comments, in order to inform the team about the seriousness of the issue in various communities. The purpose of the commission was not to investigate any specific land claims but rather to recommend governmental policies to address the ancestral land issue within existing post-apartheid legislation. The interim report was submitted to

52 RoHC File HC-MD-CIV-ACT-CON-2020/02424: Founding Affidavit, Willem Basson – 26 November 2020.

53 RoHC File HC-MD-CIV-ACT-CON-2020/02424: Plaintiff's Plea to Defendants' Counterclaim – 27 July 2021.

54 Government Notice no. 59, *Government Gazette of the Republic of Namibia* (no. 6858 of 15 March 2019).

FIGURE 94 Basson's Cattle near Girtis
PHOTO: L. LENGGENHAGER, 2020

Geingob in December 2019 and the final report in July 2020, one month after Sean Gilbertson initiated eviction proceedings against the river-folk along the Orange River.

The 797-page report of the Ancestral Land Commission is – with a few caveats – a remarkably conservative document. The report reiterated that it was not possible to address all ancestral land claims, especially those that overlapped in the precolonial period. It declared that all Namibians had suffered land dispossession under colonialism and apartheid, though certain communities within the so-called Police Zone had perhaps suffered more. In this context, the commissioners concluded that restorative justice would need to be a multifaceted project involving land (when possible), as well as a significant focus on 'renaming of places', 'recognition of heroes and heroines', 'erection of monuments', 'reburial of victims of genocide', and other symbolic gestures towards affected communities.[55] The team declared that the ancestral land question had become such a potent issue because land reform in Namibia 'is perceived

55 Republic of Namibia, *Final Report of the Presidential Commission of Inquiry into Claims of Ancestral Land Rights and Restitution* (Windhoek, 24 July 2020), pp. XXVIII–XXXI.

to be painfully slow', and most Namibians felt neglected in this process.[56] They therefore recommended some form of legislation that could investigate, regulate and mediate ancestral land claims, to move from general understandings of the relationship between colonial conquest and land dispossession towards specific understandings of how this affected particular communities.[57]

The Ancestral Land Commission also published more than one hundred pages of testimonies – excerpted and full – from people across Namibia explaining the main land issues in their communities. Although the Bondelswarts Traditional Authority were not visited by the commission, they were invited to submit a written statement about their ancestral land holdings, as well as what restitution and reconciliation would look like for them. The traditional authority's proposals were quite radical. They demanded the return of lands stretching from the Aussenkehr grape irrigation estates to the |Ai-|Ais/Richtersveld Transfrontier Park and across the properties owned by the tourism company, Gondwana, as well as the assets of Sandfontein Lodge and Nature Reserve (Pty) Ltd. If land transfer were not possible, then they would demand a 51 per cent shareholding in all of these landowning companies and 30 per cent benefit sharing from mining licences.[58] The Bondelswarts' claim might have been the most specific – referring to precise farms and shareholding relationships – but most traditional authorities submitted claims with a similar ethos.

Like its forebears, the commission understood the ancestral land question as an all-or-nothing endeavour, which had made it easy to avoid addressing specific claims. This is perhaps what made Willem Basson III's claim of *Vetustas* so unique: it represented a type of ancestral land claim that had not been considered by the commission at all, but one that – given recent judgments – will likely become more common as the years go on. When South Africa took over Namibia after the First World War, it introduced the Roman-Dutch legal system from the Cape as Namibia's common law. This has made post-apartheid Namibian law a hybrid system, involving Roman-Dutch, English, German and customary legal systems operating simultaneously.[59] Many aspects of civil law hold onto Roman-Dutch legal principles, from which *Vetustas* arises.

---

56 Ibid, p. XXVIII.

57 Ibid, p. 240. This would also involve the question of generational farmworkers, an often-ignored group.

58 Ibid, pp. 133–134. Individual submissions by Bondelswart members to the commission (likely via SMS) contain a similar specificity about land parcels that are claimed by the traditional authority, giving credence to the belief that the Bondelswarts view land restitution as more central to post-apartheid reconciliation. See ibid., appendix L.

59 For some context, see D.P. Zongwe, 'Researching Namibian Law and the Namibian Legal System', *GlobaLex* (December 2020).

FIGURE 95 The Basson family's kraals, built of reeds and driftwood, near Girtis
PHOTO: L. LENGGENHAGER, 2020

As a legal principle, which dates to the juristic *Digest* (Pandects) commissioned by Byzantine Emperor Justinian I in 530 AD, *Vetustas* – sometimes called 'immemorial usage' – refers to a right held against another person, property or institution, which has existed continuously since 'time immemorial'. According to the nineteenth-century Dutch jurist, Joël Emanuel Goudsmit:

> When any state of things had endured so long a time that its origin dated back to a period to which the memory of man did not extend, there was a legal presumption that such origin had been legitimate, and the parties were dispensed from furnishing proof that it was so [...] The memory of man, however, in this matter, was not restricted to that which persons themselves remembered, but extended to things stated to the existing generation by that which had preceded it.[60]

Goudsmit emphasised that although *Vetustas* rights did not require documentary proof, per se, it did necessitate that a 'double direction' [*dubbele richting*] state of affairs existed. On the one hand, it had to be shown that the current generation understood that a particular right, use or practice had existed for so long that neither they nor their forebears could precisely point to when it began. On the other hand, there also had to be no clear evidence that another

60 [*'Heeft een toestand zoo lang geduurd, dat zijn aanvang ligt buiten het bereik van's menschen geheugen, zoo word zijn wettige oorsprong vermoed en het bewijs daarvan ontbeerlijk ... Het menschengeheugen bepaalt zich echter niet bloot tot eigen waarneming, maar strekt zich ook uit tot mededeelingen den nog levenden door hunne voorgangers gedaan'*]. J.E. Goudsmit, *Pandecten-Systeem* (Leiden, 1866), pp. 195–196.

state of affairs regularly existed. This existence of a given use or right since 'time immemorial' implies and assumes a lawful origin, and it then falls upon the other party to prove that the right has an unlawful origin.[61]

Even before it was formalised in the *Digest*, the principle of *Vetustas* was invoked in the Roman empire as early as the second century AD as sufficient justification for the existence of a legal servitude over a particular property, especially as it related to extracting water, grazing and watering livestock, and channelling water to crops.[62] Similar to right-of-way and bridleway customary laws in England and Scotland, *Vetustas* entered Roman-Dutch canon as a historically grounded usufruct right to reasonably use (or traverse) property belonging to another person.

In 2018, a landmark case in South Africa opened the door for *Vetustas* to impact on jurisprudence related to land reform and ancestral land in Namibia, which follows RSA courts for legal precedent. On the private commercial farm Grootkraal, near Oudtshoorn in the Eastern Cape, stands a small United Congregational Church and its primary school. After the death of the farm owner, his descendants sought to evict the school from the property and prohibit community access to the site. The 'Community of Grootkraal' laid a counterclaim seeking a public servitude over that portion of the property for religious, educational and cultural purposes, based on the principle of *Vetustas*.[63] Their case was successful, and a public servitude was established, enabling community members to continue to use and access the school and church.

The importance for the broader question of ancestral land rights is how the Supreme Court of Appeals handled the historical aspects of the case. According to the judgment:

> The Community is not a formal body, nor is it capable of exact definition. It is said to consist of those individuals who have historic and family ties with the Grootkraal area, where they and their forebears have lived

61 L.T.C. Harms and J.H. Hugo, *Amler's Precedents of Pleadings*, Third Edition (Durban, Butterworths, 1989), p. 300. See also J.M. Otto, 'Vetustas, Onheuglike Tye en die Witpad by Langebaan', *Fundamina*, 19, 1 (2013), pp. 48–60.

62 C. Möller, 'Time as an Argument in Roman Water Law', *Water History*, 15 (2023), pp. 67–80. *Vetustas* rights need not result in a praedial servitude akin to Way of Necessity (*via ex necessitate*), on the grounds that the right is not held against one property by other – such as a landlocked private property against another with road access – but rather is tied to continuity of community use. See, C.G. van der Merwe, 'Louisiana Right to Forced Passage Compared with the South African Way of Necessity', *Tulane Law Review* 73, 4 (1999), pp. 1363–1414.

63 *Community of Grootkraal v. Kobot Business Trust* (1219/2017) [2018] ZASCA 158 (28 November 2018).

> and worked for many generations. By way of example, Ms Kiewits and Ms Tiemie are the fifth generation of their families to have lived and worked in the Grootkraal area. Other deponents claimed a similarly lengthy connection with the area and the property, and one deponent said that she was the ninth generation to have lived there. They and their parents and children attended the school, and they have longstanding connections with the church that has existed on the property, so they say, for nearly 200 years. The members of this Community are largely drawn from the Coloured sector of the population and are historically disadvantaged. They include farmworkers, artisans, domestic workers at local resorts and people working in various capacities at the Cango Caves. Although people come and go from the area, there is obviously a core of people having close family and working relationships with one another and a connection to both the school and the church on the property.[64]

The acceptance of the 'Community of Grootkraal' as an entity capable of entering into legal proceedings concerning ancestral land rights was groundbreaking. The fact that the community was mostly Coloured implies that the 'community' could not have logically existed as such prior to, say, 1652, because (following the logic of *Vetustas*) Coloureds as a population group did not exist prior to that date. Nevertheless, the judgment held that 'at some uncertain date, between 1820 and the building of the church in the late nineteenth century, a Christian community was established at Grootkraal, with connections possibly to Dyssselsdorp, but definitely to Oudtshoorn'.[65] This imprecision qualified as being outside of living memory and within the definition of 'since time immemorial', and the historical continuity of the community could be proven based on missionary records, which the court consulted.[66] Thus, a potentially heterogenous 'community' of disadvantaged peoples could be historically considered outside the auspices of ethnic categories or traditional authorities.

In Namibia, prior to the Orange River case at hand, legal motions that related to ancestral land, broadly defined, were bound within the traditional authorities system, and *Vetustas* had never been invoked for this purpose. The only other major ancestral land case in Namibia – claims made by members of the Hai||om San community over Etosha National Park – collapsed in 2019

64 Ibid, para. 4.

65 Ibid, para. 38.

66 For an analysis of the case, see P. de Vos, 'Little-Used Legal Doctrine Secures Informal Land Rights of the Grootkraal Community', *Constitutionally Speaking* (29 November 2018).

when the Hai||om Traditional Authority invoked the Traditional Authorities Act to oppose the motion. The applicants – being eight registered members of the traditional authority – argued that the Hai||om Traditional Authority itself was conflicted and could not in good faith represent their community, but the High Court of Namibia would not hear the case on the grounds that the traditional authority wanted to be in the driver's seat.[67]

However, the precedent that the *Grootkraal* case set regarding *Vetustas* potentially changes future ancestral land claims. Although *Vetustas* does not imply the transfer of land ownership or title deeds – and does not necessarily need to result in a legal servitude – it does allow for the potential use of or traversing another's property based on immemorial use.[68] Whereas the *Grootkraal* case involved a church and its school grounds, the original purpose of *Vetustas* in Roman-Dutch canon revolved around water, grazing and transhumance. Especially for pastoralist communities in Namibia and South Africa, this element could prove crucial, because land ownership is often of less importance than secured access to grazing, water and trek-paths.

Willem Basson's invocation of *Vetustas* was based on the fact that he and his community of Bondelswarts near the Orange River had consistently traversed Gilbertson's and Morkel's farms to access other Nama communal areas since time immemorial, and they sought a public servitude to reflect this.[69] They elaborated by describing themselves as members of the Bondelswarts Traditional Authority, though comprising a community specific within this ethnic group. It was they who filed the counterclaim, not the traditional authority.

> The defendants are all members of the Bondelswart (!Gami‡nûn) people, a traditional community duly recognised as such and pursuant to, inter alia, Section 1 of the Traditional Authorities Act 25 of 2000. All of the members of the !Gami‡nûn share a common historical tradition, language, culture, religion, ethnic identity, law and customs. The defendants are among those individuals comprising the !Gami‡nûn people who have direct, historic, cultural, and family ties with the Warmbad area of

67 *Tsumib v Government of the Republic of Namibia* (A 206/2015) [2019] NAHCMD 312 (28 August 2019). The applicants were also represented by the Legal Assistance Centre. For more information, see W. Odendaal, *'Beggars on our own Land …': Tsumib v. Government of the Republic of Namibia and its Implications for Ancestral Land Claims in Namibia* (Basel, Basler Afrika Bibliographien, 2024).

68 See 'The doctrine of Vetustas – Communities may secure formal legal rights to use and occupy land owned by somebody else', *Fasken Institute* (1 April 2019).

69 RoHC File HC-MD-CIV-ACT-CON-2020/02424: Founding Affidavit, Willem Basson – 26 November 2020.

FIGURE 96 Basson's goats drink from the Orange River near Girtis
PHOTO: L. LENGGENHAGER, 2020

> Southern Namibia and the Orange River in the vicinity of the plaintiffs' farms, where they and their forebears have lived and worked for many generations since time immemorial ... The !Gamiǂnûn people as a collective indigenous grouping in Namibia have historically possessed, occupied, and used a specific territory, including a 'trek-path' in the nature of a public servitude to access the Orange River, its islands and inner banks by traversing the plaintiffs' farms prior to and after the plaintiffs' farms were registered. The defendants and other !Gamiǂnûn people have continued to, inter alia, utilise the natural resources along the inner banks and islands of the Orange River as their predecessors have done, in accordance with their customs, norms and laws since time immemorial.[70]

The counterclaim is clear that: (1) the river-folk constitute a distinct community of Bondelswarts with specific connections to the Orange River (that is, a specific grouping of Bondelswarts); (2) they and their ancestors have continued to

70 RoHC File HC-MD-CIV-ACT-CON-2020/02424: First to Fourth Defendants' Counterclaim – 7 May 2021 (para. 4–9).

make use of trekpaaitjies that stretch from the Orange River north to Warmbad and other Nama communal areas; and (3) the public roads that traverse Sandfontein, Homsrivier, Norechab, Girtis and other properties belonging to the plaintiffs were built in part on top of these ancient trek-paths. They therefore claim that the principle of *Vetustas* applies for the purpose of achieving a public servitude for the Bondelswarts as a group to cross the properties that form the ORKCA conservation zone.

One of the current authors (Moore) was brought on by the Legal Assistance Centre in late 2022 to submit expert witness reports about the history of the Orange River region and the precolonial, colonial and postcolonial history of the Bondelswarts. These reports are included among the court filings and discovery. We are not legal practitioners, but it is our professional opinion that – given the requirements to prove *Vetustas* for the *Grootkraal* case – the case of the river-folk has merit. The data presented throughout this book reveals that despite losing most of their lands by legal fiat during the German and South African colonial eras, the Bondelswarts/river-folk by the Orange River maintained a connection to their ancestral grounds. They continued to combine subsistence stock-keeping with wage labour for White farmers on both sides of the river, or even at Tantalite Valley mine. During the apartheid years of the Odendaal Plan, the Bondelswarts never allowed their communal lands to be fully seized by the government, and both traditional leaders and ordinary river-folk themselves maintained these transhumant and pastoral connections between formal communal lands and the Orange River. Apartheid-era legislation made crossing these commercial farms more difficult but not completely impossible. This relationship with their ancestral lands began beyond living memory of any individual person today or of previous known generations, and it continues up to the present day, despite the actions of capitalist farmers and 'conservationists'.

Sean Gilbertson and Pete Morkel were having none of this, and their responses to Basson's counterclaim were revealing of the ways in which post-apartheid land ownership and 'conservation' relies on and indeed honours the colonial past. Gilbertson's plea rejected all claims of *Vetustas* or ancestral land rights:

> The plaintiffs specifically deny that the defendants and/or the !Gamiǂnûn traditional people have any legitimate claim to the plaintiffs' farms or portions thereof, irrespective of the fact whether they at any stage in history had direct or indirect ties to the plaintiffs' farms or not … The plaintiffs further deny that the defendants and/or the !Gamiǂnûn traditional

> people are in law entitled to use or trespass upon the plaintiffs' farmland and/or in law have a legitimate claim to a historical 'trek-path' which traverses the plaintiffs' farmland.[71]

Gilbertson – via his legal team – argued that the Bondelswarts had communal lands elsewhere, and that the ORKCA properties were protected by title deeds from the colonial era, which Basson did not have. Gilbertson denied the existence of and/or relevance of public rights of way or roads across his properties and challenged the Bassons to submit proof thereof to show that historical 'trek-paths' existed, that historical ties existed and that these belonged to the river-folk and the Bondelswarts community. Furthermore, the plea declared that if historical ties or trek-paths were indeed discovered, his team would declare (1) that the Bassons had not 'kept to the parameters' of such a servitude, and (2) that the Bassons had used this servitude to 'commit various crimes such as the illegal hunting of game and/or poaching of wild animals belonging to the plaintiffs'.[72]

Despite denying the relevance of these roads, trek-paths and rights of way, Gilbertson's legal team quickly filed an application with the Roads Board in Karasburg to have farm roads 217, 284 and 311 closed.[73] These public roads run from the Orange River across Homsrivier, Sandfontein, Sandfontein West, Girtis, Norechab and other properties up towards Warmbad and the Bondelswarts communal areas. Given that there was an ongoing case concerning these rights of way, the legality of the application itself is unclear to us. In time, Gilbertson's team admitted to filing the application to close the roads, but they continued to deny the relevance of the roads for the purpose of *Vetustas* as well as the existence of Bondelswarts ancestral connections to the area, full stop.[74] When confronted by the LAC attorneys about applying to have the roads closed after the fact, Gilbertson's legal team 'had nothing really useful to say about it, except looking a bit sheepish'.[75]

71 RoHC File HC-MD-CIV-ACT-CON-2020/02424: Plaintiff's Plea to Defendant's Counterclaim – 27 July 2021.

72 Ibid, para. 3.5.3. This claim that the Bassons secretly use their liminal status to engage in crime is often made by Pete Morkel and Adriaan Mulder; see E. Hattingh, 'Rivier van Wettelosheid', *Republikein* (27 October 2021). In reality, most if not all of these irregularities are committed by individuals who cross over from Witbank on the South African side. William Rooi, interview with Bernard C. Moore and Luregn Lenggenhager (Girtis farm, 11 November 2021).

73 'Application that a Portion of Farm Road 217 and Farm Roads 284 and 311 be Closed: District of Karasburg', *Government Gazette of the Republic of Namibia* (no. 7841 of 1 July 2022).

74 RoHC File HC-MD-CIV-ACT-CON-2020/02424: Sean T. Gilbertson: Plaintiff's Witness Statement – 21 February 2023.

75 Peter Watson (LAC), personal communication with Bernard C. Moore (27 April 2023).

Gilbertson and Morkel implied that kraals belonging to Basson and/or the river-folk existed exclusively on Girtis and Kambreek, seeming to ignore that between these properties lies another farm, Hartebeesmund.[76] As shown in Figure 80, Hartebeesmund is one of the last examples of Crown Land in Namibia (it was surveyed as an Orange River commercial farm during the 1920s and 1930s, but the South West Africa Administration was unable to find a buyer for the property at all throughout the twentieth century).[77] Some White farmers in the area took out short-term grazing licences to move animals to Hartebeesmund during extreme droughts, but the hyper-mountainous landscape of the property rendered it unsaleable. In the Windhoek Deeds Office Archives, the Hartebeesmund file is completely blank and empty. Nevertheless, as shown in chapter 7, the Bassons and other river-folk continued to live on Hartebeesmund from time to time, and this has continued into the present day. Although most of the structures that house the river-folk and their workers are located near to Girtis and Kambreek farms, these also exist near the Crown land of Hartebeesmund, which still belongs to the Ministry of Lands.[78] This is confirmed by satellite images, as shown in Figure 93.

Hartebeesmund is a 6,500-hectare property surrounded on all sides by properties belonging to Sean Gilbertson and Ian Craig's companies, and it lies within the privately owned Orange River-Karoo Conservation Area. It belongs to the Namibian government and is occasionally traversed by river-folk. It appears highly improbable that the Ministry will part with its lands and sell it off to private investors now or in the future. Furthermore, ORKCA and its constituent members have admitted that they do not own the Orange River waters or have properties that actually extend to the waters themselves, nor have they yet acquired any properties on the South African side of the river. Despite believing that private land acquisition through deed of sale or custodianship agreements were key to transfrontier conservation, ORKCA lacks ownership or control of the key resource that it wishes to 'rewild', namely the Orange River itself.

In his most recent affidavit to the court, Willem Basson III hinted that it may not actually be the goal of these landowners to 'own' the Orange River. In the same way that Basson and the river-folk invoked *Vetustas* not to gain

76 RoHC File HC-MD-CIV-ACT-CON-2020/02424: Sean T. Gilbertson: Plaintiff's Witness Statement – 21 February 2023.

77 NAN LAN 338 File 714/2: Inspeksie-Rapport: Hartebeesmund and Eselruh – 1 January 1928.

78 William Rooi, interview with Bernard C. Moore and Luregn Lenggenhager (Girtis Farm, 11 November 2021); Monika Basson, interview with Bernard C. Moore and Luregn Lenggenhager (Warmbad, 9 November 2021).

control of Sandfontein, Girtis and other farms – but rather to beneficially use or traverse the properties – ORKCA does not need to 'own' the river to use it.

> The consortium of the three plaintiffs plan to take the river frontage from the general population use, in order to exercise exclusive rights over it for themselves and their guests from time to time. They do not want to see us nor our cattle in the area because it spoils their perception of ambience for themselves and their guests to see livestock or herders attending to them. Rather, they seek to claim as part of their exclusive playground, river frontages devoid of other people.[79]

•••

The Norechab Sand & Tent Club adopted the motto 'Space is the Ultimate Luxury' as early as the year 2000 when Sean Gilbertson, Heye Daun and Finn

FIGURE 97 Aerial view of Gerrit Luttig's old farmhouse on Pelladrift Farm. the house was constructed near to pre-existing river-folk kraals along the north bank of the Orange River, near to point N on the map in Figure 93. This location is almost impossible to access from the Namibian side of the river.
PHOTO COURTESY OF ROSITA SMEENK AND MAURITS MATERS, USED WITH PERMISSION

79 RoHc File HC-MD-CIV-ACT-CON-2020/02424: Willem Basson, Witness Statement – 8 December 2022.

Behnken purchased Norechab Farm. Later, Morkel, Craig and their supporters together created ORKCA as a purportedly 'conservationist' organisation. It is hard to believe that these aesthetic concerns have not continued, perhaps in modified form. Although ORKCA/Oana is technically not involved in this court case – reiterated several times to us by Ian Craig and Andreia Pawel – their vision of conservation likely depends in part on the successful eviction of Black Namibian pastoralists from the Orange River area by their business partners. To attain both the 'luxury' sought by the Norechab Club and the 'conservation' goals of ORKCA, they do not actually need to own the river: it must be emptied, however.

If the Orange River were offered to the conservationists, it is likely that they would buy it. However, they are not fighting the river-folk over who *owns* the Orange River but rather who has the right to be at and *use* the Orange River. Exclusive use for these wealthy conservationists is as good as ownership, and this is what the river-folk are fighting against. In this sense, the ancestral land claims advocated by the river-folk via Willem Basson – based on the principle of *Vetustas* – is a historically grounded justification for their *presence* on lands which colonialism and apartheid formally stripped from them.

Furthermore, because the court case is ultimately a Namibian one, it does not address the fact that the Bondelswarts were (and are) a transnational community with ancestral lands on both sides of the Orange River. The case is ongoing, and we do not wish to make firm predictions. Nevertheless, for the river-folk and for those who are seeking to further the project of ancestral land restitution in Namibia, this court case is promising as a step to revitalise the land question in Namibia.

One of the key tools in 'rewilding', for advocates from Namibia to Kenya to Holland to Montana, is wildlife corridors and conservation easements that allow game to move between 'core' conservation zones. In many ways, what the Bassons and other river-folk seek is not so different. By legitimising ancestral trek-paths through servitudes or other legal mechanisms, *Vetustas* claims would create what we call *corridors for humans*, allowing individuals to move between cores of human settlement in a safe manner to facilitate ancestral transhumance both for the purpose of agricultural production and to maintain cultural connections. Ancestral land claims, like 'rewilding' projects, seek to recreate and secure how things operated in the past.

The biggest difference between *Vetustas* and 'rewilding' (as presented by ORKCA) is that the latter can be distinctly anti-human, or at least anti-poor, in worldview. The data presented throughout this book has shown that to bring the landscapes of southern Namibia and the Orange River back to what they were like centuries before would ultimately necessitate a coexistence and conviviality between humans and non-humans. All historical sources point to this. Yet, 'rewilding' as displayed by ORKCA and its constituent members avoids this, ignores this and even condemns this. Private nature conservation as epitomised by Sandfontein, Oana and ORKCA is a political and economic project that fundamentally values charismatic wildlife and title deeds more than human beings. That is, to say, human beings who did not arrive at the lodge by private Cessna aeroplane. This case in southern Namibia reveals a disturbing trend whereby the poor are increasingly and inherently viewed as threats to the 'purity' of the conservation landscapes of the rich, threatening to destroy their project for the sake of mundane things like food production or cultural connections to ancestral grounds. When examining private nature conservation projects – especially of the 'rewilding' sort – observers should ask themselves the following questions: *Ancestral land for whom*? *Ancestral land for what*? Doing so allows us to distinguish between private nature conservation projects that are bona fide about *conservation* and those that are more about being *private*.

CHAPTER 15

Weidner's Ghost

'I just don't understand why Sean Gilbertson and Heye Daun won't let the mining companies build a pipeline from the Orange River to Tantalite Valley', said a despondent Lisias Pius as he took a sip of tea in his favourite Windhoek café. 'They made all their money in mining, and now they won't allow us to be more responsible with our water. It doesn't make sense'.[1] Tantalite Valley has been reopened for mining, and efforts are being made to build on and expand Peter Weidner's original operations. However, the expansion of mining is capped by the amount of water available, and in an arid region pipelines are key. Lisias Pius is one of the fast-rising Black businessmen within the mining industry in Namibia, a sector dominated by Whites. He has successfully targeted, applied for and been granted at least eight exclusive prospecting licences (EPLs) across Namibia over the past ten years.[2] Some of these are simply speculative, others have borne fruit, particularly the areas surrounding Tantalite Valley mine, which became available in 2012 around the same time that Ian Craig purchased KumKum and Pelladrift.

'I was at the University of Namibia studying IT and computer science. I actually knew very little about mining at the time', admitted Pius. 'But I had a friend who worked at the Ministry of Mines and Energy who called me one day in 2012, and he urged me to apply for this EPL way out there in ǁKaras region. Apparently no one had applied for it yet'. This was EPL 5047, which covers the entirety of Umeis, Kinderzitt and Norechab farms, with the exception of the small 450-hectare area that formed Peter Weidner's old mining licence at Tantalite Valley. The licence granted the twenty two year old Pius 26,321 hectares of land where he had the right to prospect for and run tests on 'base and rare metals, industrial metals, and precious metals', covering most minerals apart from uranium and precious stones.[3] In Namibia, mining resources are the property of the central government – not the landowner – which means that although the EPL-holder must negotiate a land-use agreement with the landowner, the latter has no inherent right to the minerals. 'I didn't know much about geology or mining engineering at the time, so I enrolled in a course at the university,

1 Lisias Pius, interview with Bernard C. Moore (Windhoek, 2 August 2022).

2 See, MME: Exclusive Prospecting Licence Master List (12 July 2022).

3 MME File 14/2/1/4/2/5047: Exclusive Prospecting Licence no. 5047: Lisias Pius (18 December 2012).

FIGURE 98 Storms approach southern Namibia on the road to Tantalite Valley
PHOTO: B.C. MOORE, 2022

and I spent the next few years trying to locate investors who would come to Namibia to work with me to prospect and open up a mine'.[4]

Lisias Pius was not the only party interested in opening up mining in the area around Tantalite Valley. Although his licence did not include Weidner's original mine itself, the famous pegmatites on Tantalite Valley were going to get a second life. In 2010, the United States Dodd-Frank Act was passed by Congress. While its main purpose was to overhaul financial regulation in the aftermath of the Great Recession, buried deep within Section 1502 of the Act was a requirement that mining companies and manufacturers that do business with the USA disclose whether any 'conflict minerals' (gold, tungsten/wolfram and, crucially, tantalum) have originated from the Democratic Republic of the Congo or neighbouring nations.[5] Because the DRC and Rwanda were two of the largest global producers of tantalum at the time, the price of ore from these nations plummeted as companies scrambled to rebuild and monitor

4 Lisias Pius, interview with Bernard C. Moore (Windhoek, 2 August 2022).

5 International Energy Agency, 'Dodd Frank Wall Street Reform and Consumer Protection Act' (4 November 2022) https://www.iea.org/policies/16713-dodd-frank-wall-street-reform-and-consumer-protection-act

their supply chains. The global capacitor and semiconductor market – one of the main uses of tantalum in the twenty-first century – was feeling the effects of this Act, and it sought to source tantalum from 'non-conflict' areas like Namibia.

This is what drove the Welsh company Kazera Global plc (formerly, Kennedy Ventures plc) to enter into negotiations over 2014–2015 to buy Weidner's old Tantalite Valley mine.[6] As mentioned in chapter 9, the Australian company Magnum Mining and Exploration Ltd had gone into business in 2007 with Steen Severin's former engineer, Peter J. Cox, to buy the land, the mining licence and the equipment from Severin. Cox had been acquainted with Kazera directors Giles Clarke and Nick Harrison since at least 2012, through their collaborative work on the board of Welsh mining company Ironveld plc,[7] and it was likely he who urged his British colleagues to invest in Tantalite Valley. Kazera started negotiations with Magnum Mining, and by the end of 2015 it had finalised purchase of the latter's interest in Tantalite Valley for ZAR 15 million.[8] Through a complex arrangement of holding companies, Kazera took ownership of the mining licence, equipment and buildings, while Peter Cox and Sean Gilbertson's companies owned the land on which the mine stood.[9]

6 Kennedy Ventures plc, 'Preliminary Announcement of Unaudited Results for the Year ended 30 June 2014' (14 November 2014). Kennedy Ventures plc, *Annual Report 2016* (16 December 2016), p. 29.

7 Ironveld plc, *Annual Report and Financial Statements for the Year ended 30 June 2022* (28 December 2022), p. 11.

8 Magnum Mining and Exploration Ltd, *Annual Report for the Year ended 31 December 2015* (31 March 2016), p. 4.

9 In 2015, Kazera (then still called Kennedy Ventures) bought out Magnum's interest in the mine, via Kazera's Namibia-domiciled holding company African Tantalum (Pty) Ltd. At this time, African Tantalum was still 25% owned by Cox's company Warmbad Investment Holdings (Pty) Ltd, though Kazera bought out this remaining percentage in 2020 in exchange for GBP 30,000 and 3% interest in Kazera Global plc. This rendered African Tantalum a 100% subsidiary of Kazera. Mining Licence no. 77, representing Tantalite Valley mine, is owned by Tameka Shelf Company Four (Pty) Ltd, which is a 100% subsidiary of African Tantalum. The equipment and the buildings are owned by Namibia Tantalite Investments (Pty) Ltd, also a 100% subsidiary of African Tantalum. Kazera holds no financial interest in the land – Umeis and Kinderzitt farms – which is controlled by a different corporate structure. Warmbad Investment Holdings owns 95% of Kindermeis Holdings (Pty) Ltd, and the remaining 5% is allegedly held in trust for the Bondelswarts community. Kindermeis in turn owns 70% of Tantalite Valley Estates (Pty) Ltd, the legal owner of the land. The remaining 30% was acquired by Sean Gilbertson's BVI company, Arianna Investments Ltd in 2022. See Kennedy Ventures plc, *Annual Report 2016* (16 December 2016), p. 29; 'Kazera Global plc acquired the remaining 25% stake in African Tantalum Limited', *Market Screener* (25 June 2020); Peter J. Cox, interview with Bernard C. Moore (telephone, 15 January 2023).

Kazera's decision to buy Tantalite Valley likely helped to stimulate interest in Lisias Pius's EPL, which surrounded Kazera's mining licence. At a 2016 mining expo in South Africa, Pius met Jurie Wessels and Philip le Roux, who were intrigued with the mining possibilities in southern Namibia. Pius began applying for, and was granted, additional EPLs on KumKum and Pelgrimsrust farms,[10] but the focus would be on EPL 5047, near the main mine on Tantalite Valley. Together, the three of them founded Orange River Pegmatite (Pty) Ltd, to which they transferred the prospecting licences.[11] Orange River Pegmatite is an 80 per cent subsidiary of Arcadia Minerals Ltd, which is a Guernsey-domiciled mining company listed on the Australian Stock Exchange.[12]

Arcadia, like Kazera, may be interested in the tantalum mineralisation in the area, but its broader goals involve the lepidolite and spodumene pegmatites that are interspersed throughout the Pofadder Shear Zone. Both these pegmatic ores are important sources of lithium. Although mining lithium via pegmatites is more labour intensive and less cost effective than sourcing lithium from brines (as is common in South America), the global growth in demand for electric vehicles has led to a massive surge in lithium prices, making pegmatites valuable again.[13] Along these lines, Arcadia now describes itself as a 'diversified battery explorer' company and frames its hunt for lithium and tantalum as a means of responding to the 'electric vehicle revolution'.[14] Interestingly, the directors speak of lithium mineralisation in terms of the quantity

10 See, MME File 14/2/4/1/6940: Exclusive Prospecting Licence 6940, Lisias Pius – 1 October 2018. MME File 14/2/4/1/7295: Exclusive Prospecting Licence 7295, M. Rosario – 13 May 2019. Very little prospecting has been done on the KumKum/Pelgrimsrust EPLs and the project remains 'inherently speculative in nature'. See, J. Hattingh, 'Independent Geological Report on the Nickel-Copper-PGE Sulphide Hosted Mafic-Ultramafic Deposits at KumKum and Surroundings', prepared for Orange River Pegmatite (Pty) Ltd (March 2021).

11 BIPA File 2018/0020: Annual Return (CM-23) – 10 February 2021.

12 Arcadia Minerals Ltd also features Wessels, Le Roux and Pius on its board of directors. Arcadia's largest shareholder is the South African diversified contractor company SPH Kundalila (Pty) Ltd, which is itself a subsidiary of the South African infrastructure company Raubex Group Ltd, such that the latter has almost 29% voting power in Arcadia. Raubex Group Ltd, *Annual Financial Statements for the Year ended 28 February 2023* (9 May 2023), p. 101. ASX, 'Arcadia Minerals Ltd: Notice of Change of Interests of Substantial Holder, Raubex Group Ltd' (Form 604) – 4 July 2023.

13 Arcadia also holds EPLs in Namibia's Hardap region, where it is investigating lithium brines as well.

14 Jurie Wessels, Arcadia Minerals Ltd, interview with *ProActive Investors* (1 September 2022). https://www.proactiveinvestors.com/companies/news/991601 (Accessed 30 June 2023).

of cars that could be powered from their deposits, rather than simply giving figures in million tonnes of lithium ore.[15]

Neither Arcadia nor Kazera have done substantial mining on their properties, and most of the work they have done is merely prospecting and identifying the quality of the deposits. Not all of this is circumstantial, however. Arcadia's drilling programme revealed an average tantalum pentoxide mineralisation grade higher than in any other mine globally, including Greenbushes in Australia, which is the most productive in the world.[16] But while the *grade* of Arcadia's deposits is impressive, the heterogeneity of the Pofadder Shear Zone means that the actual *quantity* is not comparable with Greenbushes' deposits, which are more than one hundred times as large, if of a lower grade.[17] Economies of scale matter in mining, and while Arcadia's tantalum and lithium deposits appear larger than Kazera's, it is unclear if they are economically viable outside of boom years, when lithium, microchips and semiconductors are in high demand. Nevertheless, on 3 June 2022, the Ministry of Mines and Energy approved a fifteen-year mining licence #223, which would be granted to Arcadia's subsidiary Orange River Pegmatite.[18]

In the months after Arcadia was granted an official mining licence, Chinese investors quickly took notice. Chinese refineries and factories already account for nearly sixty per cent of all lithium processing globally, and these large conglomerates are increasingly seeking longer-term contracts with mines, often buying mining licences across Africa to control the supply.[19] This applies not just to lithium but also other important battery, microchip and semiconductor minerals like cobalt, nickel, graphite and tantalum. In September 2021, Kazera entered into an offtake agreement deal with Jiujiang Jinxin Nonferrous Metals Co. Ltd, which agreed to buy the entirety of Kazera's tantalum production for two years.[20] In June 2022, it managed a similar agreement concerning lithium with Hebei Xinjiang Construction Co., which then decided in December 2022

15 Philip le Roux, CEO Arcadia Minerals, interview with *Assay TV* (7 February 2023) https://youtu.be/jczSZBT-fmU

16 Arcadia Minerals Ltd, *Annual Report for the Year ended 30 June 2022* (28 September 2022).

17 Arcadia Minerals Ltd, 'Investor Presentation: 1–2–1 Mining Indaba' (Cape Town, May 2022).

18 Arcadia Minerals Ltd, *Annual Report for the Year ended 30 June 2022* (28 September 2022), p. 8.

19 H. Dempsey and J. Cotterill, 'How China is Winning the Race for Africa's Lithium', *Financial Times* (3 April 2023).

20 IAWB: Kazera Global plc, 'Kazera-Align Update, October 2022'. https://web.archive.org/web/20221103145753/https://kazeraglobal.com/wp-content/uploads/2022/10/Kazera-Align-Update-October-2022.pdf

to simply buy out Kazera from Tantalite Valley entirely, for USD 13 million in cash.[21] It was expected that the mine would officially turn over to Hebei after all paperwork and payments were made, by the end of 2023.[22] Hebei Xinjian made a similar offer to Arcadia Minerals, declaring that it would buy all its tantalum and lithium ore.[23] Hebei would also provide USD 7 million in funds to construct concentrate plants on Arcadia's new mining licence area in return for a share of the profits.[24]

From the very beginning, the problem that Kazera and Arcadia faced in getting their prospecting operations off the ground was the question of water. Mining requires a tremendous amount of water, always. The more mechanical and/or chemical a process is – as opposed to being labour intensive – the more water it will require. From the 1950s to the 1970s, Peter Weidner primarily used Pleitz jigs and other rudimentary sorting machines to conduct preliminary processing of the pegmatite ore. This necessitates water – which Weidner obtained from Umeis's boreholes – but much less than is required for hydrometallurgical refining or even froth flotation machinery. Mining operators are left in a quandary. Use more labour in a less-effective separation method? Or use scarce water sources to improve yields and reduce transport costs by shipping partially refined ore? Lepidolite has quite a lot of mica in it, and the cleavage process of mica is horizontal sheeting, so it is much harder to crush and process via low-water gravitational separation. One generally needs to set up a metallurgical plant to extract lithium efficiently, but that requires more water.[25]

Kazera had briefly entered into production in 2017, only to quickly shut down its operations when borehole waters fell too low.[26] It secured additional borehole rights on Arus Farm to the south-east but additional sources needed

21 Kazera Global plc, *Annual Report for the Year Ended 30 June 2022* (15 March 2023).

22 Kazera Global plc, 'Tantalum and Lithium Project' https://kazeraglobal.com/portfolio/tantalite-valley-mine/ (Accessed 30 June 2023).

23 H. Mulholland, 'Arcadia Minerals receives letter of intent for potential offtake from Swanson Tantalum and Lithium Project', *Mining.com.au* (29 November 2022).

24 M.Plan International, 'Competent Persons Report on the Swanson Tantalum and Lithium Definitive Feasibility Study', prepared for Arcadia Minerals Ltd (29 May 2023), p. 2. The letter of intent was signed on 23 March 2023 between Hebei and Arcadia, formalising the offtake agreement, whereby Hebei Xinjiang would take over 38% of the shares in Orange River Pegmatite (Pty) Ltd.

25 Sabine Anderson (née Severin), interview with Bernard C. Moore (telephone, 2 January 2023). Peter J. Cox, interview with Bernard C. Moore (telephone, 25 March 2023).

26 IAWB: Kazera Global plc, 'Kazera-Align Update, October 2022' https://web.archive.org/web/20221103145753/https://kazeraglobal.com/wp-content/uploads/2022/10/Kazera-Align-Update-October-2022.pdf

to be identified.[27] In 2018, Kazera hired a team of consultants to investigate possible ways to build a pipeline from the Orange River to Tantalite Valley. Expanding production needed more water, and the most sustainable way was to pump it from the only perennial resource in the region. The consultants proposed that Kazera could build the pipeline either along the D206 road from the Velloorsdrift border post (longer and more expensive) or else due south from Tantalite Valley across Sandfontein and Homsrivier farms to the river (shorter and cheaper to build and operate).[28] With permission from the landowners of Umeis,[29] Kazera applied to the Ministry for an environmental clearance certificate, which was granted.[30] Sean Gilbertson, however, stopped the project in its tracks, refusing Kazera the rights to build the pipeline over his land. This rendered the project unaffordable, and it was scrapped.

'Look', said Peter Cox, 'Sean does not believe in the mine, and he doesn't want the disturbance which building a pipeline would cause'.[31] It appears that the other ORKCA partners are on the same page as Gilbertson, and that the only option for Kazera (and now its successors) would be to build the pipeline an extra forty five kilometres to the south-east. 'Having tracked the Tantalite Valley project since 1992, I don't believe it will deliver the hopes so many hold for it, unless there is a MAJOR uptick in tantalum prices', said Gilbertson. 'As for the current "lithium" hype on and around our projects, time will tell, but I consider their prospects to be thin'.[32] Although it was Kazera that originally proposed the pipeline to the river, it is likely that Arcadia could have negotiated with them for use of the Orange River water, rather than boreholes, so both companies would benefit.[33] Nevertheless, the lack of a pipeline meant that Arcadia was forced to negotiate use of the already weak borehole system

27 Johannes Gertze, interview with Bernard C. Moore (Arus Farm, 16 October 2022).

28 KPM Environmental Consulting, 'Environmental Management Plan for the Abstraction of Water from the Orange River to Namibia Tantalite Mine' (30 April 2021). This report is from its application for a renewal licence, but the 2018 EMP is nearly identical.

29 Caroline McCleod, Tantalite Valley Estates (Pty) Ltd to Environmental Commissioner, Ministry of Environment, Forestry and Tourism 'Letter of Consent: Tameka Shelf Company Four (Pty) Ltd' – 25 March 2021.

30 Teofilus Nghitila, Environmental Commissioner MET to CEO, African Tantalum (Pty) Ltd 'Environmental Clearance Certificate for the Proposed Abstraction of Water from the Orange River to Namibia Tantalite Mine at Umeis' – 9 March 2018.

31 Peter J. Cox, interview with Bernard C. Moore (telephone, 15 January 2023). Sean Gilbertson wrote to us in 2024 that he had offered 'terms' [unspecified to us] to Kazera to guide the building of the pipeline, but such 'terms' were not accepted by the Welsh company.

32 Sean T. Gilbertson to Bernard C. Moore (7 January 2023), letter in the authors' possession.

33 Lisias Pius, interview with Bernard C. Moore (Windhoek, 2 August 2022).

at Warmbad village. It is in the process of building a forty-kilometre pipeline costing at least USD 500,000.[34]

One of the groups that is vocally supporting the pipeline and the redevelopment of Tantalite Valley are the river-folk. Their spokesman Willem Basson wrote a petition to then-Namibian President Hage G. Geingob, explaining the situation that his people face at the Orange River, linking much of their problems to Sean Gilbertson's exclusive view of land tenure and perceived expansionist ambitions. In perhaps exaggerated language, Basson claimed that 'Mr. Gilbertson also alleges that the entire Orange River boundary belongs to him too. He has said to us that he is going to buy the entire Orange River and there is nothing we can do about it'.[35] Furthermore, the petition identified Gilbertson's refusal to build the pipeline as emblematic of a general refusal for development to take place in a manner that could benefit Black Namibians. While the miners and landowners at Tantalite Valley are not necessarily friends of the river-folk,[36] Basson draws on a long history of his people's cooperation and engagement with Tantalite Valley, and it is clear why he and his family hold affinity for it.

Finally, and perhaps most importantly, the construction of a pipeline to the Orange River would inevitably involve the government parastatal corporation, NamWater, because the river banks are state lands. Constructing the pipeline would mean the creation of a public servitude from the Orange River at least as far as the public road to Tantalite Valley mine. This would provide access for the river-folk to continue practising their transhumance pastoralism, from 'time immemorial' into the future. For private conservationists, the economic and aesthetic concerns of having African pastoralists nearby may weigh more heavily than the potential ecological damage that would come from mines abusing groundwater resources. For some of these 'conservationists', the Earth can wait; private property is at stake.

•••

34 M.Plan International, 'Competent Persons Report', pp. 58, 270–4. Peter Cox's companies will receive approximately USD 1,420 per month in servitude fees for this.

35 WBPA: Willem Basson III to President of the Republic of Namibia 'Petition to Reclaim Land' – 4 June 2021.

36 Indeed, Tantalite Valley Estates has recently joined the calls for evicting the river-folk, as well as interdicts against trekking animals. LAC BCF: Shikongo Law Chambers (on behalf of Tantalite Valley Estates (Pty) Ltd) to Willem Basson III 'Trespassing on Private Property: Farm Umeis no. 110' – 10 May 2023.

FIGURE 99 Night descends on Sandfontein farm, 2023.
PHOTO COURTESY OF ROSITA SMEENK AND MAURITS MATERS, USED WITH PERMISSION

When the Norechab Sand & Tent Club was founded, after Sean Gilbertson and his mates purchased their first farm in southern Namibia, it is unlikely that Sean or any of them believed that they would end up being part of one of the largest private conservation initiatives in Africa. They also would likely have been surprised that they and their companies would become involved in one of the first ancestral land cases in Namibian history. For these young, fast-rising elites within mining finance, the Norechab Club did not represent any desire to become large-scale, apartheid-style agricultural capitalists so maligned globally. They started the twenty-first century at Tantalite Valley, and they likely had little intention of repeating the original sins that colonialism had wreaked upon this landscape. After all, they did not want to make money from the land; they just wanted space: as their motto goes, 'Space is the Ultimate Luxury'.

The alleged innocent escapism of the Norechab Sand & Tent Club morphed into something many might view as rather sinister, not because of the moral deviance of the individuals involved, per se. Rather, it was because the political economy of southern Namibia permitted it and facilitated it. This regional economy was shaped by a century of violence, genocide, land dispossession and apartheid racism. Sean's desire to visit his properties in southern Namibia and, as he once put it, 'listen to the silence of the wind', could easily strike one as some form of apolitical catharsis. Sean's rise within globalCOAL and Gemfields brought him the stresses of board rooms, mergers and acquisitions, and it would be reasonable for him to desire to return to something pure and 'natural'. Likewise, Sean's self-professed understanding that conservation is necessary to rein in the damage that we have done to our planet (see chapter 2) likely helped him justify his land acquisitions along those lines. But Sean Gilbertson, Ian Craig and Pete Morkel each grew up in colonial and/or apartheid societies. They chose to join forces in Namibia – another ex-apartheid society – which afforded them a familiar social and legal system where land accumulation could take place. These individuals had the resources to accumulate many title deeds in a country where such deeds afford exclusive use over entire landscapes.

This book has shown that institutional, economic and legal foundations present in southern Namibia made it conducive for private nature conservation projects. These processes are based on more than a century of historical conditioning, but there still is nothing inevitable about land being used in a certain way. The recent reopening of Tantalite Valley mine is testament to this, as is the deep desire for the river-folk and other Bondelswarts to continue to make use of their ancestral lands for small-scale farming. That being said, the rise of private nature conservation has been a radical shift in how landowners relate to their property. It is unlikely that Sandfontein or ORKCA will ever become highly profitable initiatives, but this is probably not the goal. From this viewpoint, private nature conservation represents landowners who deliberately choose to use their property in one of the *least* remunerative ways, and who gradually strip away all the infrastructure (such as fencing) that made the farms profitable for livestock. However, this does not mean that these unprofitable conservation operations are removed from the capitalist mode of production. Foundations that fund these sorts of schemes – like The Nature Conservancy, Fauna & Flora International, Wilderness Foundation Africa – are integral to large corporations that funnel profits to what their directors deem worthy causes. And nature conservation, especially private nature conservation, is especially worthy. The rising carbon finance market, which ORKCA seeks to tap into, represents another way in which these lands are bound to these broader global processes.

When conservationists argue in favour of 'rewilding', they are making a claim about the past. When they do so with title deeds gripped tightly in hand, they are making a claim about the future. Nature conservation *cannot* be decoupled from the land question. For all the fencing that the conservationists have removed from KumKum, Pelladrift, Sandfontein and other properties, transhumant pastoralism has become only more difficult for the river-folk. The actions of the Orange River-Karoo Conservation Area and its constituent members have given a strong impression that the poor are simply not welcome. Nature conservation and 'rewilding', as is practised here, represents an equal or even more exclusive use of the land than their apartheid-era sheep-farming predecessors. In many ways, this explains the difficulty the river-folk have in viewing the conservationists as a 'break' in how land is used; they only see continuity.

Private nature conservation is a political claim about how land should be used and who should be welcome on that land. One need not be advocating for large-scale, intensive, commercial livestock-farming, irrigation or mining to argue that we must reconsider how nature conservation should operate: new approaches and new models are needed. In embryonic form, these exist in Namibia and beyond, and they must be taken seriously. For all the critiques that we have levelled against the Northern Rangelands Trust in Kenya and community based natural resource management (CBNRM) more broadly, these initiatives are proof that the exclusion of Indigenous pastoralists from 'conservation zones' is not the only way forward. The pastoralism of the river-folk is compatible with wildlife conservation and even limited approaches to 'rewilding', but it appears that rewilding as proposed by ORKCA and many others has no place for people like the river-folk.

It is beyond the scope of this book to propose a precise solution to the conflagration that we see in the twenty-first century alignment of capitalist interests with certain conservationists. We do, however, suggest considering what political ecologists Büscher and Fletcher call 'convivial conservation': a 'post-capitalist approach to conservation that promotes radical equity, structural transformation, and environmental justice, and so contributes to an overarching movement to create a more equal and sustainable world'.[37] We are at a crossroads, where authoritarian right-wing governments increasingly erode the small gains we have made towards legalising environmental protection, and mainstream conservationists increasingly claim that the ecological damage that we have done to our earth can be saved only by opening conservation to market forces. This is unsustainable. 'Convivial conservation' is based on the simple premise of a frugal good life, and it incorporates five key elements.

First, there should be less emphasis on 'protected areas' that restrict human access and use and more emphasis on 'promoted areas' where specific elements of biodiversity, for example, are emphasised as important, while still allowing humans to exist on and use the land in a responsible way.

Second, conservationists should focus less on 'saving' a form of non-human nature and more on celebrating and appreciating the ways in which humans live within nature. Nature is not something we can escape into but something

37 B. Büscher and R. Fletcher, 'Towards Convivial Conservation', *Conservation & Society*, 17, 3 (2019), pp. 283–296.

we must consciously live responsibly with, and we must structure our societies accordingly.

Third, 'touristic voyeurism' as a means to promote and fund conservation is not ecologically or socially sustainable at all. The commodification of nature to be enjoyed in brief touristic experiences is escapist, and we should instead focus on longer-term 'democratic engagement' with these areas with a focus on social and ecological justice.

Fourth, we must transition from 'spectacular' to 'everyday' iterations of ecological engagement, appreciating the mundane (like grass or goats) as much as the extreme (like rhinoceros or giraffe).

Fifth, and finally, conservation must move away from a technocratic expert-oriented form of top-down management of nature towards a more democratic management of nature.[38] All five of these overlapping emphases necessitate a move away from looking at nature conservation through the lens of the profit motive and, *especially*, of land ownership.

It is very hard to imagine a form of convivial conservation compatible with private land tenure as described throughout this study. To open ORKCA or Sandfontein to conviviality necessitates them surrendering exclusive control and management over their property, being both the land asset and the wildlife asset. In many ways, Büscher and Fletcher remind us, convivial conservation requires 'historic reparations' that acknowledge the often (neo)colonial histories of conservation and land tenure. If local communities have been dispossessed of their land in the name of conservation, or where that dispossession has facilitated 'conservation', reparative steps need to be taken. This may mean that local communities receive their land back, or at least are offered a meaningful kind of joint-ownership and joint-management.[39] But, as we have mentioned before, human issues – such as dispossession and violence – do not factor into most conservationists' definitions of biodiversity gains or losses. It is very hard to imagine Sean Gilbertson, Ian Craig or Pete Morkel and their teams ascribing to convivial definitions of conservation, because it may be the case that they value their title deeds and the privileges these bring more than their purported ecological goals.

Private actors in Namibia and elsewhere have increasingly captured nature conservation, turning it into an industry which depends upon maintaining colonial and capitalist structures. As long as conservation operates along this disastrous status quo, the only option that previously disadvantaged Namibians have is to counter the rewilder's imprecise non-human historical baselines

38 Ibid, pp. 286–288.

39 Ibid, p. 291.

with the historical precision of their ancestral ties to the land. Willem Basson III's insistence on the river-folk's immemorial use (*Vetustas*) of the trek-paths that connect the Orange River to the Bondelswarts communal areas is an example of this. And for those historians who wish to contribute towards building both social and ecological equity in this world, it is our duty to help demonstrate these longstanding connections to the land.

APPENDIX 1

Companies Addressed Extensively in the Text

Association Reference Codes

BG: Brian Gilbertson
C-F: Carbon Finance
F: Fabergé
G: Gemfields
O: ORKCA
R: Rembrandt
S: Sandfontein
SG: Sean Gilbertson
SS: Steen Severin
T-V: Tantalite Valley

Name of entity	Domiciled country	Reference codes
African Tantalum (Pty) Ltd	Namibia	T-V
Arcadia Minerals Ltd	Guernsey	T-V
Arianna Investments Ltd	British Virgin Islands	S, SG, T-V
Autumn Holdings Asset Inc	British Virgin Islands	BG, F, G, SG
BHP Group Ltd	Australia	BG, SG, T-V
Charbonnages de France	France	SS
Chiebella (Edms) Bpk	Namibia	R
Compagnie Financière Richemont S.A.	Switzerland	R
Distillers Korporasie Bpk	South Africa	R
Fabergé Ltd (*formerly: Project Egg Ltd*)	Cayman Islands	BG, F, G, SG
Fairbairn Trust Ltd	Jersey	F, G, SG
Gemfields Group Ltd	Guernsey	F, G, SG
Gemfields plc (*formerly: Gemfields Resources plc*)	England (UK)	F, G, SG
Gemhouse Inc.	Canada	G, SG
Gencor Ltd	South Africa	BG
General Electric Co.	New York (USA)	T-V
Gigajoule Ltd	England (UK)	SG
Global Coal Ltd	England (UK)	SG
Global Coal Market Ltd (globalCOAL)	England (UK)	SG
Great River Game Ranch (Pty) Ltd	Namibia	S, SG, SS, T-V
Greentop International Inc	British Virgin Islands	G, SG

 | DOI:10.1163/9789004712454_017

Name of entity	Domiciled country	Reference codes
Hagura Mining Ltd	British Virgin Islands	G, SG
Hagura Mining Ltd	England (UK)	G, SG
Heibei Construction Group Corp. Ltd	China	T-V
Johannesburg Consolidated Investments Ltd	South Africa	BG
Kagem Mining Ltd	Zambia	G, SG
Kazera Global plc (*formerly: Kennedy Ventures plc*)	Wales (UK)	T-V
Kindermeis Holdings (Pty) Ltd	Namibia	SS, T-V
Krinera Group SA	Panama	BG, G, SG
KumKum Adventure Tourism (Pty) Ltd	Namibia	O
Lukin-Heydebreck Holdings (Pty) Ltd	Namibia	O, S, SG
Marwilben Farming cc	Namibia	O, S, SG
Magnum Mining & Exploration Ltd (*formerly: Magnum Gold NL*)	Australia	T-V
Namaqua Ltd	England (UK)	O
Namibia Tantalite Investments (Pty) Ltd	Namibia	T-V
Namura Mineral Resources (Pty) Ltd	Namibia	S
Native, A Public Benefit Corporation	Delaware (USA)	C-F
Orange River Pegmatite (Pty) Ltd	Namibia	T-V
Oude Meester Groep Bpk	South Africa	R
Pallinghurst Resources LLP	England (UK)	BG, F, G, SG
Pella Refractory Ores (Pty) Ltd	Namibia	T-V
Pelladrift (Edms) Bpk	Namibia	O, R
Reinet Investments SA	Luxembourg	R
Rembrandt Groep Bpk	South Africa	R
Remgro Ltd	South Africa	R
Renova Holdings Ltd	Russia	BG, F
R.G. Niemöller (Edms) Bpk	South Africa	T-V
Rox Ltd	Cayman Islands	BG, F, G, SG
Rox Conduit Ltd	Cayman Islands	BG, F, G, SG
Sandfontein & Homsrivier Properties (Pty) Ltd (*formerly: Tantalite Mining & Exploration (Pty) Ltd*)	Namibia	S, SG, SS

Name of entity	Domiciled country	Reference codes
Sandfontein Lodge & Nature Reserve (Pty) Ltd	Namibia	O, S, SG
Severin Minerals Corporation (Pty) Ltd	South Africa	SS
Severin Mining & Development (Pty) Ltd	South Africa	SS
Silikadan A/S	Denmark	SS
Soils for the Future LLC	Delaware (USA)	C-F
Southern Sphere Mining & Development (Pty) Ltd	South Africa	SS, T-V
Steen Severin A/S	Denmark	SS
Tameka Shelf Company Four (Pty) Ltd	Namibia	T-V
Tantalite Valley Estates (Pty) Ltd (*formerly: Tantalite Valley Minerals (Pty) Ltd*)	Namibia	O, SG, T-V
Tantalite Valley Mining (Pty) Ltd	Namibia	SS, T-V
Utah International Inc	California (USA)	T-V
Warmbad Investment Holdings (Pty) Ltd	Namibia	T-V
Xemplar Energy Corporation	Canada	S

APPENDIX 2

Foundations and Non-Profits Addressed Extensively in the Text

Name of entity	Domiciled country
Africa Parks Foundation	South Africa
Lewa Wildlife Conservancy	Kenya
Natural State Ltd, 501(c)(3)	Delaware (USA)
Northern Rangelands Trust	Kenya
Oana Community Development Project (Pty) Ltd, sec. 21	Namibia
Ol Pejeta Conservancy Ltd	Kenya
Orange River-Karoo Conservation Area (Pty) Ltd, sec. 21	Namibia
Peace Parks Foundation, NPC	South Africa
The Nature Conservancy, 501(c)(3)	Virginia (USA)
Wilderness Foundation Africa	South Africa
Verra, 501(c)(3)	Washington, DC (USA)

APPENDIX 3

Archival Abbreviations

BAB	Basler Afrika Bibliographien
PA.24	Personenarchiv: Hubertus Graf zu Castell-Rüdenhausen
BArch	Bundesarchiv Berlin-Lichterfelde
BIPA	Business and Intellectual Property Authority of Namibia
BSB	Bayerische Staats-Bibliothek
CCEW	Charity Commission for England and Wales
EAKS	Erfdeel Argief en Kultuursentrum (Windhoek)
GNPA	R.G. Niemöller Private Archive (Pofadder)
IAWB	Internet Archive Wayback Machine
KAB	Western Cape Archives & Records Service (Buro-KAB)
ACLT	Agricultural Credit & Land Tenure
MOOC	Master of the Orphan Chamber
KBN	Karakul Board of Namibia (Windhoek)
LAC	Legal Assistance Centre (Windhoek)
BCF	Basson Context File
MME	Ministry of Mines and Energy of the Republic of Namibia
M/S	Museet for Søfarts, Denmark
NAN	National Archives of Namibia
A.486	Bondelswarts Commission of Inquiry
ADG	Administrator General (Governmental Affairs)
AGR	Agricultural Branch
ANA	Administration for Namas
AP	Official Publications
ADM	Secretary for the Protectorate
BAC	Chief Bantu Affairs Commissioner (Windhoek)
CRN	Coloured, Rehoboth and Nama Affairs (Windhoek office)
IMW	Inspector of Mines and Works
KGR	General Rehabilitation Commission
LAB	Land and Agricultural Bank
LAN	Lands Branch
LKE	Magistrate at Keetmanshoop
LKG	Agricultural Credit and Land Tenure
LKW	Magistrate at Karasburg
LUKS	Odendaal Commission Liaison Committee
NTB	Division of Nature Conservation

PLA	Local Authorities
SWAA	South West Africa Secretariat (A-Series)
SWAS	South West Africa Secretariat (AS-Series)
SAG	South African Administrator-General
WAT	Water Affairs Branch
ZBU	Zentralbureau des kaiserlichen Gouvernements
NASA	National Archives of South Africa (Pretoria)
BAO	Bantu Affairs and Development
NTS	Native Affairs
TES	Treasury
VWN	Social Welfare Division
NAUK	National Archives of the United Kingdom (Kew)
CO	Colonial Office
NLN	National Library of Namibia
NSS	Namibia Scientific Society
ROHC	Registrar of the Master of the High Court of Namibia
TBK	Western Cape Archives and Records Service (Buro-TBK)
KUS	Coloured Affairs
UKCHA	United Kingdom Company House (Cardiff)
WBPA	Willem Basson Private Archive (ǂGâbes)
WDO	Windhoek Deeds Office, Ministry of Land Reform

Bibliography

Agence France Presse, *AFP Editorial Standards and Best Practices* (22 June 2016).

D. Anderson, *Histories of the Hanged: The Dirty War In Kenya And The End Of Empire* (New York, W.W. Norton, 2005).

S.R. Archer, T.W. Boutton and K.A. Hibbard, 'Trees in Grasslands: Biogeochemical Consequences of Woody Plant Expansion', in E.D. Schultze et al. (eds.) *Global Biogeochemical Cycles in the Climate System* (New York, Academic Press/Elsevier, 2001), pp. 115–137.

Aussenkjer Plase (Pty) Ltd, *The Potential for Agricultural Development of the Aussenkjer-Noordoewer Area* (Stellenbosch, 1985).

S. Bailey, 'Gilbertson's Pallinghurst Raising $1 Billion for Resource Fund', *Bloomberg News* (28 May 2007).

'Bedreigde jagluiperds geskiet, vleis vir werkers gegee', *News24* (3 July 2012).

A.P.J. Beris CMM, *From Mission to Local Church: One Hundred Years of Mission by the Catholic Church in Namibia* (Windhoek, RC Archdiocese, 1996).

A.P.J. Beris CMM, *Making the Desert Bloom (vol. 2)* (Windhoek, RC Diocese of Keetmanshoop, 2003).

A.P.J. Beris CMM, *Making the Desert Bloom (vol. 4)* (Windhoek, RC Diocese of Keetmanshoop, 2014).

B. Berman and J. Lonsdale, *Unhappy Valley: Conflict in Kenya and Africa (Vol. 1)* (Oxford, James Currey, 1992).

'Bernardus Swartbooi is Fired', *Namibian Sun* (13 December 2016).

B. Bersaglio and F. Cleaver, 'Green Grab by *Bricolage*: The Institutional Workings of Community Conservancies in Kenya', *Conservation and Society*, 16, 4 (2018), pp. 467–480.

M.M. Betsill et al., 'Philanthropic Foundations as Agents of Environmental Governance: A Research Agenda', *Environmental Politics*, 31, 4 (2022), pp. 684–705.

M.J. Betts, 'Namibia's No Man's Land: Race, Space, and Identity in the History of Windhoek Coloureds under South African Rule, 1915–1990' (PhD thesis, University of California, Los Angeles, 2009).

'BHP Chief in Shock Resignation', *CNN* (5 January 2003).

'BHP now a "Moderate Sized International Company": Takeover Deal for Utah Completed', *The Canberra Times* (4 April 1984).

C. Birkby, *Thirstland Treks* (London, Faber & Faber, 1936).

M. Biwa, '"Weaving the Past with Threads of Memory": Narratives and Commemorations of the Colonial War in Southern Namibia' (PhD Thesis, University of the Western Cape, 2012).

D.I. Bleiwas, *Estimated Water Requirements Gold Heap Leach Operations* (U.S. Geological Survey, 2012).

D. Bradley and A. McCauley, *A Preliminary Deposit Model for Lithium-Cesium-Tantalum (LCT) Pegmatites* (Reston, US Geological Survey, 2013).

N. Breed, *From Ox Cart to Email: The Kenya Story of Delia Craig* (Sussex, Book Guild, 2011).

M. Bollig, *Shaping the African Savannah: From Capitalist Frontier to Arid Eden in Namibia* (Cambridge, Cambridge University Press, 2020).

J.L. Bolton, *Alfalfa: Botany, Cultivation, and Utilisation* (London, Leonard Hill Ltd., 1962).

G. Borgström, *The Hungry Planet: The Modern World at the Edge of Famine* (New York, Macmillan, [1965] 1972).

R.W. Bovey, *Mesquite: History, Growth, Biology, Uses, and Management* (College Station, Texas A&M University Press, 2016).

E. Brandt, 'Tantalite Mine Opens on Back of Favourable Market', *The Namibian* (1 August 2001).

E. Brandt, 'Tantalite Mine in South Closes "Temporarily"', *The Namibian* (9 January 2002).

'Brian Gilbertson: The Industry's Godfather', *Global Mining Observer*, 109 (2014).

B. Bravenboer, *Karakul: Gift from the Arid Land, 1907–2007* (Windhoek, Karakul Board of Namibia, 2007).

S. Broad and J. Jenkins, 'Gibbons in their Midst? Conservation Volunteers' Motivations at the Gibbon Rehabilitation Project, Phuket, Thailand', in K.D. Lyons and S. Wearing (eds.) *Journeys of Discovery in Volunteer Tourism: International Case Study Perspectives* (Wallingford, CABI, 2008), pp. 72–85.

D. Brockington, *Fortress Conservation: The Preservation of the Mkomazi Game Reserve, Tanzania* (Oxford, James Currey, 2002).

D.M. Browning and S.R. Archer, 'Protection from Livestock Fails to Deter Shrub Proliferation in a Desert Landscape with a History of Heavy Grazing', *Ecological Applications*, 21, 5 (2011), pp. 1629–1642.

K.F.R. Budack, "n Volkekundige Studie van die Tses Reservaat met Besondere Verwysing na die Geskiedenis en die Inter-Etniese Verhouding van die Bewoners' (MA thesis, University of Pretoria, 1965).

A. Buller, *The Value of a Whale: On the Illusions of Green Capitalism* (Manchester, Manchester University Press, 2022).

B. Büscher, *Transforming the Frontier: Peace Parks and the Politics of Neoliberal Conservation in Southern Africa* (Durham, Duke University Press, 2013).

B. Büscher and R. Fletcher, 'Towards Convivial Conservation', *Conservation and Society*, 17, 3 (2019), pp. 283–296.

B. Büscher and M. Ramutsindela, 'Green Violence: Rhino Poaching and the War to Save Southern Africa's Peace Parks', *African Affairs*, 115, 458 (2016), pp. 1–22.

'Businesses for Sale', *The Economist* (17 January 2004).

B. Butler, 'Caymans Court Lambastes former BHP Chief Brian Gilbertson', *Sydney Morning Herald* (28 July 2014).

J. Carruthers, *The Kruger National Park: A Social and Political History* (Pietermaritzburg, University of Natal Press, 1995).

J-P. Ceppi, 'Portait: Ils ont largué les amarres Franka Severin: Croqueuse de pépites du Tyrol à l'Afrique du Sud, elle a été comptable, ou starlette, avant de chercher de l'or au pays de l'apartheid', *Libération* (17 August 1996).

M. Cioc, *The Game of Conservation: International Treaties to Protect the World's Migratory Animals* (Athens, Ohio University Press, 2009).

E. Claassens, 'Rupert se Droom Nagmerrie vir Boere', *Landbou Weekblad* (21 November 2008).

H. Connon, 'Mining a Very Rich Seam', *The Guardian* (10 July 2004).

A. Cooper and M. Tusenius, *International Investment in South Africa* (Washington DC, Investor Responsibility Research Center, 1987).

M. Creamer, 'Russia's Renova Mulling Uplift of South African Manganese Business', *Mining Weekly* (2 December 2011).

W. Cronon, 'The Trouble with Wilderness, or, Getting back to the Wrong Nature', *Environmental History*, 1, 1 (1996).

S. Counsell, *Blood Carbon: How a Carbon Offset Scheme Makes Millions from Indigenous Land in Northern Kenya* (Survival International, March 2023).

T. Dedering, *Hate the Old and Follow the New: Khoekhoe and Missionaries in early-Nineteenth Century Namibia* (Stuttgart, Franz Steiner Verlag, 1997).

D.H. de Jager and J.W. von Backström, *The Sillimanite Deposits in Namaqualand near Pofadder* (RSA Department of Mines, Geological Survey, 1961).

W.J. de Kock, 'Ekstraterritorial Vraagstukke van die Kaapse Regering (1872–1885) met Besondere Verwysing na die Transgariep en Betsjoeanaland' (DPhil thesis, Stellenbosch University, 1948).

B. de la Bat, 'An uneasy marriage' *African Wildlife*, 27 (likely 1979).

H. Dempsey and J. Cotterill, 'How China is Winning the Race for Africa's Lithium', *Financial Times* (3 April 2023).

J. Dempsey and P. Bigger, 'Intimate Mediations of For-Profit Conservation Finance: Waste, Improvement, and Accumulation', *Antipode*, 51, 2 (2019), pp. 517–538.

A. Deneault, *Legalizing Theft: A Short Guide to Tax Havens* (Halifax, Fernwood Publishing, 2018).

P. de Vos, 'Little-Used Legal Doctrine Secures Informal Land Rights of the Grootkraal Community', *Constitutionally Speaking* (29 November 2018).

U. Dieckmann, *Hai//om in the Etosha Region: A History of Colonial Settlement, Ethnicity, and Nature Conservation in Namibia* (Basel, Basler Afrika Bibliographien, 2007).

J. Dlamini, *Safari Nation: A Social History of the Kruger National Park* (Johannesburg, Jacana Media, 2020).

C.J. Donlan et al., 'Pleistocene Rewilding: An Optimistic Agenda for Twenty-First Century Conservation', *The American Naturalist*, 168, 5 (2006), pp. 673–674.

A. Ducros and P. Steele, *Biocredits to Finance Nature and People: Emerging Lessons* (London, IIED, 2022).

S.M. Durant, N. Pettorelli and J.T. du Toit, 'The Future of Rewilding: Fostering Nature and People in a Changing World', in *Rewilding* (Cambridge, British Ecological Society, 2019), pp. 413–425.

J.T. du Toit, 'Pleistocene Rewilding: An Enlightening Thought Experiment', in N. Pettorelli, S.M. Durant, J.T. du Toit (eds.) *Rewilding* (Cambridge, British Ecological Society, 2019), pp. 55–72.

P. du Toit, *The Stellenbosch Mafia: Inside the Billionaires' Club* (Cape Town, Jonathan Ball, 2021).

R. Eden, 'Prince William's former "Fiancée" Jecca Craig calls off wedding', *The Telegraph* (11 October 2009).

C.H. Ellersgaard et al., 'Say My Name? Anonymity or not in Elite Interviewing', *International Journal of Social Research Methodology*, 5 (2022), pp. 673–686.

C.W Erichsen, 'Forced Labour in the Concentration Camp on Shark Island', J. Zimmerer and J. Zeller (eds.), *Genocide in German South West Africa: The Colonial War (1904–1908) in Namibia and its Aftermath* (Monmouth, Merlin Press, 2003), pp. 84–99.

P.L. Errington, *Of Predation and Life* (Ames, Iowa State University Press, 1967).

A.M. Evans, *An Introduction to Ore Geology* (Second Edition) (Oxford, Blackwell, 1987).

'Fabergé Seeks Partnership with Alrosa', *Rough Polished* (28 May 2008).

J. Farrell, *Billionaire Wilderness: The Ultra-Wealthy and the Remaking of the American West* (Princeton, Princeton University Press, 2020).

R. Felli, 'On Climate Rent', *Historical Materialism*, 22, 3–4 (2014), pp. 251–280.

T.R. Filley et. al., 'Chemical Changes to Nonaggregated Particulate Soil Organic Matter Following Grassland-to-Woodland Transition in a Subtropical Savanna', *Journal of Geophysical Research: Biogeosciences*, 113, G3 (2008), pp. 1–11.

H. Fincker (ed.), *Franka: Une Femme Affranchie: Chronique d'une vie* (Paris, Éditions Balland, 2018).

T. Fischer and H. Knuth, 'Grün getarnt', *Die Zeit* (19 January 2023).

T. Fischer and H. Knuth, 'Phantom Offsets and Carbon Deceits', *Die Zeit* (19 January 2023).

T. FitzJohn,'25 Years of Mkomazi', *The Horn* (2014), pp. 22–23.

R. Fitzner (ed), *Deutsches Kolonial-Handbuch, 1901* (Berlin, Hermann Paetel, 1901).

N. Fletcher, 'British Energy is Hot Nuclear Favourite', *The Guardian* (8 January 2008).

D. Foreman and B. Haywood (eds.), *Ecodefense: A Field Guide to Monkeywrenching* (Chico, Abzug Press, 1993).

M. Forster, 'Der Bau einer neuen Schule in Gaidip', *Das Licht* (1930).

G. Fowlds and G. Spence, *Rewilding Africa: Restoring the Wilderness on a War-Ravaged Continent* (Johannesburg, Jonathan Ball, 2022).

G.R. Fox, 'Images, Uncertainties, and Success Stories: The Politics of Conservation in Laikipia North, Kenya' (PhD thesis, McGill University, 2018).

G.R. Fox, 'The 2017 Shooting of Kuki Gallmann and the Politics of Conservation in Northern Kenya', *African Studies Review*, 61, 2 (2018), pp. 210–236.

G.R. Fox, 'Maasai Group Ranches, Minority Land Owners, and the Political Landscape of Laikipia County, Kenya', *Journal of Eastern African Studies*, 12, 3 (2018), pp. 473–493.

B. Fuller and G. Eiseb, 'The Commercial Farm Market in Namibia: Evidence from the First Eleven Years', Institute for Public Policy Research Working Paper no. 15 (November 2002).

S. Gilbertson, letter to the editor of *The Economist* (3 August 2002).

P. Gilolmo Lobo, 'In the Wake of the Namibian Second National Land Conference', *Agrarian South* (29 January 2019).

A.L. Giraud, 'Energy in France', *Annual Review of Energy*, 8 (1983), pp. 165–191.

G. Giuliani and L.A. Groat, 'Geology of Corundum and Emerald Gem Deposits: A Review', *Gems & Gemology* (Winter 2019), pp. 464–489.

G. Giuliani et al., 'Emerald Deposits: A Review and Enhanced Classification', *Minerals*, 9, 2 (2019).

'Global Coal's Gilbertson on New Online Coal Market', *Bloomberg News* (5 June 2001).

J.E. Goudsmit, *Pandecten-Systeem* (Leiden, 1866).

M.L. Gravesen, *The Contested Lands of Laikipia: Histories of Claims and Conflict in a Kenyan Landscape* (Leiden, Brill, 2020).

L.G. Green, *To the River's End* (Cape Town, Howard B. Timmins, 1948).

P. Greenfield, 'Revealed: More than 90% of Rainforest Carbon Credits by Biggest Certifier are Worthless, Analysis Shows', *The Guardian* (18 January 2023).

P. Greenfield, 'Biggest Carbon Credit Certifier to Replace its Rainforest Offset Scheme', *The Guardian* (10 March 2023).

B.L. Gunsallus and F.P. Uswald, 'Kyanite and Related Minerals', *Minerals Yearbook, 1954* (United States Geological Survey, 1954), pp. 645–650.

F. Habashi, 'A Short History of Hydrometallurgy', *Hydrometallurgy*, 79 (2005), pp. 15–22.

H.H. Hahn, 'Die Rol van Dr. J. Theophilus Hahn in Suider-Afrika' (MA thesis, Stellenbosch, 1993).

A. Hammill and C. Besançon, 'Measuring Peace Park Performance: Definitions and Experiences', in S.H. Ali (ed.) *Peace Parks, Conservation and Conflict Resolution* (Cambridge, MIT Press, 2007).

L.T.C. Harms and J.H. Hugo, *Amler's Precedents of Pleadings*, Third Edition (Durban, Butterworths, 1989).

E. Hattingh, 'Rivier van Wettelosheid', *Republikein* (27 October 2021).

M.W. Hayward et al., 'Top-Down Control of Ecosystems and the Case for Rewilding: Does It All Add Up?' in N. Pettorelli, S.M. Durant, J.T. du Toit (eds.) *Rewilding* (Cambridge, British Ecological Society, 2019), pp. 325–354.

D. Henrichsen, *Herrschaft und Alltag im vorkolonialen Zentralnamibia: Das Herero- und Damaraland im 19. Jahrhundert* (Basel, Basler Afrika Bibliographien, 2011).

L. Hill, 'Uranium Explorer's Shares Jump on Rio Bid Speculation', *Engineering News* (RSA) (7 January 2008).

L. Hill, 'Uranium Explorer Xemplar Appoints President, *Mining Weekly* (4 December 2008).

W. Hillebrecht, 'The Nama and the War in the South', in *Genocide in German South West Africa: The Colonial War of 1904–1908 in Namibia and its Aftermath* (Monmouth, Merlin Press, 2004), pp. 143–158.

K. Ho, *Liquidated: An Ethnography of Wall Street* (Durham, Duke University Press, 2009).

P. Holz, *The Pella Group Story* (Karasburg, Pella Group, c. 1969).

L. Hughes, *Moving the Maasai: A Colonial Misadventure* (New York, Palgrave, 2006).

P.H. Hugo, 'Follikulêre Ondersoek na die Verskillende Patroonvormende Eienskappe van Karakoellammers' (MSc thesis, UVS: Departement van Kleinveekunde, 1982).

A.I. Ivanov, *Alfalfa* (Russian Translations Series no. 63) (Rotterdam, A.A. Balkema, [1980] 1988).

H. Jäckel, *Die Landgesellschaften in den Deutschen Schutzgebieten* (Jena, Gustav Fischer Verlag, 1909).

S. Jamieson and R. Savage, 'Prince William Joins Champagne Reception Celebrations for Ex-Girlfriend Jecca Craig's Wedding', *The Telegraph* (26 March 2016).

C.L. Jardine, 'The Proposed Gariep Transfrontier Conservation Area: Conservation Overlay Evaluation of the South African Section' (MPhil thesis, University of Cape Town 1998).

P. Jepson and C. Blythe, *Rewilding: The Radical New Science of Ecological Recovery* (Cambridge, MIT Press, 2022).

B. Jewsiewicki, 'African Historical Studies, Academic Knowledge as a 'Usable Past', and Radical Scholarship', *African Studies Review*, 32, 3 (1989), pp. 1–76.

D. Johns, 'History of Rewilding: Ideas and Practice', in N. Pettorelli, S.M. Durant and J.T. du Toit (eds.) *Rewilding* (Cambridge, British Ecological Society, 2019), pp. 12–33.

T. Johnson, 'Global Uranium Supply and Demand', *Council on Foreign Relations* (14 January 2010).

E. Joubert, *On the Clover Trail: The Plight of the World's Rhinos* (Windhoek, Gamsberg Macmillan, 1996).

E. Joubert, P.A.J. Brand and G.P. Visagie, 'An Appraisal of the Utilisation of Game on Private Land in South West Africa', *Madoqua: Tydskif vir Natuurbewarings- en Woestynnavorsing in SWA*, 13, 3 (1983).

D. Jørgensen, 'Rethinking Rewilding', *Geoforum*, 65 (2015), pp. 482–488.

P. Kaapama, 'The Ancestral Land Debate prior to and during Namibia's Second National Land Conference: The Redistribution-Recognition Dilemma', *Anthropology Southern Africa*, 42, no. 3 (2019), pp. 232-246.

L. Kalvelage et al., 'Territorialising Conservation: Community-Based Approaches in Kenya and Namibia', *Conservation & Society*, 19, no. 4 (2021), pp. 282-293.

T. Kanogo, *Squatters and the Roots of Mau Mau* (London, James Currey, 1997).

A.B. Kelly, 'Conservation Practice as Primitive Accumulation', *Journal of Peasant Studies*, 38, 4 (2011), pp. 683–701.

A.B. Kelly and N.L. Peluso, 'Frontiers of Commodification: State Lands and their Formalization', *Society and Natural Resources*, 28 (2015), pp. 473–495.

J. Klemann, 'Mariensiedlung am Oranjefluß', *Das Licht* 31 (1938), p. 6.

G.P. Klinghardt, 'Missions and Social Identities in the Lower Orange River Basin, 1760–1998' (PhD thesis, University of Cape Town, 2005).

R. Knight, *Unified List of United States Companies Doing Business in South Africa* (New York, The Africa Fund, 1990).

A. Knobel, 'Rethinking Limited Liability: Beneficial Ownership Transparency to Reform the Liability System', Tax Justice Network Working Paper (November 2021).

A. Knobel, 'Complex Ownership Structures: Addressing the Risks for Beneficial Ownership and Transparency', Tax Justice Network Working Paper (February 2022).

R. Kößler, *In Search of Survival and Dignity: Two Traditional Communities in Southern Namibia under South African Rule* (Windhoek, Gamsberg Macmillan, 2005).

B. Kötting, 'Namibia's Black Rhino Custodianship Programme', *Conservation Frontlines*, 2, 2 (2020).

C.W. Lambert, 'Granitic Melt Transport and Emplacement along Transcurrent Shear Zones: Case Study of the Pofadder Shear Zone in South Africa and Namibia' (MSc thesis, Stellenbosch, 2013).

B. Lau, *Southern and Central Namibia in Jonker Afrikaner's Time* (Windhoek, National Archives of Namibia, 1987).

R.P. Lejano, W. Shan Kan and C.C. Chau, 'The Hidden Disequities of Carbon Trading: Carbon Emissions, Air Toxics, and Environmental Justice', *Frontiers in Environmental Science*, 8 (2020), pp. 1–6.

L. Lenggenhager, *Ruling Nature, Controlling People: Nature Conservation, Development, and War in Northeastern Namibia since the 1920s* (Basel, Basler Afrika Bibliographien, 2018).

L. Lenggenhager, 'Onseepkans: Irrigation, Removals and Resistance in the Borderlands of Namibia and South Africa', in L. Lenggenhager et al. (eds.) *The Lower !Garib/Orange River: Pasts and Presents of a Southern African Border Region* (Bielefeld, Transcript Verlag, 2023), pp. 151–164.

L. Lenggenhager and M. Ramutsindela, 'Property Killed a Peace Park Dream: The Entanglement of Property, Politics, and Conservation along the Gariep', *Land Use Policy*, 105 (2021), pp. 1–9.

L. Lenggenhager and A. Rosengarten, 'Flood Lines and Borderlines: Livestock Farming and Evictions Resistance at the !Garib/Orange River in Southern Africa', *Environment and Society: Arcadia* (Summer 2020).

C. Le Roux, 'Oral History Research Ethics: Should Anonymity and Confidentiality Issues be Dealt with on their Own Merit?' *Africa Education Review*, 12, 4 (2015), pp. 552–566.

P.J. le Roux, 'Oestrus Synchronisation in Karakul Ewes' (DSc thesis, University of Pretoria: Dept. of Sheep and Wool Science, 1974).

B.-O. Linnér, 'The World Household: Georg Borgström and the Postwar Population-Resource Crisis' (PhD thesis, Linköping University, 1998).

S. Likela, 'Top Government Officials Benefit from Resettlement Farms', *The Namibian* (25 September 2018).

L. Louw and M.G. Lambrechts, 'Grape Based Brandies: Production, Sensory Properties, and Sensory Evaluation', in J. Piggott (ed.) *Alcoholic Beverages: Sensory Evaluation and Consumer Research* (Sawston, Woodhead Publishing, 2012), pp. 281–298.

H. Lovell, H. Bulkeley and D. Liverman, 'Carbon Offsetting: Sustaining Consumption?' *Environment and Planning A*, 41 (2009), pp. 2357–2379.

T. MacAlister, 'Gilbertson Carpeted at Vedanta', *The Guardian* (5 July 2004).

D. MacKenzie, 'Making Things the Same: Gases, Emission Rights, and the Politics of Carbon Markets', *Accounting, Organizations, and Society*, 34 (2009).

J.M. MacKenzie, *The Empire of Nature: Hunting, Conservation, and British Imperialism* (Manchester, Manchester University Press, 1988).

N.A. Mancheri et al., 'Resilience in the Tantalum Supply Chain', *Resources, Conservation, and Recycling* 129 (2018), pp. 56–69.

J. Masson, *Jakob Marengo: An Early Resistance Hero in Namibia* (Windhoek, Out of Africa, 2001).

C. Matthews, 'Rio Tinto Linked to Offer for Xemplar', *The Namibian* (9 January 2008).

J. Mbaria and M. Ogada, *The Big Conservation Lie: The Untold Story of Wildlife Conservation in Kenya* (Auburn, Lens&Pens Publishing, 2017).

M. McCullers, 'Lines in the Sand: The Global Politics of Local Development in Apartheid Era Namibia, 1950–1980' (PhD thesis, Emory University, 2012).

R.L. McCulley et al., 'Soil Respiration and Nutrient Cycling in Wooded Communities Developing in Grassland', *Ecology*, 85, 10 (2004), pp. 2804–2817.

J. McCulloch, *South Africa's Gold Mines and the Politics of Silicosis* (Oxford, James Currey, 2012).

D. McKay, 'Gilbertson Eyes Russian Diamonds', *News24* (19 August 2008).

M. McKittrick, *Green Lands for White Men: Desert Dystopias and the Environmental Origins of Apartheid* (Chicago, University of Chicago Press, 2024).

H. Melber, 'Colonialism, Land, Ethnicity, and Class: Namibia after the Second National Land Conference', *Africa Spectrum*, 54, 1 (2019), pp. 73–86.

G. Miescher, *Namibia's Red Line: A History of a Veterinary and Settlement Border* (New York, Palgrave, 2012).

C. Möller, 'Time as an Argument in Roman Water Law', *Water History*, 15 (2023), pp. 67–80.

B.T.O Mokopakgosi, 'German Colonialism in Microcosm: A Study of the Role of Concessionaire Companies in the Development of German Colonialism' (PhD thesis, SOAS University of London, 1988).

A.C. Moore, 'The Petrography and Regional Setting of the Tantalite Valley Complex, South West Africa', *Transactions of the Geological Society of South Africa*, 78 (1975), pp. 235–249.

B.C. Moore, 'Smuggled Sheep, Smuggled Shepherds: Farm Labour Transformations in Namibia and the Question of Southern Angola, 1933–1975', *Journal of Southern African Studies*, 47, 1 (2021), pp. 93–125.

B.C. Moore, 'Swimming Upstream: From "Poor Whites" to "Coloureds" along South Africa's Lower Orange River', in L. Lenggenhager et al. (eds.) *The Lower !Garib/ Orange River: Pasts and Presents of a Southern African Border Region* (Bielefeld, Transcript Verlag, 2023), pp. 119–144.

B.C. Moore, 'Land, Labour, and Karakul in Namibia, 1910s-1960s' (PhD thesis, Michigan State University, 2025).

M. Morkel, 'Wildlife Conservation and Pete Morkel' (self-published, December 2016).

M. Morkel, *Diamonds on the Soles of His Feet* (self-published, 2019).

P.W. Morkel, *The Morkels: Family History and Family Tree* (self-published, 1961).

M. Mostafanezhad, *Volunteer Tourism: Popular Humanitarianism in Neoliberal Times* (Burlington, Ashgate, 2014).

S. Mosweu, 'Environmental Factors and Management Options for Prosopis Invasive Species: A Case Study in Botswana's Kgalagadi District' (PhD thesis, North West University, 2012).

S. Mosweu et al., 'Modification of Soil Properties by Prosopis L. in the Kalahari Desert, South-Western Botswana', *Open Journal of Ecology*, 3, 2 (2013), pp. 145–150.

A. Mukpo, 'Carbon Credits from Award-Winning Kenyan Offset Suspended by Verra', *MongaBay* (21 March 2023).

H. Mulholland, 'Arcadia Minerals receives letter of intent for potential offtake from Swanson Tantalum and Lithium Project', *Mining.com.au* (29 November 2022).

L. Mundell, 'Carry on Kraaling', *Farmer's Weekly* (25 July 1973).

A. Mureva et al., 'Soil Organic Carbon Increases in Semi-Arid Regions while it Decreases in Humid Regions Due to Woody-Plant Encroachment of Grasslands in South Africa', *Scientific Reports*, 8, 15506 (2018).

Namibia Press Agency, *NAMPA Editorial Guide and Style Book* (Windhoek, John Meinert, 2015).

L. Neubert, *The Karakul Industry: Policy Options for Independent Namibia* (Lusaka, United Nations Institute for Namibia, 1989).

R.P. Neumann, *Imposing Wilderness: Struggles over Livelihood and Nature Preservation in Africa* (Berkeley, University of California Press, 1998).

A.R. Newby-Fraser, *Chain Reaction: Twenty Years of Nuclear Research and Development in South Africa* (Pretoria, Atomic Energy Board, 1979).

R.V. Nghitevelekwa, *Securing Land Rights: Communal Land Reform in Namibia* (Windhoek, University of Namibia Press, 2020).

M. Nicol, *Hydrometallurgy: Volume 1, Theory* (Amsterdam, Elsevier, 2022).

M. Nicol, *Hydrometallurgy: Volume 2, Practice* (Amsterdam, Elsevier, 2022).

K.G. Nustad, *Creating Africas: Struggles over Nature, Conservation, and Land* (London, Hurst, 2015).

'Obituary: Anna Merz', *The Telegraph* (27 May 2013).

W. Odendaal, *'Beggars on our own Land ...' : Tsumib v. Government of the Republic of Namibia and its Implications for Ancestral Land Claims in Namibia* (Basel, Basler Afrika Bibliographien, 2024).

'Offshore with Sharks', *Noseweek* (June 2010), pp. 28–30.

V. Ogle, 'Archipelago Capitalism: Tax Havens, Offshore Money, and the State, 1950s–1970s', *American Historical Review*, 122, 5 (2017), pp. 1431–1458.

Orange River Vineyard Investments (Pty) Ltd et al., 'Aussenkehr Agricultural Developments: Environmental Assessment Scoping Report' (November 2018).

J.M. Otto, 'Vetustas, Onheuglike Tye en die Witpad by Langebaan', *Fundamina*, 19, 1 (2013).

M. Paterson, 'Governing Mobilities, Mobilising Carbon', *Mobilities*, 9, 4 (2014), pp. 570–584.

N. Penn, *The Forgotten Frontier: Colonist and Khoisan on the Cape's Northern Frontier in the 18th Century* (Athens, Ohio University Press, 2005).

C.L. Pettersson, *In the Footsteps of Mr. Anderson: Milestones in Swedish-Namibia Relations* (Parkwood, Ulwazi, 2008).

G.A. Polis and D.R. Strong, 'Food Web Complexity and Community Population Dynamics', *The American Naturalist*, 147, 5 (1996), pp. 813–846.

B. Pool, 'Die Geskiedenis van die Afrikaner-Oorlams in die Tyd van Jonker Afrikaner, 1790–1861' (PhD thesis, Stellenbosch, 1995).

M. Ramutsindela, *Transfrontier Conservation in Africa: At the Confluence of Capital, Politics, and Nature* (Wallingford, CABI, 2007).

Reichs-Kolonialamt, *Die deutschen Schutzgebieten in Afrika und der Südsee:* [Vols. 1909–1913] (Berlin, Königliche Hofbuchhandlung, 1911–1014).

C. Rickard, 'A Borderline Case that Never Was', *Legal Brief* (21 October 2015).

M.E. Ritchie et al., 'Increased Pastoralist Livestock Mobility is Associated with Large-Scale Rangeland Restoration and Soil Carbon Sequestration', *agriRxiv* (2024).

N. Roos, 'Work Colonies and South African Historiography', *Social History*, 36, 1 (2011), pp. 54–76.

A. Rosengarten, 'Entangled Networks: Ethnicity, Mobility, and Exchange in the Lower !Garib/Orange River Region in the Late-18th Century', in L. Lenggenhager et. al. (eds.), *The Lower !Garib/Orange River: Pasts & Presents of a Southern African Border Region* (Bielefeld, Transcript Verlag, 2023), pp. 27-50.

W. Rutishauser, 'The Orange River Boundary and the Ongoing Dispute between Namibia and South Africa', L. Lenggenhager et al. (eds.) *The Lower !Garib/Orange River: Pasts and Presents of a Southern African Border Region* (Bielefeld, Transcript Verlag, 2023), pp. 211–224.

S.A. Farm Consultants (Pty) Ltd, *Proposals for the Restoration of Profitability in of Farming in SWA/Namibia* (Pietermaritzburg, January 1983).

C. Schetter, K. Mkutu and M. Müller-Koné, 'Frontier NGOs: Conservancies, Control, and Violence in Northern Kenya', *World Development* 151 (2022), pp. 1–12.

H-M. Schindler (ed.) *S.W.A. Handboek* (Windhoek, South West Agency Co., 1972).

G.O.J. Schmid (ed.) *S.W.A. Handboek* (Windhoek, South West Agency Co., 1969).

B.A. Schreiber, 'Anna Merz', *Encyclopaedia Britannica* (2015).

J.C. Scott, *The Art of Not Being Governed: An Anarchist History of Upland Southeast Asia* (New Haven, Yale, 2009).

K. Sena, *Due Diligence Report Concerning the Report 'Stealth Game' Published by the Oakland Institute* (The Nature Conservancy, 2022).

B. Sergeant, *Brett Kebble: The Inside Story* (Cape Town, Struik, 2006).

B. Sergeant, *The Kebble Collusion* (Auckland Park, Jacana Media, 2012).

G.A. Sessions and S.D. Sessions, *Utah International: A Biography of a Business* (Ogden, Weber State University Press, 2002).

S. Shah, 'Gilbertson Quits Vedanta for $50 Million Russian Job', *The Independent* (7 July 2004).

R. Sherbourne, *Guide to the Namibian Economy, 2017* (Windhoek, Institute for Public Policy Research, 2017).

R. Sherbourne, *Guide to the Namibian Economy, 2022* (Windhoek, IPPR, 2022).

J.B. Shetler, *Imagining Serengeti: A History of Landscape and Memory in Tanzania from the Earliest Times to the Present* (Athens, Ohio University Press, 2007).

J.E. Shingley and A.R. Kampf, 'Gem-Bearing Pegmatites: A Review', *Gems & Gemology* (Summer 1984), pp. 64–77.

J.M. Simon OSFS, *Bishop for the Hottentots: African Memories, 1882–1909* (New York, Benziger Brothers, 1959).

J.M. Simon OSFS, *African Memories: Part Two, 1909–1932* (Keimoes, RC Diocese of Keimoes-Upington, 2006).

F.T. Sisco and E. Epremian (eds), *Columbium and Tantalum* (New York, John Wiley & Sons, 1963).

M. Soulé and R. Noss, 'Rewilding and Biodiversity: Complementary Goals for Continental Conservation', *Wild Earth* (Fall 1998), pp. 18–28.

J. Song, 'An Incomplete Guide to Biodiversity Credits', *Conservation Finance Network* (14 February 2023).

S.J. Sparks, 'Apartheid Modern: South Africa's Oil from Coal Project and the History of a South African Company Town' (PhD thesis, University of Michigan, 2012).

M. Spierenburg, C. Steenkamp and H. Wels, 'Enclosing the Local for the Global Commons: Community Land Rights in the Great Limpopo Transfrontier Conservation Area', *Conservation and Society*, 6, 1 (2008), pp. 87–97.

M. Spierenburg and H. Wels, 'Conservative Philanthropists, Royalty, and Business Elites in Nature Conservation in Southern Africa', *Antipode*, 42, 3 (2010).

M. Spierenburg & S. Brooks, 'Private Game Farming and its Social Consequences in Post-Apartheid South Africa: Contestations over Wildlife, Property and Agrarian Futures', *Journal of Contemporary African Studies*, 32, no. 2 (2014), pp. 151–172.

E.L.P. Stals, *Môrewind oor die Karasberge: 'n Kultuurhistoriese Verkenning van die Karassreek van die laat Negentiende Eeu* (Pretoria, Protea Boekhuis, 2009).

J.B. Stevens, P.S. van Heerden, P. Reid, A. Liebenberg, E. Hagedorn and G. de Kock, *Technical Learner Guide: Irrigated Crop and Fodder Production (Part 8)* (Republic of South Africa, Department of Agriculture, Forestry and Fisheries [Water Research Commission], 2012).

Stealth Game: 'Community' Conservancies Devastate Land and Lives in Northern Kenya (The Oakland Institute, 2021).

St. George, 'Pricing Nature: Can Biodiversity Credits Propel Global Conservation?' *Yale Environment 360* (6 April 2023).

S. Sullivan, 'Banking Nature? The Spectacular Financialisation of Environmental Conservation', *Antipode*, 45, 1 (2013), pp. 198–217.

S. Sullivan et al. 'Historicising Black Rhino in Namibia: Colonial-Era Hunting, Conservation Custodianship, and Plural Values', Future Pasts Working Papers no. 13 (December 2021).

Y. Suzuki, *The Nature of Whiteness: Race, Animals, and Nation in Zimbabwe* (Seattle, University of Washington Press, 2017).

J.-C. Svenning et al., 'Science for a Wilder Anthropocene: Synthesis and Future Directions in Trophic Rewilding Research', *Proceedings of the National Academy of Sciences*, 113, 4 (2016), pp. 898–906.

J. Swanepoel, 'In the Land of the Jackals: Postcolonial Aridity in Southern Namibia' (PhD thesis, University of the Free State, 2023).

'Swartbooi Resigned, not Fired – Kapofi', *New Era* (14 December 2016).

R. Swilling, 'Sandfontein Lodge: Far South of Namibia', *Travel News Namibia* (undated, likely 2012).

J.W.N. Tempelhoff, 'Omgewingslagoffers of Armblankes? Water en Welvaartskepping in die Era van Hertzog (1924–1939)', *Generaal J.B.M. Hertzoggedenklesing XXXV* (21 September 2006).

'The Bushpilot Training Course – An African Adventure', *African Pilot* (January 2007).

The Journal of Hendrik Jacob Wikar (1779), ed. E.E. Mossop (Cape Town, Van Riebeeck Society, 1935).

The Journals of Brink and Rhenius, ed. E.E. Mossop (Cape Town, Van Riebeeck Society, 1947).

J.L. Thompson and J. Day, 'Towards a Model of Entrepreneurial Behaviour: An Evaluation of the History of Fabergé from 1842 to 2017', *Journal of Management History*, 29, 1 (2023), pp. 134–155.

P. Thompson and R. Macklin, *The Big Fella: The Rise and Rise of BHP-Billiton* (Sydney, William Heinemann, 2009).

E. Tjirera, 'Namibia's Intractable Land Question', *Current History*, 122, no. 844 (2023), pp. 167-171.

H. Türk, 'From Oil to Coal? The International Energy Agency and International Coal Policy since the End of the 1970s', in L. Bluma et al. (eds.) *Boom-Crisis-Heritage: King Coal and the Energy Revolutions after 1945* (Berlin, De Gruyter Oldenbourg, 2021), pp. 81–92.

United Nations Institute for Namibia, *Namibia: Perspectives for National Reconstruction and Development* (Lusaka, UNIN, 1986).

US Department of Agriculture, 'Exports', *Journal of Commerce* (1988), p. 12B.

O. van Beemen, *Im Namen der Tiere: Wie eine NGO große Teile Afrikas beherrscht* (München, C.H. Beck, 2024).

M. van den Akker, *Monument of Nature? An Ethnography of the World Heritage of Mt. Kenya* (Leiden, African Studies Centre, 2016).

A. van der Colf, 'Stigting en vroeë geskiedenis' in *Historiese Oorsig van Karasburg* (Karasburg, 1956).

C.G. van der Merwe, 'Louisiana Right to Forced Passage Compared with the South African Way of Necessity', *Tulane Law Review* 73, 4 (1999), pp. 1363–1414.

A. van der Westhuizen, 'Obituary: Hugh Jenner-Clarke', *Geobulletin* [Geological Society of South Africa] 60, (June 2017), pp. 31–34.

A.J.A. van Niekerk, ''n Studie van Teeltprobleme by Karakoelskape met Spesiale Verwysing na die Niemöllerkudde' (DSc thesis, University of the Free State, 1972).

L. van Sittert, 'Bringing in the Wild: The Commodification of Wild Animals in the Cape Colony/Province c. 1850–1950', *Journal of African History*, 46 (2005), pp. 269–291.

L. van Sittert, 'Writing on Skin: The Entangled Embodied Histories of Black Labour and Livestock Registration in the Cape Colony, c.1860–1909', *Kronos* 40 (2014), pp. 74–98.

L. van Vuuren, *In the Footsteps of Giants: Exploring the History of South Africa's Large Dams* (Water Research Commission of South Africa, 2012).

A-M. van Wyk, 'South Africa's Nuclear Programme and the Cold War', *History Compass*, 8/7 (2010), pp. 562–572.

E.R. Varley, *Sillimanite* (New York, Chemical Publishing Company, 1968).

Vioolsdrift-Noordoewer Joint Irrigation Authority Scheme, *Water Management Plan* (Orange-Senqu River Commission, March 2014).

R.A. Virginia and W.M. Jarrell, 'Soil Properties in a Mesquite-Dominated Sonoran Desert Ecosystem', *Soil Science Society of America Journal*, 47, 1 (1983), pp. 138–144.

K. Visser and K. du Pisani, 'Die Vestiging van Skanskopeilande-Besproeiingsnedersetting as 'n Mylpaal in Selfwerksame Armoedeverligting', *Historia*, 57, 1 (2012), pp. 103–126.

R.A. Voeltz, *German Colonialism and the South West Africa Company, 1884–1914* (Athens, Ohio University Press, 1988).

C. Vogel, 'Russian Industrialist Buys Fabled Forbes Fabergé Eggs', *New York Times* (4 February 2004).

H.O. von Schöllenbach, *Die Besiedelung Deutsch-Südwestafrikas bis zum Weltkriege* (Berlin, Dietrich Reimer Verlag, 1926).

A. von Wietersheim, 'Im Jagdrevier zwischen Kalahari und Namib (1984)', in *Vom Schutzgebiet bis Namibia 2000* (Windhoek, Klaus Hess Verlag, 2000).

J. Walker, 'Gencor Pops Champagne on Achieving its Ambitious Dream', *Business Times* (22 June 1997).

S. Walker, 'Russian Billionaire Feathers his Fabergé Nest Egg', *The Independent* (10 December 2010).

M. Wallace, *A History of Namibia: From the Beginning to 1990* (London, Hurst, 2011).

C. Walsh, 'A Big Moment in the Life of Brian', *The Guardian* (2 September 2006).

K. Ward, 'For Wilderness or Wildness? Decolonising Rewilding', in N. Pettorelli, S.M. Durant, J.T. du Toit (eds.) *Rewilding* (Cambridge, British Ecological Society, 2019), pp. 36–54.

K. Wedekind, *Impfe und herrsche: Veterinärmedizinisches Wissen und Herrschaft im kolonialen Namibia, 1887–1929* (Göttingen, Vandenhoeck & Ruprecht, 2021).

F. Wehrl, *Mission am Oranje: Geschichte der Oblaten-Mission der Vikariate Keimoes und Keetmanshoop nach Briefen, Tagebüchern und Visitationsberichten* (Eichstätt, Franz Sales Verlag, 1994).

P. Weidner, *The World Owes You Nothing* (Cape Town, c. 1969).

V. Wessels and S. Bailey, 'Gilbertson, Investec Start $1.5 Billion Resource Fund', *Bloomberg News* (28 May 2007).

B.J. Wigley, W.J. Bond and M.T. Hoffmann, 'Bush Encroachment under Three Contrasting Land-Use Practices in a Mesic South African Savannah', *African Journal of Ecology*, 47 (supp. 1) (2009), pp. 62–70.

K.M. Woodhouse, *The Ecocentrists: A History of Radical Environmentalism* (New York, Columbia, 2018).

A. Zaby OMI, 'Die Riemvasmakers: Geschichte und Kultur einer Gemeinschaft', *Namibiana*, 4, 1 (1982).

D.P. Zongwe, 'Researching Namibian Law and the Namibian Legal System', *GlobaLex* (December 2020).

G. Zucman, *The Hidden Wealth of Nations: The Scourge of Tax Havens* (Chicago, University of Chicago Press, 2015).

Index

www.ingramcontent.com/pod-product-compliance
Lightning Source LLC
Chambersburg PA
CBHW071413230926
51846CB00001B/72
* 9 7 8 9 0 0 4 7 1 2 4 4 7 *